What
Is Water?

The Nature|History|Society series is devoted to the publication of high-quality scholarship in environmental history and allied fields. Its broad compass is signalled by its title: nature because it takes the natural world seriously; history because it aims to foster work that has temporal depth; and society because its essential concern is with the interface between nature and society, broadly conceived. The series is avowedly interdisciplinary and is open to the work of anthropologists, ecologists, historians, geographers, literary scholars, political scientists, sociologists, and others whose interests resonate with its mandate. It offers a timely outlet for lively, innovative, and well-written work on the interaction of people and nature through time in North America.

General Editor: Graeme Wynn, University of British Columbia

NATURE|HISTORY|SOCIETY

What Is Water?

The History of a Modern Abstraction

JAMIE LINTON

FOREWORD BY GRAEME WYNN

UBC Press • Vancouver • Toronto

20 19 18 17 16 15 14 13 12 11 10   5 4 3 2 1

Printed in Canada with vegetable-based inks on FSC-certified ancient-forest-free paper (100% post-consumer recycled) that is processed chlorine- and acid-free.

Library and Archives Canada Cataloguing in Publication

Linton, Jamie
    What is water? : the history of a modern abstraction / Jamie Linton.

(Nature/history/society, 1713-6680)
Includes bibliographical references and index.
ISBN 978-0-7748-1701-1 (bound); 978-0-7748-1702-8 (pbk);
978-0-7748-1703-5 (e-book)

    1. Water. 2. Hydrologic cycle. 3. Water – History. 4. Water – Environmental aspects. 5. Water – Social aspects. I. Title. II. Series: Nature, history, society

GB665.L55 2010   553.7   C2009-906381-6

Canadä

UBC Press gratefully acknowledges the financial support for our publishing program of the Government of Canada (through the Canada Book Fund)), the Canada Council for the Arts, and the British Columbia Arts Council.

This book has been published with the help of a grant from the Canadian Federation for the Humanities and Social Sciences, through the Aid to Scholarly Publications Programme, using funds provided by the Social Sciences and Humanities Research Council of Canada.

UBC Press
The University of British Columbia
2029 West Mall
Vancouver, BC V6T 1Z2
www.ubcpress.ca

Contents

Illustrations

Tables

Making Waves

by Graeme Wynn

Beginning a book as Jamie Linton does this one, with the claim that "water is what we make of it," is an act of provocation. Just as a sudden gust of wind ruffles reflections on the mirror surface of a still pond, so this assertion disturbs the seemingly self-evident truth that water is water. There are dangers in such a strategy. Readers who take the existence of this most common of the earths' natural substances for granted may well be puzzled by Linton's opening words. Those who think about water only as fresh or salt, hard or soft, hot or cold, dirty or clean might concede individual responsibility for heating or chilling small quantities of it and accept collective human blame for water pollution, yet insist that water is fundamentally the same as it was when "the Spirit of God moved upon the face of the waters."[1] Those who know that water is a colourless, transparent, tasteless, scentless compound of oxygen and hydrogen – H_2O in its intermediate state between ice and vapour – might doubt the veracity of the author. Others, equally misguided, might see this sentence as the beginning of yet another convoluted and ultimately irrelevant exercise in academic hair splitting.

Like the effects of the perturbing wind, the implicit challenge of these first words may dissipate, but it cannot be ignored. Even as reflections paint the calming waters anew, those who watched their precursors disappear appreciate their fragility, sense their imperfection, and know them differently. So, once made, the assertion that water is what we make of it complicates our thinking about it. Just as a stone thrown into a lake spreads ripples outward across its surface, so Linton's provocation sends intellectual

shock waves hammering into pervasive ways of understanding and defining water, invites reflection on the ways in which people have thought
about water in the past, and heightens awareness of the consequences that
will flow from what we make of water in the future. These are not small
matters.

Water is everywhere these days, at least metaphorically and represen-
tationally. It is in the news: too little or too much, it is the centrepiece of
stories of crisis and disaster precipitated by droughts, floods, or tsunamis,
and impure, it is a source of disease and death. It is the topic of countless
magazine articles: read about "Situational Waste in Landscape Watering,"
about the Great Salt Lake as America's Aral Sea, or about the economics
and psychology of bottled water, on which Americans spent more in 2007
than they did on iPods or movie tickets, an astonishing $15 billion. It has
spawned a groaning shelf of books: it appears as blue gold, liquid assets,
an uncooperative commodity, a precious resource, and a disconcertingly
diminished drop at the centre of the last oasis. It is on film and DVD:
see *Waterlife*, winner of the 2009 Canadian Hot Docs Special Jury Prize
for the story of the last huge supply of fresh water on earth, the Great
Lakes, or Deepa Mehta's film *Water*, which is not really about water at
all! It is even in museums: a major (US$ 3 million plus) exhibition "Water:
H_2O = Life" opened at the American Museum of Natural History in 2007,
offering "the fascinating story of water's influence on Earth and, simultaneously, a cautionary tale of growing demands on an essential and
limited resource."[2]

Making a distinctive contribution to this hubbub is no easy task, but
Linton rises to the challenge by tracing the development of particular ways
of thinking about water in the twentieth century and pondering their
consequences. He does so with a singular vision and an unusual voice,
both rooted in personal experience. Trained first in political philosophy
and international development studies, Linton was an environmental
policy researcher and environmental activist with a special interest in
aquatic ecosystems and water issues before beginning the doctoral work
in geography from which this book is derived. Reflecting its author's
varied training and his engagement on the front lines of debates about
water in the 1990s, when it was widely asserted that the world was on the
brink of unprecedented and calamitous water scarcities, *What Is Water?* is
an extended essay making the argument that this "water crisis" owed (and
owes) its existence and rhetorical power to the ways in which modern
Westerners think about water.

The crux of Linton's case lies in an insight offered by the philosopher and social critic Ivan Illich in the 1980s: in developing the idea of water as a scientific abstraction, modern society disenchanted that "ineffable stuff called water," robbed it of its history, and made it almost impossible for its members to know the waters of "the deep imagination" (75). Put slightly differently, this suggests that premodern societies typically lived with the reality of various waters diversely known, whereas modern societies essentialize water to the point where it is extracted from the social contexts of human experience and treated as an invariant essence, be that essence H_2O or a "resource." In more formal terms yet, this is to say that Linton's argument draws its impetus from the identification of an epistemological revolution in the way that people knew and represented water, a revolution that presumed and established a fundamental separation between the natural and the social realms and thus, sometime in the nineteenth century, robbed water of its social nature.

There is much to think about in all of this. Reflecting his geographical training and ongoing debate in the discipline of geography about what is commonly described as socionature, Linton first outlines the theoretical approach he uses to develop his understanding of the "essential relations between water and society" (20) and to underpin his basic point that water (and nature) are far less natural than they have generally been taken to be by those educated in the modern Western tradition. He draws together notions of relational dialectics and hybridity to offer a way of "analyzing both the history of water and how the idea of water articulates with its material and representative forms to produce this history" (41). This discussion is far less difficult than this stark summary suggests; taken in full it provides a thoughtful engagement with diverse literatures and an intriguing platform for reflecting more broadly upon human-environment relations.

Although a historical perspective is central to Linton's argument, his discussion of the past is neither chronological nor comprehensive. It begins retrospectively, noting recent expressions of dissatisfaction with the ways in which water has been conceptualized and represented in the industrialized world through the twentieth century, to establish both the need for critical perspective on this way of seeing – treating water as an abstraction – and the importance of examining, once more, "the institutional, social and political dimensions of the water-society nexus" (48). His argument then proceeds to consider a series of "moments" that might be regarded as "separate, independent, or self-sufficient," but which are here understood through the lens of relational-dialectical analysis to "actually produce each other in mutually constitutive processes" (27). In this central section of his book, Linton reflects in turn upon the ways in which premodern conceptions of water differed from those framed through the practices of modern science, engineering, and state-building. He compares the hydrological cycle (a scientific concept) with older ideas about the circulation of water and examines its importance as the predominant mode of thinking about the ways in which water flows through the hydrosphere. He also shows how articulations of the hydrological cycle with government agendas in mid-twentieth-century America facilitated unprecedented manipulation and control of rivers. The projection of the hydrologic cycle on a global scale yields the further abstraction that Linton describes as "global water," a necessary ingredient of the full-blown water crisis that emerged in the 1990s. Finally, *What Is Water?* takes the turn suggested by its title, to explore the philosophical commitments that underpin the abstraction "modern water" and to argue that the idea of a global water crisis emerges from the incommensurability of "global water" and a second, equally gross, abstraction, "world population." The conclusion then outlines a new approach to thinking about water in relation to both social and hydrological circumstances.

Readers with very different interests will be intrigued by the rich array of arguments, ideas, interpretations – and provocations – in this book. Those preoccupied by theoretical considerations and larger questions about the nature-society binary, as raised by Bruno Latour, Erik Swyngedouw, David Harvey, and others, will find reflexive engagement with the ideas of these writers and a substantive treatment of some of the questions their works raise, at least as they are evinced by water, every instance of which, as Linton has it on page 36, "combines nature and society, the properties of H_2O, the material practices of people, and the effects of discourse."

Those interested in modernity and especially its promise of "human emancipation through the domination of nature,"[3] the culmination of which James C. Scott has described and interrogated as "high modernism,"[4] will likewise find much to mull over in these pages. Big dams holding back enormous rivers are almost archetypal products of what Scott sees as large-scale, authoritarian planning intended to realize grand utopian schemes. In the minds of many twentieth-century politicians and planners, these dams were icons of progress and development, veritable "temples of modernity" as Indian prime minister Jawaharlal Nehru famously described them in the 1950s.[5] As parts of complex engineering works for the generation of electricity, they helped turn rivers into organic machines.[6] As devices for controlling flows of water, changing the seasonal rhythms of rivers running to the sea, and redefining their purpose according to new metrics of "efficiency," they were integral elements of a pervasive water management paradigm that more than doubled per-capita rates of water withdrawal from rivers, lakes, and aquifers during the twentieth century. In Linton's account, their escalating costs, both ecological and economic ("the next water project costs twice as much as the last" quip those who plan them), are forcing water managers in many parts of the world to think anew about the challenges they face when planning new developments.

In this context, the story of modern water management follows its own intriguing cycle, moving through a series of supply-side emphases – from resource development, through water management, to sustainable resource management – to a recent demand-side emphasis on managing consumption. There are echoes in this cycle of the distinction that Samuel Hays draws between the conservation and environmental movements in the United States. In Hays' view, the early twentieth-century conservation movement grew out of and reflected producers' concerns about the finitude of resources and promulgated a "gospel of efficiency" in resource use. By contrast, he sees post–Second World War environmentalism as a consumer-led movement that reflected growing public anxiety about the quality of human life on earth.[7] Both environmental historians and resource managers might find a good deal to think about, and with, in these parallels.

Thanks to the work of Donald Worster, the fact that the state became "an agency for [the] conquest" of water in the early twentieth-century United States is hardly a revelation,[8] but Linton's discussion of these developments (Chapter 7) is worth reading for two reasons: first for its treatment of the role played by W.J McGee, sometimes described as "the chief theorist of the conservation movement,"[9] in defining water as a resource; and second for its discussion of the ways in which this definition depended upon developing the capacity to measure and inventory water (in effect, upon making water legible, in Scott's terms).[10] Developing a quantitative view of water was part of the process that enabled science, in the words of German philosopher Martin Heidegger, "to pursue and entrap nature as a calculable coherence" (183). It allowed estimates of the stock of water, identified the limits of supply, implied the prospect of scarcity, and facilitated the exercise of allocative power over what became, in this context, a finite resource.

These moves were foundational for the development of resource management and, as Linton has shown elsewhere, they were codified and extended by the work of the economist and geographer Erich W. Zimmerman between 1933 and the early 1950s.[11] Most often remembered for his aphoristic comment that "resources are not, they become," Zimmerman was no harbinger of a social constructivist view of the world but a firm believer in the divide between (civilized) humans and nature. He thought of nature as "neutral stuff" and the word "resource" as an abstraction, referring not "to a thing or substance but to a function which a thing or substance may perform or to an operation in which it may take part."[12] From this perspective, nature was either (economically) useful to modern humans or it was meaningless. It follows from this that once something is identified as a resource, it is open to exploitation "to yield the highest return" to capital or society (more abstractions) in order to further "the promotion of real wealth."[13] Thus questions of access to the resource are reduced to the language of calculation and technique, political agendas are hidden behind a veil of bureaucratic competence, difficult questions of social justice (who defines such terms as resource, return, wealth?) are swept aside, and resource managers are empowered to make what they take to be the best of the situation.

These developments were important because they framed ways of thinking that led inexorably to "the gloomy arithmetic of water" (203) – that there is simply not enough to go around, that "all land-bound life has to share one ten-thousandth of the planet's water" (194) – and thus produced the notion of a general water crisis. For proponents of this way of thinking, the crisis is largely a consequence of runaway demand. The earth's supply of fresh water is finite, and the world is running dry because its human population continues to grow. But this crisis is, Linton insists, manufactured in another sense. The people factored into these calculations and facing water crises are, in his fine and ironic phrasing, "one dimensional, consuming, procreating, biological units, whose relation with water is as fixed and determinate as the statistical methods by which they are made known" (198).

The noble essence of Linton's position is that people are more than Malthusian figures, and that water is much more to them than a cluster of molecules or a resource defined in functional terms. A couple of decades ago, amid much debate about the development of bulk water exports from Canada to the United States, Chief Kathy Francis of the Klahoose Nation of British Columbia offered a moving counterpoint to the arguments of those who saw water as something "to be captured or tamed, put in containers ... and transported far away to be used or sold for money." Her people, she said, saw water very differently:

A creek, which to a non-native person may be seen simply in terms of flow rates and acre-feet per year, may have a special name and spiritual significance. It may be a private bathing place for special ceremonies or initiation rites, or in some cases be owned by a particular individual or family. It not only physically and spiritually cleanses people, but it also cleanses the earth and, eventually the sea to which it inevitably flows, if left alone.[14]

Here is the heart of Linton's argument: "that phenomena [as the English geographer and proponent of socionature Noel Castree has written] do not have properties in themselves but only by virtue of their relationships with other phenomena."[15] Linton's answer to the question *What Is Water?* is that water becomes what it is in relation to other things and processes; it is what we make of it.

Provocation indeed. This simple observation is a call to arms. It signals rejection of utilitarian, managerial attitudes toward nature, undergirded by catch-phrases such as "sound ecology is good economics" and sustained by the conviction that "environmental planning can make the most of nature's resources so that human resourcefulness can make the most of the future."[16] In its refusal to accept definitions of water as a commodity, it echoes American conservationist Aldo Leopold's call for an ethic in which humans see themselves as part of a "community of interdependent parts."[17] And in the process it drives waves of dissent against the very foundations of what the Indian anthropologist and human rights activist Shiv Visvanathan, reacting against the managerial underpinnings of the Report of the World Commission on Environment and Development, entitled *Our Common Future,* once described as "Mrs Bruntland's disenchanted Cosmos."[18]

NOTES

1 Genesis 1:2.
2 Endter-Wada et al., "Situational Waste in Landscape Watering"; Bedford, "The Great Salt Lake as America's Aral Sea?"; Fishman, "Message in a Bottle"; Barlow and Clarke, *Blue Gold*; Boberg, *Liquid Assets*; Bakker, *An Uncooperative Commodity*; De Villiers, *Water*; Postel, *The Last Oasis*; McMahon, dir., *Waterlife* (see also: http://waterlife.nfb.ca); Mehta, dir. *Water*; Schmit, "Water: H_2O = Life."
3 Kaika, *City of Flows*, 12.
4 Scott, *Seeing Like a State.*
5 Williams and Mawdsley, "Postcolonial Environmental Justice."
6 White, *The Organic Machine.*
7 Hays, *Conservation and the Gospel of Efficiency*; Hays, *Beauty, Health and Permanence.*
8 Worster, *Rivers of Empire.*

9 See Helms, "The Early Soil Survey."
10 McGee, "Water as a Resource."
11 Linton, "The Social Nature of Natural Resources."
12 Zimmerman, *World Resources and Industries*, 7.
13 Ibid., 17.
14 Francis, "First They Came and Took Our Trees."
15 Castree, *Nature*, 224.
16 Conable, "Address to the World Resources Institute," 6, 3.
17 Leopold, *A Sand County Almanac*, 204.
18 United Nations World Commission on Environment and Development (Gro Harlem Brundtland, chair), *Our Common Future*; Visvanathan, "Mrs Bruntland's Disenchanted Cosmos."

Preface

In the 1990s, I worked as a freelance writer and researcher specializing in water issues. There was a lot to write about, as problems like water scarcity, water pollution, and inadequate water services for billions of people were then (as now) giving rise to concerns about a global water crisis. In 1997, I put together a small book on water issues that was published by the Canadian Wildlife Federation and titled *Beneath the Surface: The State of Water in Canada.* The book was a snapshot of the contemporary health of aquatic ecosystems throughout Canada. On virtually every page, along with data on water quality, the hydrology of rivers, and the status of freshwater biota, there was information and comment about people: the legacy of activities such as mining on water quality and of land clearance on wetland ecosystems; the effects of large dams on the diet of First Nations peoples in the North; the impacts of agricultural and industrial practices on rivers; the influence of recreational fishing on the species composition of the Great Lakes. In a book devoted to "The State of Water in Canada," I found it unavoidable to include a long chapter at the beginning that dealt with the history of human-water relations. The concluding chapter asked, "Are things getting better or worse?" and dealt largely with the water policy of the federal government.

I learned from producing that first book the difficulty of writing, talking, and even thinking about water without involving people in the story. The state of water always reflects, in one way or another, the state of society. And yet perhaps the greatest hydrological accomplishment in the modern world has been to construct an idea of water as something apart from the broader social contexts in which it occurs. Water has been made known as an abstraction – as H_2O, the stuff that flows through the hydrologic cycle. This book provides a history of this abstraction and a critique of the kind of management thinking that flows from it.

Among the many people who have helped bring this book to light, there are several whom I wish to thank in particular: Graeme Wynn, the Nature | History | Society series editor at UBC Press, has been an inspiration and mentor as well as a source of suggestions and ideas along the way. Randy Schmidt, the acquisitions editor with whom I have worked throughout the project, has provided invaluable assistance and much-appreciated doses of humour when needed. Laraine Coates and the editing and production team at UBC Press have made it a pleasure to bring this book out while doing an excellent job of making it as presentable as possible. Although I cannot thank them personally, the comments of the three anonymous reviewers engaged by the Press have greatly improved the original manuscript. I am grateful to several colleagues for having reviewed parts of the text or otherwise having helped to improve it, especially Andrew Baldwin, Bruce Braun, Mike Brklacich, David Brooks, Sean Carey, Simon Dalby, Alex Loftus, Bill Nuttle, and Iain Wallace. I also wish to thank all those involved in the Department of Geography and Environmental Studies at Carleton University (Ottawa) and the Department of Geography at Queen's University (Kingston) for providing me with a solid and convivial academic home in which to carry out my research and writing. I am also very pleased to acknowledge the Social Sciences and Humanities Research Council (of Canada) (SSHRC) for supporting my study habits in recent years. My deepest thanks of all go to Deb Vuylsteke for her assistance and support throughout this project and to our children, James Jules and Samantha, for the same.

PART I
Introduction

I

Fixing the Flow:
The Things We Make of Water

W ater is what we make of it. This is not a particularly novel assertion. The philosopher and historian of religions Mircea Eliade wrote that water "is *fons et origio,* the source of all possible existence ... it will always exist, though never alone, for water is always germinative, containing the potentiality of all forms in their unbroken unity." Everyone knows that we can't exist without water. But neither can water, as *fons et origio,* exist without us. We give to water that which enables it to realize its potential. All by itself, water is supremely fluid, fluctuating, fleeting. We mix language, gods, bodies, and thought with water to produce the worlds and the selves we inhabit. *Encyclopaedia Britannica* reports that "the body of a normal man weighing 65 kilograms (about 145 pounds) contains approximately 40 litres (about 42 quarts) of water." "We made every living thing of water," says the Qu'ran. "Water has a nearly unlimited ability to convey metaphors," declares the social critic Ivan Illich. Indeed, almost anything can be distilled into a watery metaphor. But then we can always (re)turn to water as a means of dissolving the very things we have made of it: "Only in contemplation of [water] do I achieve true self-forgetfulness and feel my own limited individuality merge into the universal," writes Thomas Mann. "Yes, as everyone knows, meditation and water are wedded forever," wrote Herman Melville in his classic work on the human condition.[1]

If we were to ascribe to water an essential nature, this might best be described as its legendary fecundity. Something of this essence can be gleaned from ancient cosmology, particularly theories about the origin of the universe. People in ancient Egypt, Babylonia, and Greece conceived of water as the fundamental substance out of which everything came into being.[2] Thales of Miletus (a seventh century BC Greek thinker often considered the first philosopher in the Western tradition for having postulated a reason for the manifold of nature) declared that it was water out of which the entire world took form. The Judeo-Christian tradition bequeathed a similar sense of the fecundity of water, even though it required a transcendent mind (God) to bring things to life: "In the beginning," declares the first sentence of the Old Testament, the universe consisted of formless, dark waters. Then, during the course of the first six days, "a wind from God swept over the face of the waters" to produce light, dry land, vegetation, animals, people, and every other thing that populates this worldly realm.[3]

These ideas and accounts of the origin of the universe reveal a kind of truth about water that doesn't translate easily into modern language. We are more used to thinking of – and representing – water as a fixed thing rather than a principle or process out of which things occur. One purpose of this book is to consider the difficulties that arise when we lose sight of water's essential fecundity and consider its essence in more fixed, material terms. Certainly, water is among the least cooperative of things when it comes to being contained in words and in deeds. Water is what we make of it, but it seldom stays that way for long. When we do contrive to slow down the flow for long enough to substantiate it in language, represent it in numbers, or confine it in Euclidian spaces, water transforms and slips into impermanence; reservoirs rise and fall, winter comes along and the stuff turns to ice, sublimates, and gets spirited away on the first available breeze. Even H_2O, that pregnant compound that emerged from the eighteenth-century laboratory of French chemist Antoine Lavoisier is shockingly promiscuous – it goes and bonds with practically everything once it escapes the lab![4]

We will be considering water primarily as a process rather than a thing. The "water process" is that out of which every specific instance of water gets abstracted, including scientific representations such as H_2O. On this view, things such as H_2O do not constitute the fundamental reality of water but, rather, are fixations that occur at the nexus of the water process and the social process of producing and representing scientific knowledge. The stability of such representations of water, moreover, is contingent on these social processes. Every instance of water that we can think of occurs as a product of the water process and various kinds of social processes and practices. It is in this sense that we discuss the *social nature of water* – not that society produces water per se, but that every instance of water that has significance for us is saturated with the ideas, meanings, values, and potentials that we have conferred upon it.[5]

The water process has a remarkable capacity to connect things. The ancient Romans didn't go to the public baths primarily to wash their private bodies of the dirt of the city; they went to the baths to cement a civic bond. "Cleanliness was a shared civic experience," writes sociologist Richard Sennett, "and a public bath was the most popular building a ruler could erect. The baths mixed the enormous diversity of the city together in a common nakedness."[6] But just as it has the capacity to dissolve things in a common solution, water is able to undo the world completely, even if only to allow it to be put together again in some new and improved way: The significance of the biblical flood was that it ended after forty days and forty nights, allowing things to get established on a sounder footing thereafter. The significance of baptism by submersion is in receiving the initiate as born again upon his or her emergence from beneath the surface. Dissolving and resolving ourselves in water, we partake in a dialectical process of solvent and solution in which emergent and diverse capacities of water, and people, come into relevance. It is this dialectic that I want to explore here, ultimately with the aim of arguing that society and water can be understood to make each other, a process by which both water and society are changed.

I direct this argument to a broad audience. Specifically – as might have been inferred from the liberal use of the pronoun "we" in the paragraphs above – it is addressed to those who, like me, understand themselves to be the heirs and critics of a broad tradition described as Western thought. This tradition may be considered as both the cause and effect of narratives of intellectual development through the various epochs of Western history, described by the philosopher Richard Tarnas as "ancient and Classical Greece, the Hellenistic era and imperial Rome, Judaism and the rise of Christianity, the Catholic Church and the Middle Ages, the Renaissance, Reformation, and Scientific Revolution, the Enlightenment and Romanticism and onward to our own compelling time." The argument that this is a tradition "whose sum and consequence we all bear within ourselves" will be relevant to most readers,[7] and I would add that it is an argument only strengthened by the vehemence with which we may reject the tradition. In the broadest sense, then, this book is about the idea of water in Western thought. More specifically, it is about how a particular idea of water has attained widespread prominence in recent times, and about the implications and consequences of this idea.

To speak of Western thought is to invoke a wide variety of (often completely opposite) ideas and ways of seeing; in fact, it is impossible to characterize Western thought as a whole. Nonetheless, at any point in history it may be said that certain presumptions about things and about how it is possible to gain knowledge of them predominates. The historian of ideas, R.G. Collingwood, described three key phases in the history of the idea of nature in Western thought: ancient Greek, Renaissance, and modern.[8] Taking a very different approach, the social theorist and philosopher Michel Foucault described a sequence of "regimes of knowledge" or "epistemes" dating from the sixteenth century that he identified as Renaissance, Classical, and Modern.[9] Although Foucault's "archaeology" of these epistemes somewhat eclipsed the history of ideas approach, he nevertheless retained the notion that in any given period certain ways of thinking predominate "in the mainstream of a culture such as ours." Foucault's aim was to excavate these epistemes, "to rediscover on what basis knowledge and theory become possible."[10] In short, although people in every society have simultaneously entertained wonderfully different concepts and ways of thinking, one may nevertheless speak of *predominant* ideas that pertain in any given time and place, and analyze these ideas in ways that consider the reasons for, as well as the effects of, their predominance.

For most of the twentieth century, we generally took water for granted. We held a firm understanding of what water was and what it meant, as well as a certain faith in its material abundance. Today, of course, this is no longer the case. Various factors – reflected in growing concerns about a global water crisis, water scarcity, water pollution, the uneven geographic and social distribution of access to water and to water services, and the potentially disastrous hydrological consequences of climate change – suggest that we no longer take water for granted in a material sense. Rather than fecundity or potentiality, we are now more likely to associate water with scarcity, pollution, war, and crisis. It could be said that "Water and Dreams" – the title of Gaston Bachelard's meditation on the place of water in the Western psyche – has given way to the generalized "water nightmare" of a recent newspaper headline.[11]

No longer taking water for granted in a material sense, we have also begun to *think* about water in a different way. Water is now more complicated than it seemed in the mid-twentieth century. In modern times, water has most commonly been thought of as a resource that could be considered and managed in abstraction from the wider environmental, social, and cultural context(s) in which it occurred. Today, however, water is complicated by (and co-implicated with) these contextual circumstances. It has become far more difficult to think about water in the abstract: We can no longer ignore water's *ecological* dimensions, such as its importance for sustaining healthy terrestrial, as well as aquatic, ecosystems.[12] Nor can we ignore water's *cultural* dimensions, as in the myriad ways that water articulates with people to produce different meanings and different kinds of relationships.[13] And we cannot ignore water's *political* dimensions, as marked by the distribution of economic benefits and affordances associated with particular modes of water governance.[14] These various dimensions have always been present in water, of course. Now, however, the ecological, cultural, and political aspects of water present themselves to us in ways that challenge and defy our abstract understanding of water's nature.

This complication of water, I argue, is at the root of what is often called the water crisis. The water crisis is often understood in terms of things such as water scarcity and water pollution and is presented as something that manifests physically, often at a global scale. There is no question that people in different parts of the world face enormous and urgent problems associated with water. It is a stark fact, for example, that over a billion people lack access to safe sources of drinking water and over two billion people lack adequate sanitation services. But to gather these issues under the rubric of the "global water crisis" diverts attention from the political and social circumstances that produce such problems and frames their solution in predominantly technical and hydrological terms. This book develops the argument that our idea of water needs to be complicated by the fact that in every instance, water bears the traces of its social relations, conditions, and potential. We are indeed facing a water crisis, and this crisis stems from the fact that we can no longer presume a simple identity for water as well as from the facts of water scarcity and water pollution.

In an effort to describe the futility of identifying the fixed truth of things, the philosopher Friedrich Nietzsche described the world as "an unstable foundation of running water."[15] If pinning down the world is so difficult, how much more difficult it must be to fix the flow that Nietzsche invoked to illustrate the very principle of inscrutability. But we do have a penchant for fixing things, and water is no exception. This habit of pinning things down perhaps owes something to our biological inheritance, which predisposes us, like other primates, to "see the environment as a collection of things rather than merely as a pattern."[16] But our tendency to perceive discrete things that relate to one another as separate entities is greatly aided by how we are *taught* to see. How have we been taught to see water? How have we managed to pin down water as a "thing" (for example, a compound of hydrogen and oxygen denoted by the chemical formula "H_2O") despite its rather obvious processual nature? In the chapters that follow, I describe how we have produced the idea of water as an abstraction and have upheld this abstraction as water's essence. "Modern water" is the term I use to describe this way of knowing water and the habits of thought and practice that it helps induce. One virtue of modern water is that it is not complicated by ecological, cultural, or social factors. This has made it relatively easy to manage. Another virtue of modern water is its universality – all waters, in whatever circumstances they may occur, are reducible to this abstraction. A third virtue is its naturalness – not only may all waters be reduced to H_2O but the product of this reduction is understood to constitute water's essence, its basic nature.

Today, we take these virtues quite for granted. After all, our modern understanding of water is perfectly true in its consistency; as a basis for managing water for human society it has been immeasurably beneficial, helping improve health and standards of living for most people, especially in the modern, developed world. However, what we now presume to be the basic nature of water is actually a remarkable accomplishment, one that can be traced through the history of relations between water and people, including the myriad uses to which water has been put, the attachment of various meanings to water, the social differences and conflicts that have been mediated by water, and the ways that water has been made known to (and by) philosophers and scientists. Foucault developed an approach for studying the history of things that are not usually considered to have a history – things like sexuality, conscience, sentiments, and instincts. He called this approach "genealogy."[17] To some extent, this book falls in the genealogical tradition in its investigations of something – water – which, despite the substantial literature in water history, described in Chapter 3, *itself* isn't usually considered to have a history.[18] I have retained the word "history," because Foucault's genealogy is concerned primarily with investigation of the discourses that produce particular forms of human subjectivity – especially how discourses have the effect of objectifying the human subject – rather than aspects, or ideas, of nature per se. Thus, Part 2 of the book deals with the "history" of modern water.

The word "modern" is used here in two related senses. First, it denotes an approximate period of time. Because I have located the more salient historical developments of modern water in the Scientific Revolution (approximately the seventeenth century) and the Enlightenment (approximately the eighteenth century), this may be considered the beginning of the modern period for our purposes. This period extends to the present albeit in attenuated form since the last third or so of the twentieth century, when a growing awareness that a basic rupture in Western thought was underway began to be reflected by changes in the ways people thought about water.[19] Second, "modern" corresponds to a particular way of understanding things. Maria Kaika, a geographer who has investigated the social and material dimensions of flows of water in urban environments, describes modernity in terms of "a new, forward-looking world view and a new set of social expectations." The most significant feature of this world view, as Kaika and others have pointed out, is a sense of the capacity for "human emancipation through the domination of nature."[20] This sense in turn is associated with an intellectual manoeuvre – the Cortesian claim that the human mind is separate from nature, and therefore properly knows it as an object. Thus, modern water implies a way of knowing water that has become dominant, along with the idea that this way of knowing water reveals its true essence.

To speak of the domination of a particular idea or understanding requires some explanation: At any given time and place, different concepts and understandings of water prevail. Even from moment to moment, I might regard water as a source of inspiration, then as a natural resource, and then again as a medium of social relations. Nevertheless, it is possible to speak of a *hegemonic* construction of water in that there is one way of knowing water that has attained general predominance in my own place and time.[21] Because the uses to which water is put, the people in whom authority is vested for its control, and the distribution of benefits that accrue from its allocation are central to any social order, it can be said that such order is maintained through discourses, laws, and tacit as well as formal rules that reinforce certain ideas and meanings of water in any given society. These ideas and meanings, moreover, get fixed in a material sense, as in the concrete engineering and infrastructural works that materialize hydrosocial relations in different places and times.

In a memorable phrase, David Harvey describes how certain kinds of social and natural relations – or connections – become consolidated through "an ecological transformation which requires the reproduction of those relations to sustain it."[22] Thus, all these things – ideas, meanings, laws, concrete fixtures, management techniques – hang together in a way that makes the hegemony of an idea seem natural, at least until confronted with a problem or contradiction that reveals it to be held in place by a web of powerful but ultimately changeable relations.

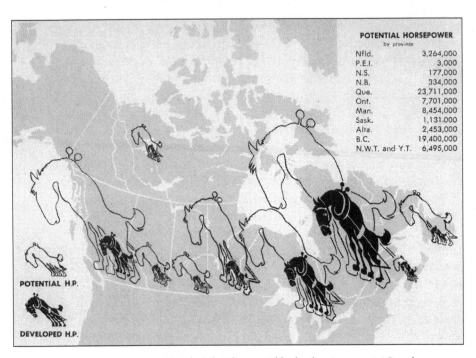

POTENTIAL HORSEPOWER	
by province	
Nfld.	3,264,000
P.E.I.	3,000
N.S.	177,000
N.B.	334,000
Que.	23,711,000
Ont.	7,701,000
Man.	8,454,000
Sask.	1,131.000
Alta.	2,453,000
B.C.	19,400,000
N.W.T. and Y.T.	6,495,000

POTENTIAL H.P.

DEVELOPED H.P.

FIGURE 1.1 Developed and potential hydroelectric power in Canada
J.E. Robbins, ed., *Encyclopedia Canadiana* (Ottawa: Grolier Society of Canada, 1957), 5:214.

Let me give an example from Canada. One of the dominant ways of fixing water (conceptually and materially) in Canada over the past century has been bound up with its capacity to assist people in the production of electricity. This consolidation of water's identity is particular to a certain kind of society, namely one in which people subscribe to ideas of technological advancement, economic development, and centralization of social power (i.e., the state). Perhaps because of its sense of advancement, development, and centralization, our society – at the height of its modernist vision – has thought it only reasonable to project this meaning onto the waters of a vast geographical space, thus representing the entirety of Canada's rivers as a quantity of hydropower just waiting to be harnessed (see Figure 1.1). The materialization of such meaning in the form of hydroelectric dams constitutes an important chapter in late twentieth-century Canadian history, sometimes referred to as "the postwar era of dams and diversions"[23] (see Figure 1.2.). Meanwhile, the dams have literally helped consolidate this meaning of water, just as the dams themselves have been fixed in a growing

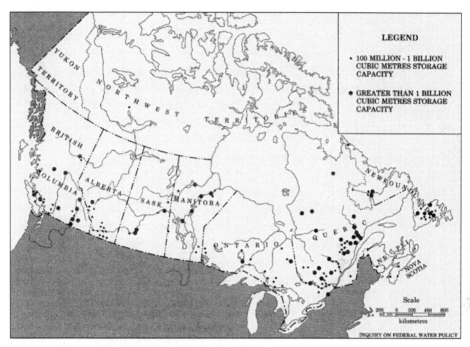

FIGURE 1.2 "Largest dams in Canada," 1985
Most of these dams were built since 1950 and more than 80 percent serve the purpose of generating electricity. | Pearse, Peter, F. Bertrand, and J.W. MacLaren, *Currents of Change: Final Report, Inquiry on Federal Water Policy* (Ottawa: Government of Canada Inquiry of Federal Water Policy, 1985), 32.

entanglement of transmission lines, laws and regulations, transformers, industrial growth, consumer demand for electricity, and discourses that produce nationalist sentiments through the control of northern rivers.[24]

Thus, a particular kind of identity, representation, and material form of water can get caught and held within a web of social and hydrological relations. Such webs can impose a kind of inertia, or "sclerosis" upon the water in question, and upon the society that produces it.[25] This is what I mean by hegemony, and it provides an example of what can be described as the hegemony of modern water. Such ways of imagining, representing, and materializing water might be considered hegemonic when alternative kinds of water are made out to be less real or less legitimate, or when they become so overshadowed that they are made invisible.

A hegemonic construction may enter a period of crisis when alternatives are recognized as being just as real and just as valid. With respect to water, alternative identities and ideas become more apparent as we consider the waters of marginalized groups of people.[26] When government agencies built the dams in the decades following the Second World War, the views of water held by First Nations peoples inhabiting the affected areas of the vast physiographic region known as the Canadian Shield were considered of so little importance that they were left out of the calculations that resulted in the harnessing of rivers for electricity. These alternative waters became visible to the dominant society only when the people who give them life and meaning managed to place them(selves) directly in our line of vision.

The southward journey of the *odeyak* from the shores of James Bay in 1990 provided such a moment of recognition. The odeyak was a hybrid canoe/kayak built by the Cree and the Inuit of northern Quebec.[27] These peoples wished to demonstrate their opposition to the plans of Hydro-Québec (the public hydroelectric utility in the province of Quebec) to build a series of hydro dams on the Great Whale River, located approximately in the region represented by the head of the largest, black horse in Figure 1.1. The dams would have flooded vast territories and drastically altered the ecology of the river and, along with it, the livelihoods and lifeways of the people for whom it was an integral part of their cultures. The construction of the odeyak by residents of Whapmagoostui and Kuujjuaraapik – respectively the Cree and Inuit communities that would have been most heavily impacted by the dam – on the shores of James Bay symbolized the solidarity of the two communities. The craft was taken to Ottawa and from there was paddled by members of the two communities to the Hudson River and eventually to New York City in the spring of 1990. This voyage was intended to bring the Great Whale River, as it was known to these peoples, to the very doorsteps of the people whose consumption of electricity threatened to transform it into a foreign object. As one of the organizers of the voyage put it, "I think you can do that by having the people who live at the mouth of the river build a large paddling canoe to become an ark, a symbol of their way of life, their culture."[28] The campaign was an important contribution to a significant historical change of direction: the premier of Quebec announced a halt to the Great Whale Project in 1994, mainly because Hydro-Quebéc's prospective customers in the northeastern United States – now more aware of the implications of their actions on the waters of northern Quebec – backed away from initial commitments to purchase electricity from the project.

A host of water-related problems, together with glimpses of alternative waters such as were represented by the odeyak, is beginning to change our view of rivers and waters generally. Water is becoming much more complicated than it was in the 1950s and 1960s, when hydro-horsepower seemed to map so naturally onto Canadian territory. Nevertheless, the way we see water remains rather fixed and simplified in our textbooks, in the speeches of our politicians, and in the physical infrastructure of a large portion of the hydrosphere surrounding us. Overcoming the hegemony of modern water involves changes in how we think about water, as well as how we represent, manage, distribute, value, and use it, for all these are closely related.

We can see how particular kinds of water can be held fast in recursive webs of social and natural processes. Because such fixations – like the identity of water as a resource for producing hydroelectricity – are the product of mixing water with social processes, they perform a kind of political work in the sense that they strengthen some social relations while making it difficult for others to establish or sustain themselves. To treat water as an economic resource allows some people to use it as a means to whatever ends they may have the economic and technological capacity to effect. Thus, alternative, potential meanings and relations with water may be ignored or shunted aside, along with the people for whom such meanings and relations are constitutive of life and livelihood. The business of fixing water, in other words, is hardly just an intellectual performance; in each instance, it allows for certain hydrosocial realities while making it difficult or impossible for others to spring to life. The meanings of water that get fixed in any particular time and place can therefore be seen as a function of the relative power of different social actors.

Modern Water

In subsequent chapters, we examine how the modern tendency to fix water got established in scientific discourse, was eventually made invisible to society at large, and thus became hegemonic. Our aim is to make modern water visible in a way that will help us reconstitute waters in different ways. A few introductory remarks are in order.

First, modern water is an intellectual achievement with far-reaching consequences for human society and the environment. By relating the history of this achievement – and thus considering the extent to which it involves human and cultural, rather than purely natural, phenomena – we do not discount its value to modern society. Much of what follows may be taken as a critique of the hydrological sciences because it puts the practice of these sciences in a social and cultural context. However, it is readily acknowledged that the intellectual abstraction of water as a natural quantity has been of primary importance to modern economic development, urbanization, agriculture, transportation, public health, and flood control, to name just a few modern necessities that have been made possible by developments in hydrological science. The question of whether water science might even provide "the basis of civilization" has been raised by scholars working in a variety of fields.[29] Such a contention only underscores the importance of modern water and of critical examination of the knowledge practices that have made it possible.

Modern water can be defined as the dominant, or natural, way of knowing and relating to water, originating in western Europe and North America, and operating on a global scale by the later part of the twentieth century. A moment's reflection will reveal that people do indeed relate to water in many different ways. However by "modern water" I mean to highlight the particular way of knowing and understanding what water is that dominates and pervades modern discourse. In essence, modern water is the presumption that any and all waters can be and should be considered apart from their social and ecological relations and reduced to an abstract quantity. Modern water's historicity is suggested by geographer Derek Gregory when he writes that, toward the end of the nineteenth century, "a new discourse of hydrology and hydraulic engineering emerged which translated 'nature' into mathematical formulae. In these there would be no place for 'local' knowledge, and the hydraulics of irrigation channels and the mechanics of dam construction could be made the same the world over."[30]

The notion of translating "'nature' into mathematical formulae" suggests the predominantly scientific discourse in which modern water has incubated and proliferated. Through this discourse, all water is made known as an abstract, isomorphic, measurable quantity that may be reduced to its fundamental unit – a molecule of H_2O – and represented as the substance that flows in the hydrologic cycle. The scientific nature of (modern) water shows up quite clearly in our standard definitions of the term:

> water 1a: the liquid that descends from the clouds as rain, forms streams, lakes, and seas, issues from the ground in springs, and is a major constituent of all living matter and that when pure consists of an oxide of hydrogen H_2O or $(H_2O)x$ in the proportion of 2 atoms of hydrogen to one atom of oxygen and is an odourless, tasteless, very slightly compressible liquid which appears bluish in thick layers, freezes at 0°C and boils at 100°C, has a

maximum density at 4°C and a high specific heat, contains very small equal concentrations of hydrogen ions and hydroxide ions, reacts neutrally, and constitutes a poor conductor of electricity, a good ionizing agent, and a good solvent.[31]

water 1. Colourless transparent tasteless scentless compound of oxygen and hydrogen in liquid state convertible by heat into steam and by cold into ice, kinds of liquid consisting chiefly of this seen in sea, lake, stream, spring, rain, tears, sweat, saliva, urine, serum, etc.[32]

Water – Water (H_2O) occurs in the atmosphere and above and below the Earth's surface as a liquid, solid or gas. It is continually changing state (e.g., by freezing/thawing, evaporation/condensation) and location (e.g., by gaseous, liquid or glacier flow). All water is involved in a continuous hydrologic cycle, of which evaporation into the atmosphere from oceans, lakes, rivers and land surfaces and transpiration through plant leaves may be considered the first phase. This moisture is transported, often great distances, by winds and is precipitated, as rain or snow, upon water and land surfaces.[33]

Or, to cite a highly popular "authority" on all things, the opening sentence of Wikipedia's main article on "Water" states: "Water is a common chemical substance that is essential to all known forms of life."[34] Water – in its modern guise – is thus made common to every circumstance – geographical, cultural, ecological, and historical – in which we find it. This commonality – or universality – is representable in diagrams and tables that have the effect of reducing all possible waters to a quantifiable substance (see Figure 1.3 and Table 1.1). The flow of modern water through the earth's hydrosphere is represented by means of the ever-popular hydrologic cycle (see Figure 1.4). And to provide one further illustration, Figure 1.5 represents (modern) water in what is perhaps its most abstract form.

Although we can – and shall – trace the path of H_2O back to the late eighteenth century, and the scientific hydrologic cycle to a little over a century later, modern water is suggested in the notion that H_2O and the hydrologic cycle have always existed but simply required the application of proper (scientific) method to be brought to light. Of course, the universality of modern water is true in a sense: Water didn't change its physical properties when the chemists pronounced it a compound. And it didn't behave any differently when the hydrologists put it through the hydrologic cycle. However, to equate this (way of understanding and

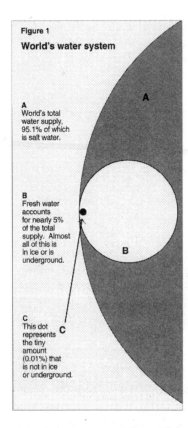

Figure 1

World's water system

A
World's total
water supply,
95.1% of which
is salt water.

A

B
Fresh water
accounts
for nearly 5%
of the total
supply. Almost
all of this is
in ice or is
underground.

B

C
This dot   C
represents
the tiny
amount
(0.01%) that
is not in ice
or underground.

◄ FIGURE 1.3 "World's water system"
"Water – Here, There and Everywhere," Freshwater
Series A-2 (Ottawa: Environment Canada, 1992), 1.
Reproduced with permission of the Department.

FIGURE 1.5 The ultimate abstraction

▼ FIGURE 1.4 "The hydrologic cycle"
National Atlas of the United States, 5 March 2003.
http://nationalatlas.gov. Reproduced with permission
of the National Atlas of the United States.

TABLE I.I

Water reserves on the Earth

Type of water	Area of distribution (km² × 10³)	Volume (km² × 10³)	Water layer (m)	Fraction of total volume of hydrosphere (%)	Fraction of fresh water (%)
World ocean	361,300	1,338,000	3,700	96.5	–
Ground water (gravity and capillary)	134,800	23,400	174	1.7	–
Predominantly fresh ground water	134,800	10,530	78	0.76	30.1
Soil moisture	82,000	16.5	0.2	0.001	0.05
Glaciers and permanent snow cover:	16,227.5	24,064	1,463	1.74	68.7
Antarctica	13,980	21,600	1,546	1.56	61.7
Greenland	1,802.4	2,340	1,298	0.17	6.68
Arctic islands	226.1	83.5	369	0.006	0.24
Mountainous regions	224	40.6	181	0.003	0.12
Ground ice of permafrost zone	21,000	300	14	0.022	0.86
Water in lakes:	2,058.7	176.4	85.7	0.013	–
Fresh	1,236.4	91.0	73.6	0.007	0.26
Salt	822.3	85.4	103.8	0.006	–
Swamp water	2,682.6	11.5	4.28	0.0008	0.03
River stream water	148,800	2.12	0.014	0.0002	0.006
Biological water	510,000	1.12	0.002	0.0001	0.003
Water in the air	510,000	12.9	0.025	0.001	0.04
Total volume of the hydrosphere	510,000	1,386,000	2,718	100	–
Fresh water	148,800	35,029.2	235	2.53	100

Source: I.A. Shiklomanov, "World fresh water resources," in *Water in Crisis: A Guide to the World's Freshwater Resources*, ed. P.H. Gleick (New York and Oxford: Oxford University Press), 13.

representing water's) essence and behaviour with the nature of water itself is to tread modern water. It might well be asked, Was water not always H_2O, even before it was recognized as such? Of course the answer to such a question must be yes. However, the significant point here is that to ask this very question – or to make the assertion that water was always H_2O – is characteristic, even definitive, of modern water. Modern water reduces *all* water to this essential substance, this homogeneous chemical compound, both spatially and temporally. Thus, all water was, is, and always will be H_2O. That we find it necessary to make this assertion is distinctly modern.

By means of its conceptual abstraction, modern water materializes modern *man's* legendary distaste for mud, muck, and swamps of all kinds.[35] Modern water has been a tremendous ally of drainage projects and the creation of hardened shorelines. And just as we like to keep it neat and separate in the physical environment, we like to keep it separate from people too. Even though it flows constantly through our bodies and our psyches, in the modern cosmos, water has been banished to the Cartesian realm of extended substance.[36] Society and modern water are externally related as two independent and intransigent categories; water may be understood as affecting society, and society may be understood as affecting water, but neither may be understood as being fundamentally (internally) changed as a result of these exchanges.

Another characteristic of modern water can be described as its deterritorialization. The philosopher Bernard Kalaora has noted how the conquest of water by means of its conceptual abstraction and technical control has broken relations that otherwise bind specific groups of people to the waters of particular territories. A corollary of the placelessness of modern water (perhaps best symbolized by the tap) is the transfer of water control to placeless discourses of hydrological engineering, infrastructural management, and economics. Kalaora describes this in terms of a "déresponsabilisation" by which we have left all the responsibility for maintaining relations with water to experts.[37] Making a similar argument, Colin Ward describes our water problems in the title of his book, *Reflected in Water: A Crisis of Social Responsibility.*[38] And as described by anthropologist Veronica Strang, water has been dematerialized, rendered "a metaphorical abstraction ... in which it ceases to be particular to any place or group." Thus, deterritorialized and dematerialized, modern water "denies the reality of local, specific human-environmental relationships and alienates the medium through which individuals can identify with a locale and its other inhabitants."[39]

Other characteristics of modern water will surface in the discussion that follows. For now, we might consider how the environmental historian Donald Worster describes the Friant-Kern Canal (a water diversion from the Sierra Mountains through the Great Central Valley of California) as a rough description of our subject:

> Quite simply, the modern canal, unlike a river, is not an ecosystem. It is simplified, abstracted Water, rigidly separated from the earth and firmly directed to raise food, fill pipes, and make money. Along the Friant-Kern Canal, as along many others like it, tall chain-link fences run on either side, sealing the ditch off from stray dogs, children, fishermen (there are no fish anyway), solitary thinkers, lovers, swimmers, loping hungry coyotes, migrating turtles, indeed from all of nature and of human life except the official managerial staff of the federal Bureau of Reclamation. Where the canal passes under highways large, ominous signs are posted: "Stay alive by staying out." The intention of the signs, of course, is to promote public safety by warning the innocent of the dangers of drowning, of being sucked into siphons by the swift current. However their darker effect is to suggest that the contrived world of the irrigation canal is not a place where living things, including humans, are welcome.[40]

This passage illustrates how the characteristics of modern water – its intellectual abstraction, scientific specification, material containment, and alienation from society and from the rest of non-human nature – hang together: the modern idea of water as an objective, homogeneous, ahistorical entity devoid of cultural content is complemented by its physical containment and isolation from people, and reinforced by modern techniques of management that have enabled many of us to survive without having to think much about it. I argue that modern water has entered a critical phase wherein each of these characteristics is recognized as untenable, or unsustainable, and that this crisis is forcing us to think about, and get involved with, water in ways to which we are little accustomed.

Downstream from Here

Chapter 2 provides an outline of the theoretical approach used in this study. One of the virtues of theory is that it helps instigate ways of seeing and understanding that suggest how things might be changed for the better. The approach described here directs attention toward the essential relations between water and society; it provides a way of seeing that allows us to recognize the sense in which water is a social product, and thus the sense in which water can be changed as a result of changes in society. The name of this theoretical approach – relational dialectics – might suggest that it involves a discussion comprehensible only to people with a PhD in philosophy. That it derives from current ideas in the discipline of geography will likely be of only minor consolation. I have made every effort to ensure that the discussion is relatively easy to follow in order to encourage those who are neither geographers nor of a theoretical bent to engage in the argument that informs everything that follows.

Part 2 (Chapters 3 to 8) offers an account of the history of modern water. Here, modern water is seen as an idea that has arisen out of historical circumstances and has contributed to history approximately since the seventeenth century. However, we approach modern water retrospectively, beginning with the present and working our way back. Water is in a critical state and this condition compels us to look for the roots of the crisis. This search – or intimations of the need for such a search – is the main focus of Chapter 3. Here it is shown how, over the past twenty or more years, researchers and writers in many disciplines have engaged in work that stresses the importance of different ways of thinking about water. We survey examples of this work and show that, when taken all together, it points to the need for a direct critique of modern water.

Because it is usually taken to be natural (or timeless), an effective critique of modern water begins by describing its history. This description begins in Chapter 4 by considering an article written by historian of science Christopher Hamlin. Hamlin has identified a transition from an empirical emphasis on diverse "premodern waters," which were regarded as heterogeneous entities exhibiting different properties and qualities, to a modern "essentialist conception of water itself," a transition that appears to have taken place throughout the industrialized world by the end of the nineteenth century.[41] A key moment in this conception of water itself was the naming and representation of water as a chemical compound of hydrogen and oxygen, and we will consider the importance of this moment to the idea of modern water. Furthermore, while agreeing with Hamlin's analysis, we will broaden the argument by describing how the consideration of what might (anachronistically) be called hydrological questions has long effected an essentialist conception of water, and how the emergence of scientific, quantitative, hydrological practice beginning in the late seventeenth century has been instrumental in giving rise to modern water.

In Chapters 5 and 6, we consider the historiography of water science through the concept of the hydrologic cycle. Along with H_2O, the development and dissemination of the concept of the hydrologic cycle represents an important contribution to the idea of abstract, modern water. This contribution, as well as its centrality to the discipline of scientific hydrology, warrants consideration of the hydrologic cycle in some depth. Three different histories of the hydrologic cycle are presented, the first two in Chapter 5 and the third in Chapter 6, each describing a slightly different thing. The first history presented (and the one most commonly told) is that offered by modern-day hydrologists. This is the history of the scientific hydrologic cycle, the one that has always existed and was known intuitively to a host of ancient savants but awaited the application of correct scientific method to be truly revealed for the natural system that it is. The second history involves the story of how natural philosophers in the tradition of natural theology constructed what I call the "sacred hydrologic cycle" to buttress their theological arguments.[42] This history would probably have remained untold had it not been for the geographer Yi-Fu Tuan, who published a unique study on the subject in 1968.[43] He shows that differences between the scientific and the sacred hydrologic cycles, and the difference between the histories that animate them, are explained mainly in terms of the distinctions between the epistemological approaches of modern hydrology and natural theology.

The third history of the hydrologic cycle, offered in Chapter 6, considers how the hydrologic cycle can be regarded as a concept that was deliberately *constructed* in, rather than *revealed* through, scientific practice. The hydrologic cycle that it describes is different from the scientific and sacred versions discussed in Chapter 5; it is a concept whose origin is specific to a particular time and place and yet has succeeded in obliterating its origins by virtue of having been planted firmly in the soil of nature by its progenitor. I call this the Hortonian hydrologic cycle, after Robert E. Horton (1875-1945), the American hydrologist who first presented it in 1931.[44]

In Chapter 7, the links between modern water, the hydrologic cycle, and the modern state are explored. The main intention here is to show how modern water and the modern state are related. Essentially, the state has materially engineered modern water as a resource, while water resources have strengthened the apparatus of the state. This relation is illustrated by showing how the hydrologic cycle was implicated in the mid-twentieth century state-sponsored project to control and regulate the major river basins of the western United States. Following on the discussion of the Hortonian hydrologic cycle in Chapter 6, attention is focused here on the history of modern water as it developed in the United States in the twentieth century. It is noted, however, that similar developments have occurred throughout the modern Western world.

Chapter 8 concludes Part 2 with the historical culmination of modern water – the abstraction and representation of the world's total hydrological stocks and flows, which I call "global water." Global water is latent in modern water, and its assessment was always implicit in the (scientific) hydrologic cycle. The most reliable methods of calculating the world's quantity of water were developed and refined by hydrologists working in the Soviet Union, where the state's needs for aggregated hydrological data were particularly demanding. These methods began to be transferred to Western hydrological practice via international collaborative efforts, the most notable of which was the International Hydrological Decade (1964-74). Although most hydrologists working in the United States and other English-speaking countries were committed to hydrological investigation on the scale of the drainage basin, "global hydrology" eventually became an accepted practice, such that by the 1990s, global water had achieved scientific credibility and growing popular currency in the West.

As modern water's culmination, global water presents a way of knowing and representing water that does not mix well with people. In Part 3 (Chapters 9 to 11), we shift from an historical to a more philosophical register to elaborate on this fundamental incompatibility. The articulation of modern water with people now becomes the focus of the argument. In Chapter 9, I argue that although it was certainly produced in relation to social (i.e., scientific) practice, modern water is nevertheless taken to be entirely independent of social relations. This fictional independence is at the core of what we will consider here in terms of the "constitution" of modern water. The constitutional metaphor is taken from the writings of philosopher Bruno Latour.[45] The main purpose of the chapter is to identify the philosophical commitments that are necessary to produce, from the ephemeral flux of the water process, water's essential, modern identity. By describing its constitution as a set of commitments that evacuate the social content of water, light is shed on the kind of difficulties that arise when attempts are made to reintroduce modern water to society.

These difficulties are the main theme of Chapter 10, which deals substantively with the water crisis, an issue that became a global concern in the 1990s. The argument is that a kind of philosophical crisis is produced when global water is juxtaposed with another abstraction: human population. As abstractions, global water and human population cannot be related dialectically, and so their juxtaposition produces an inevitable crisis of scarcity. In contrast to the way the water crisis is often presented – that is, as a measurable consequence of fixed water supplies and growing human populations – the aim here is to show how modern water itself establishes the epistemological conditions that inevitably give rise to crisis. The crisis presented in this chapter is therefore not the water crisis per se but *the crisis of modern water*.

In Chapter 11, we consider the dominant response to this water crisis. It is argued that this response has been orchestrated by a network of water experts and professionals who have managed only to strengthen modern water in the process of trying to fix it. Such is the inevitable outcome of a way of seeing that continues to regard water fundamentally as an object rather than a process capable of internalizing social relations. Responding to the crisis of modern water – rather than the water crisis – requires that we address the social circumstances that make water what it is in every particular instance.

The reconstitution of water in a way that deliberately recognizes, or restores, its social nature is the subject of the concluding chapter. Here, the term "hydrolectics" is used to describe an approach to water that redefines what it is, in relation to social as well as to hydrological circumstances. In setting out something of a program of research and political action, a distinction is made between analytical and practical hydrolectics. The overall objective of this program is to facilitate the adoption of more flexible, fluid hydrosocial relations. Together, water and people constitute great potential for changing each other in ecologically healthy and socially just ways.

2
Relational Dialectics: Putting Things in Fluid Terms

*Dialectical thinking emphasizes the understanding of
processes, flows, fluxes, and relations over the analysis
of elements, things, structures, and organized systems ...
for dialecticians hold that elements, things, structures,
and systems do not exist outside of or prior to the processes,
flows, and relations that create, sustain, or undermine
them.*

— DAVID HARVEY, JUSTICE, NATURE AND THE
GEOGRAPHY OF DIFFERENCE

*A thing cannot be understood or even talked about
independently of the relations it has with other things.
For example, resources can be defined only in relationship
to the mode of production which seeks to make use of them
and which simultaneously "produces" them through both
the physical and mental activity of the users. There is,
therefore, no such thing as a resource in abstract or a
resource which exists as a "thing in itself."*

— DAVID HARVEY, "POPULATION, RESOURCES,
AND THE IDEOLOGY OF SCIENCE"

*The doctrine of internal relations makes it impossible to
attribute "change" to any actual entity. Every actual entity
is what it is, and is with its definite status in the universe,
determined by its internal relations to other actual entities.*

— ALFRED NORTH WHITEHEAD, PROCESS AND
REALITY: AN ESSAY IN COSMOLOGY

Although the terms "dialectics" and "dialectical" are often used, they frequently carry different meanings. Here we begin with the position, suggested in the quotations that open this chapter, that dialectics sees the world as fundamentally constituted of process, relation, and change. Dialectical analysis thus seeks to make sense of how actual things and states of affairs come to be, how they endure, and how they are transformed. Dialectics provides a way of understanding the flow of history as well as the flow of water, and reveals how these flows are very closely related. I begin with a general introduction to the dialectical tradition, then move on to define the particular approach taken in this study and note how it departs somewhat from the mainstream of dialectical thinking. Even though this discussion draws heavily from ideas within the discipline of geography, it is of interest to anyone concerned with the question of the relationship between nature and society.

Although it has ancient origins, dialectics is often considered to have been brought to modern thought by the eighteenth-century German philosopher Georg Wilhelm Friedrich Hegel (1770-1831).[1] Hegel's dialectic offered a way of explaining the morphology and evolution of human society, including its intellectual, political, material, and economic aspects. Essentially, Hegel described a dynamic process by which ideas arise and are confronted – or contradicted – by opposing ideas. Out of these confrontations new ideas then arise, which are contradicted in turn, giving rise to a continuing process often described in terms of thesis, antithesis, and synthesis. According to Hegel, history is the product of this ongoing dialogue of ideas. A generation later, Karl Marx (1818-83) famously turned the Hegelian dialectic upside down (or right side up, as Friedrich Engels later put it). Marx argued that it was not the confrontation of ideas that produced stages and drove changes in society but, rather, that the material conditions of society constitute "the real ground of history," produce distinctive forms of consciousness, and establish the conditions in which particular ideas arise.[2] According to Marx, it is the contradictions arising out of material conditions – such as the social inequality produced in the capitalist production cycle – that constitute the main driving force of historical change. Out of Marx's inversion of the Hegelian dialectic has emerged the doctrine of historical materialism, by which the material (i.e., economic) conditions and contradictions of society are understood as the major determinants of the (temporal) sequence of historical events.

Over the past half-century or more, dialectical thinking has been favoured mainly by scholars working in the Marxist tradition, who seek to analyze various states of social affairs as necessarily contingent, internally contradictory, and always subject to change. Among these scholars, the geographer David Harvey has particularly influenced my approach, though there are some distinctions between his theoretical approach and the one employed here. Harvey's dialectical investigations flow from his close reading of Marx and have resulted in a rich elaboration and critique of the *spatial* dimensions of capitalism.[3] This critique has been developed into a theoretical orientation intended to explain how the course of history and the production of space are consequences of the logic and structure of capital, particularly the imperative of accumulation. Historical-geographical materialism, as this orientation is known, has been further developed by other theorists working in the Marxist tradition. Among them, Neil Smith has elaborated on the spatial unevenness of wealth and power produced by crises of capital accumulation, and Smith and Erik Swyngedouw have broadened historical-geographical materialism to theorize the production, transformation, and political nature of scale.[4]

Central to all this work is power, the power that governs the social production of space and determines the distribution of resources and wealth. Exponents of historical-geographical materialism locate this power in the economic structure of society – that is, in the capitalist mode of production, and specifically in those who command the disposition of capital. This study contributes in a preliminary way to discussions of the production of space as they apply to water.[5] However, the investigation of modern water that follows is distinguished from historical-geographical-materialist analysis by its emphasis on the role of non-structural factors in the production of modern water. There is some rationale for moving toward a more-than-structuralist position in Harvey's more recent work, but this analysis goes further than Harvey (and most of his followers) in ascribing power to cultural factors. Although Harvey has made some accommodation for the play of power in a Foucauldian sense – that is, the power of discourse – I aim to show how *ideas* can play an important role in the production of water – and of history – and to demonstrate that ideas do have power. This approach differs from the earlier history-of-ideas approach in stressing that ideas occur in relation to other processes, both human and non-human.[6] In attending to the history of the idea of modern water, the following chapters give recognition to the discursive, political, statistic (relating to the state), economic, and ecological processes by which this idea occurs – and which the idea itself helps to sustain. The lesser attention paid to the particular material conditions in which modern water arises is less neglect of the question of power than affirmation of a more catholic understanding of power, one that specifically admits of the power of ideas.

I find Harvey's approach particularly useful in considering what might be described as the organic necessity of ideas – that is, that their constitution is necessarily related to social and natural processes. Harvey describes his own method as a "relational and dialectical approach to things," as "a dialectical and relational approach," as "relational dialectics," and as a "relational conception of dialectics."[7] Emphasis on the *relational* aspect of dialectics is particularly important to this study of water. In essence, a relational approach holds that things become what they are in relation to other things that emerge through an overall process of mutual becoming. As a philosophical theme, relationality can be traced through the development of process philosophy, including ideas from the likes of Heraclitus (ca. 535-475 BC), G.W. Leibnitz (1646-1716), and the mathematician-philosopher Alfred North Whitehead (1861-1947).[8] Harvey draws from these and like-minded thinkers, who offer "strains of thought" that he describes as being parallel to Marxian dialectics.[9] To cite a classic illustration of dialectical relationality (and one to which Harvey refers), Marx described production and consumption as co-constitutive elements (he calls them "moments") of the overall production process:

> Not only is production immediately consumption and consumption immediately production, not only is production a means for consumption and consumption the aim of production, i.e., each supplies the other with its object ... but also, each of them, apart from being immediately the other, and apart from mediating the other, in addition to this creates the other in completing itself, and creates itself as the other ...
>
> The important thing to emphasize here is only that, whether production and consumption are viewed as the activity of one or of many individuals, they appear in any case as moments of one process, in which production is the real point of departure and hence also the predominant moment.[10]

Relational-dialectical analysis considers how things (or "moments") that are often understood to be separate, independent, or self-sufficient actually produce each other in mutually constitutive processes. "There is," as Whitehead asserts, "no entity which enjoys an isolated self-sufficiency of existence."[11] In the process philosophy of Whitehead, this principle of relationality applies literally to everything, as ultimately there is nothing that is not related to everything else through the cosmic process of mutual becoming.[12] Bringing this philosophy down to earth, so to speak, Harvey

(like Marx) is especially interested in describing the relational transformations that occur in specific processes characterizing capitalism, such as
the production process described in the quote above. Another example
can be drawn from Marx's description of the relationship between nature
and society, which were typically understood (at least since Cartesian
metaphysics became hegemonic) as two independent realms, but which
Marx showed to be deeply interrelated: "Man opposes himself to Nature
as one of her own forces, setting in motion arms and legs, head and hands,
the natural forces of his body, in order to appropriate Nature's productions
in a form adapted to his own wants. By thus acting on the external world
and changing it, he at the same time changes his own nature."[13]

This notion of a human-nature dialectic has been highly influential in
studies of the relationship between water and society.[14] But the idea of
something "creating the other in completing itself," to refer to Marx's
description of the relation between production and consumption, may be
applied in a variety of ways that do not necessarily involve people or social
processes. Consider the relationship between water and land, as in how
the physical landscape affects the flow of a river, which over time changes
the landscape, which changes the river, which changes the landscape, and
so on. Or, in fluvial geomorphological terms: "Bed morphology exerts a
control on the processes and it generates a hydraulic environment where
processes are initiated that will feed back onto the form ... Form influences processes that will result in a modification of the form."[15] Or, as the
breaking waves said to the sand, and the sand replied, "I would be nothing
without you."[16] Adding people to this relationship complicates but does
not fundamentally change the nature of the process. Hydrological engineers, for example, may intervene by constructing weirs in rivers and
breakwaters along shorelines. But dialectically speaking, these interventions
are moments in the overall process. Such moments affect rivers and shorelines in ways that may sustain, or change, local economies, transportation
routes, and recreational activities. Such things, in turn, might then be
considered as further moments in an overall process that does not require
an intellectual or discursive distinction between nature and society.

Although dialectics is often associated with processes relating two things,
or moments (such as production and consumption, nature and society,
river and landscape, form and process, waves and sand), dialectical thinking effectively dissolves such dualisms by considering how each term of
the binary is dependent on – and is internally related to – the other.[17] The
idea of internal relations requires further explanation. When Whitehead
asserts "the doctrine of internal relations makes it impossible to attribute

'change' to any actual entity," he means that change – which is *the* condition of the universe – does not happen *to* things but *among* things. Thus, "Every actual entity is what it is, and is with its definite status in the universe, determined by its internal relations to other actual entities."[18] The Newtonian legacy of independent objects careening about in a matrix of absolute space and time makes it somewhat difficult for us to comprehend this idea of internal relations. We often tend to think of relations *between* things – such as billiard balls colliding on a pool table – rather than *among* things – such as the organic play of subatomic particles that produces the billiard balls, the table, and the process by which they are related in a game of billiards.

The doctrine of internal relations emphasizes how things, in other words, do not relate to each other as ready-made, preformed entities (like billiard balls), nor do they emerge from these relations as independent entities. Thus, the very existence of things in a dialectical relationship presupposes the relationship, and the nature of this relationship produces internal change in the things themselves. Bertell Ollman, whose writings on dialectics have influenced Harvey, notes: "In the view which currently dominates the social sciences, things exist *and* undergo change. The two are logically distinct. History is something that happens to things; it is not part of their nature. Hence the difficulty of examining change in subjects from which it has been removed at the start."[19] One aim of relational-dialectical thinking is therefore to consider how things – like water, nature, etc. – that are usually considered in the social (and natural) sciences to be ahistorical – do indeed have a history and may therefore be historicized.

So far we have considered various examples of binary relations. However, the dialectical approach is not necessarily circumscribed by analysis of such dualistic, binary relations.[20] More complex accounts of dialectical processes may involve a plurality of heterogeneous moments comprising a variety of social and natural relations.[21] One frequently encounters such complexity in Harvey's analyses. Consider, for example, how he approaches something as apparently straightforward as a location in space: here he uses a relational and dialectical approach to complicate the usual (modern) idea of Newtonian absolute space, that is, the idea of space as an isomorphic, isometric matrix that discrete things may be said to occupy and in which they may be said to occur:

> An event or a thing at a point in space cannot be understood by appeal to what exists only at that point. It depends upon everything else going on around it (much as all those who enter a room to discuss bring with them

a vast array of experiential data accumulated from the world). A wide variety
of disparate influences swirling over space in the past, present and future
concentrate and congeal at a certain point (e.g., within a conference room)
to define the nature of that point. Identity, in this argument, means something quite different from the sense we have of it from absolute space.[22]

In what Harvey calls "relational space," not only must things be understood
in relation to other things occurring in other places and times but the
identities of the very things themselves are defined by reference to occurrences beyond, or outside, themselves. Something that appears to be very
simple in absolute space – take a bicycle, for example – is vastly complicated when considered as a thing in relational space: A bicycle congeals
out of production processes that relate things such as raw materials, human
labour, imaginative design, finance, and international trade, while at the
same time becoming ingredient to other processes affecting such things
as public health, urban transportation networks, discourses of public space,
sales of automobiles, and emissions of carbon dioxide. The same holds for
any commodity or thing which, understood in relational-dialectical terms,
both embodies and affects socio-natural processes comprising other things
occurring over more or less wide expanses of space and time.[23]

.It might now be clearer why I have found relational dialectics an insightful approach for the study of water. In the sense of its abstract, metric
identity, modern water may be considered much like absolute space. A
relational understanding of water complicates this identity by drawing
attention to the various things and circumstances that, in effect, make
water what it is. By this way of seeing – and to elaborate on a point introduced in Chapter 1 – water is not a thing but, rather, is a process of engagement, made identifiable by water's emergent properties but always
taking form in relation to the entities with which it engages. The water
held in place behind dams in northern Canada is not merely the liquid
H_2O measured in cubic metres that falls through penstocks and turbines
to generate hydroelectricity; this water is held in place by state-run power
utilities, the human labour that is extracted to produce the dams, penstocks, and turbines; abstract hydrological calculations; water management
protocols; discourses linking national identity with the generation of
hydroelectricity; networks of transmission wires; consumer expectations;
construction consortiums; and political discourses, which together have
the effect of fixing it in a particular way. This water is what it is by virtue
of events and processes that transcend the place and time of the water

itself. "To define elements relationally," Harvey notes, "means to interpret them in a way external to direct observation."[24]

Following this definition, we can say that water *itself* is constituted by its relations. To be sure, all water – at least in that part of the universe with which we are familiar – exhibits forms of behaviour that are proper to it, and these properties hold in every known instance of its occurrence. When considering water in a relational-dialectical way, however, we recognize these properties while bracketing them in order to concentrate on what might be called its relational substance. This kind of substance is not constituted prior to the relationships into which water enters but is formed in the very process of relation. In other words, when considered in a relational-dialectical sense, water *itself* clearly is an historical subject.

IDEAS AS MOMENTS IN A RELATIONAL-DIALECTICAL SCHEME

The history of water comprises an important portion (Part 2) of this study. The question of what, if anything, determines this history may be considered here from a theoretical perspective. My argument is that relational dialectics provides a means of theorizing historical processes that are not necessarily guided, or determined, by any particular moment or determining force. Moreover, I will argue that thought and ideas may be included among the forces – including material practices and discourse – that co-produce as well as emerge in historical processes.

Following Marx, Harvey has *generally* asserted the production of material existence to be the major determinant of affairs, a doctrine known as historical materialism. Historical materialism is usually interpreted in a way that allows only a derivative (i.e., superstructural) role for concepts, ideas, representations, and discourse. This interpretation follows from a strict interpretation of Marx's assertion – contra Hegel – that history determines consciousness, rather than the other way around.[25] Harvey, however, offers a more generous interpretation of Marx. Although he has never denied their fundamentally material basis, Harvey allows that concepts can arise in relation to other concepts and can contribute to the historical process as producing agents, even to the point of being considered a material force:

Concepts are "produced" under certain conditions (including a pre-existing set of concepts) while they also have to be seen as producing agents in a social situation ... Insofar as knowledge becomes a material force, the re-structuring which occurs on the conceptual plane can expand throughout the totality of society and ultimately be registered in the economic basis. Movements in the economic basis are likewise registered on the conceptual plane. But ultimately, the latter has to be related to the former if it is to be understood.[26]

Moving to the reconciliation of historical materialism with the emphasis of Foucault and others on language as the critical moment and factor in social processes, Harvey has more recently put forward a relational-dialectical schema that includes discourse as a fundamental moment of the social process. He thus recognizes the effective and affective power of discourse in society without succumbing to the kind of idealism – the "descent into discourse" – that he and many others find troubling in some poststructuralist analyses.[27] On this view, discourse stands with material practices, power, social relations, beliefs, and institutions as a relational moment in the overall social process.[28] Harvey presents a "dialectical cognitive map" of this process (see Figure 2.1) that considerably broadens the materialist interpretation of the production process and of history.

Harvey stresses that this schematic representation needs to be read in a relational-dialectical register, emphasizing that "each moment is consti-tuted as an *internal relation* of the others within the flow of social and material life."[29] Clearly, this provides a more subtle understanding of historical change than the standard binary dialectic (Hegelian or Marxian). This subtlety permits a catholic appreciation of the kinds of forces and contradictions out of which historical change emerges:

FIGURE 2.1 "'Moments' in a cognitive map of the social process"
Based on David Harvey, *Justice, Nature and the Geography of Difference* (Oxford: Blackwell, 1996), 78.

Each major theorist, it is true, tends to appeal to a particular structure of "permanences" (elements) that transfix relations between the various "moments" to give a structured order to a society. And there is often some sort of privileging of one or other moment as the locus of social change. Such a structured system seems to imprison the social process in an iron cage of reproductive circular causality (such as Marx's "logic of capital," Weber's "bureaucratic-technocratic rationality" or Spender's "prisonhouse of language") ... But these causal versions, for which plenty of textual evidence can be found, entail serious misreadings of the much more complex arguments of most major theorists. The more dialectically-minded simultaneously keep open an entirely different level of theorizing."[30]

This different level of theorizing locates potential for change at every moment of the process, or as Harvey puts it, acknowledges that "there is no moment within the social process devoid of the capacity for transformative activity." This suggests a rather unconstrained historical process that emerges as an internally relational dynamic rather than a sequence determined by any particular force or logic. "All of this," Harvey admits, "must sound, and in some respects is, profoundly at odds with traditional views of Marx's historical materialism."[31] Bruce Braun has described this theorizing as "an ontology of multiple flows and forces" that suggests "a heterogeneous socio-ecological field that has no prior or external determination."[32] In this view, there is no reason that the relational-dialectical approach that Harvey sets out cannot produce a theory of process or history free of the need for a particular sort of driving force; in this view, the driving force (if we must insist on such a thing) can be internal only to the process itself. Furthermore, there is no reason that thought and ideas, as well as discourses and concepts, should be excluded from the processes and moments of this dialectic. In sum, relational dialectics offers an approach that regards the development of ideas and concepts as well as discourses and representations as co-related moments of a dynamic process that simultaneously rests on, and produces, material and social (socionatural) relations. In the most straightforward terms relevant to this book, relational dialectics helps us see how the idea of water is internal to what people do with it, and vice versa.

Although water is what we make of it, an emphasis on these makings does not deny the "it" that constrains their possibilities.[33] We can identify the reality of H_2O in the kinds of behaviour that it presents under particular circumstances. For example, the transition from liquid water to ice and vice versa occurs at the temperature of zero degrees Celsius. Although the Celsius scale is a social construction, the liquid-solid phase transition of water is completely independent of people and constitutes a reality that definitely constrains water's (and our) possibilities. One can only imagine, for example, how historical processes might have turned in the colder latitudes were water to freeze at the temperature at which it is densest (i.e., 4 degrees Celsius). In that case, ice would have formed first at the *bottom* of lakes, rivers, and streams, requiring an entirely different strategy of survival for benthic communities living below the surface and radically affecting the economic strategies of the human communities living above. We can say therefore that water is subject to an independent reality that may have nothing to do with people. This kind of reality is usually described as a *property* of water. However, to hold this reality as a relational achievement, it should be understood that such a property is *emergent of* rather than *inherent in* water. In other words, the property does not belong to water but *becomes realized* through water's relation to processes through which its temperature rises and falls. Thus, as Tim Ingold points out,

> The properties of materials, regarded as constituents of an environment, cannot be identified as fixed, essential attributes of things, but are rather processual and relational. They are neither objectively determined nor subjectively imagined but practically experienced. In that sense, every property is a condensed story. To describe the properties of materials is to tell the stories of what happens to them as they flow, mix and mutate.[34]

Another way of considering this is to say that although water is what we make of it, we are not the only forces or actors involved in the process of realization. The history of water is a story of how people have drawn meanings, ideas, representations, and powers from water; but the story is conditioned at every turn by co-constitutive forces such as climate, season, air pressure, geomorphology, and countless other species that engage with water to make it what it is. Physical properties, as well as meanings and representations, are not essential to water, nor are they merely imposed from what is beyond. They are, rather, internalized in the things – such as ice, chemical compounds, and states of meditation – that get made of water.

Nor can we impose *any* identity or behaviour on water in the manner of a pure social construction. Our constructions – as well as our material productions – of water must be constrained by the reality of water's myriad engagements. To illustrate this point briefly, a concept such as the hydro-logic cycle (see Chapters 5 and 6) internalizes the physical properties of water, the hydrological circumstances in the places where the concept has been developed, contemporary ideas about nature, the development of hydrological science, and the uses to which water is put by society.[35] The concept of the hydrologic cycle has changed over time depending on the cultural, social, and geographical circumstances of its construction, but it has necessarily been constrained by the emergent properties of water, at least as these are defined on our planet with its particular physical characteristics. Thus, among the various – and, to modern sensibilities, rather odd – notions of the circulation of water in Western thought, none posits a version of the hydrologic cycle in which liquid surface water flows uphill against gravity, or where water vapour condenses into ice when warmed.[36]

One limitation of relational dialectics – as I have interpreted it so far – is that it does not afford an obvious place in which to locate physical realities such as the emergent physical and chemical properties of H_2O.[37] Although the representation of water as H_2O was constructed in eighteenth- and nineteenth-century scientific practice, this substance has always, and we may presume always will, change from a liquid to a gaseous state when brought to a temperature of 100 degrees Celsius at sea level. As suggested above, this fact conforms to a relational view of substance, in that the specific state of water (as solid, liquid, or gas) is relative to temperature and atmospheric pressure. These properties are nevertheless emergent of water in a way that conditions myriad hydrosocial phenomena – such as industrial processes and the cooking of foods – which enter directly into social relations. The problem of locating the place of nature in Marxian dialectical thinking has been the subject of much theoretical discussion. A number of authors, as Noel Castree points out, have "sought to re-discover nature's materiality in Marx by way of a theoretical reconstruction."[38] One outcome of this reconstruction has been a discursive and analytical shift away from concentration on social construction and the social process per se (the latter of which remains the focus of Harvey's analysis) and toward consideration of all social processes as necessarily involving non-human nature; hence, the recent circulation of the concept of socio-nature and the "socio-natural production process."[39]

As applied by Erik Swyngedouw and others, this concept (and process) marries relational-dialectical thinking with the concept of hybridity. Hybridity is consistent with relational dialectics, as both are rooted in a relational philosophy; both see things as constituted by their relations rather than existing as things in themselves.[40] The idea of the socio-natural hybrid, along with other intentionally confounding concepts such as the cybernetic organism or cyborg and "quasi-objects, quasi-subjects," comes from research in the field of "science studies," or the sociology of scientific knowledge.[41] In essence, the idea of socio-nature reflects a conviction that the separation of society from nature is a constructed or discursive ploy. As a fundamental tenet of modern epistemology, this separation has been very effective in helping to produce objective knowledge. The problem is, as Swyngedouw makes clear, that "once it became hegemonic ... [it] turned from a dominant epistemology to a dominant ontology." It turned, in other words, on the "strong belief that the world was actually ontologically split into things natural and things social."[42]

The idea of hybridity rejoins what has been driven apart by acknowledging that all things – at least so far as they enter into our consciousness, our production of knowledge, and our material practices – are at once social and natural, material and discursive. It allows for the fact of non-human nature without succumbing to the view that this constitutes nature's fundamental, or ultimate, reality: "If we ... maintain a view of dialectics as internal relations, we must insist on the need to transcend the binary formations of 'nature' and 'society' and to develop a new 'language' which maintains the dialectical unity of the process of change as embodied in the thing itself. The things are hybrids or quasi-objects (subjects and objects, material and discursive, natural and social) from the very beginning."[43]

Water is readily understood in this way. Every instance of water that one can think of – whether detected by a space probe on Mars, temporarily held behind a dam in northern Quebec, measured in cubic meters per second at the foot of a glacier, boiling on the stove, or flowing through your body as you read this – combines nature and society, the properties of H_2O, the material practices of people, and the effects of discourse. The articulation of water with human society is particularly salient in urban settings, where, as Matthew Gandy has shown, "the history of cities can be read as a history of water."[44] In a similar vein, Erik Swyngedouw has combined dialectics with hybridity to suggest a research program for studying the historical-geographical production of water as a hybrid entity.[45] Swyngedouw's studies of hydrosocial relations have been applied mainly in para-urban settings but also at larger scales.[46]

Swyngedouw develops the concept of water as a hybrid entity capturing and embodying "processes that are simultaneously material, discursive, and symbolic."[47] Such an understanding allows for investigation of the heterogeneous relations that comprise flows of water:

If I were to capture some urban water in a glass, retrace the networks that brought it there ... [quoting Latour] "I would pass with continuity from the local to the global, from the human to the non-human." These flows would narrate many interrelated tales: of social and political actors and the powerful socio-ecological processes that produce urban and regional spaces; of participation and exclusion; of rates and bankers; of water-borne disease and speculation in water industry related futures and options; of chemical, physical and biological reactions and transformations; of the global hydro-logical cycle and global warming; of uneven geographical development; of the political lobbying and investment strategies of dam builders; of urban land developers; of the knowledge of engineers; of the passage from river to urban reservoir. In sum, my glass of water entails multiple tales of the "city as a hybrid."[48]

Such an account combines "the materialist legacy" of Marx's production process (as elaborated in Smith's concept of the production of nature) with insights from poststructural theory to insist that every "representation of socio-nature is itself inevitably caught in a web of symbolic and discursive meanings."[49] This recognition of the internal (dialectical) relations between the material production of nature and its representational construction is comprehended in Harvey's dialectics of the *social process*.[50] However, for Swyngedouw, who draws on Latour's concept of hybridity, the process is understood as being social and natural from the beginning. The things or moments that are produced out of such a process are simultaneously social and natural quasi-objects or hybrids (see Figure 2.2).

This schema bears a resemblance to Harvey's relational dialectic, par-ticularly considering that "none of the component parts is reducible to the other, yet their constitution arises from the multiple dialectical rela-tions that swirl out from the production process itself. Consequently, the parts are always implicated in the constitution of the 'thing' and are never outside the process of its making."[51] This formulation allows for the in-corporation of hybridity in a way that recognizes the materiality of non-human nature in the process of its production. At the same time, it resolves a serious problem with the concept of hybridity: as deployed in science studies, hybrids come to life in networks of socio-natural relations, but not in a way that makes them subject to political criticism and change. Latour, as Noel Castree points out, "has yet to flesh out precisely what 'politics' might mean in a world of actor-networks."[52] In his synthesis of hybridity and dialectics, Swyngedouw attempts to flesh out what politics means in the world of socio-nature. In effect, the accomplishment here is

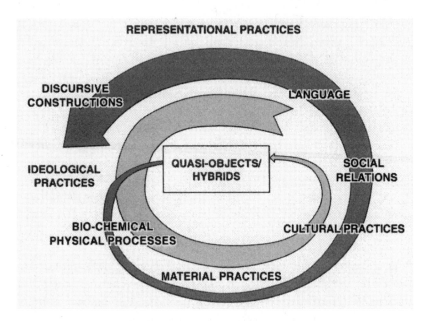

FIGURE 2.2 "The production of socio-nature"
Erik Swyngedouw, *Social Power and the Urbanization of Water: Flows of Power* (Oxford: Oxford University Press, 2004), 22. Reproduced by permission of Oxford University Press.

to place the concept of hybrids – which comprehends the sense in which nature is not socially produced or constructed – in a historical-dialectical framework – which allows for them to be historicized in narrative accounts, as well as actuated to produce historical change.

Finally, we need to allow a place for thought and ideas in the production of socio-nature. The ideas and concepts that people have formed of water in different places and times have been crucial in determining what they do with it. These ideas are not necessarily and only derived from material processes. Many emerge from rather more thoughtful engagements with water. The claim made by Herman Melville (noted in Chapter 1) that "meditation and water are wedded forever," resonates because so many people have experienced such moments of contemplation. Considering the frequency with which watery metaphors flow in language, we might say that people think *with* water as well as *about* water, and these processes are often impossible to disentangle. Of course, the ideas that people form of and with water are dependent on historical-geographical circumstances: people inhabiting deserts have tended to form ideas (and metaphors) that differ markedly from those formed by people living in humid regions. All ideas of water are hybrids in the sense that they are at once social and natural, internalizing the emergent – that is, historical and geographical – properties of H_2O along with the historical and geographical circumstances of the thinker. Nevertheless – and at the same time – the development of a water-borne idea takes on a life of its own and itself becomes a force in what we call the history of water.

Most formulations of relational dialectics allow ideas only a secondary or derivative role. This circumscription of the place of ideas is a corollary of the materialist (Marxist) foundation on which the socio-natural criticism reviewed above rests. As the basis of most contemporary dialectical thinking, materialism assigns ideas to a superstructural sphere, as phenomena that are ultimately determined by the material and metabolic processes of social production. "Clearly," as Swyngedouw summarizes it, "any materialist approach insists that 'nature' is an integral part of the 'metabolism' of social life. Social relations operate in and through metabolizing the 'natural' environment and transform both society and nature."[53]

Using the concept of metabolism as a metaphor for the production of socio-nature in effect precludes ideas from the analysis as moments in their own right.[54] Metabolism describes a material process, specifically, according to the *Oxford English Dictionary*, "the chemical processes that occur within a living organism in order to maintain life." Marx borrowed the term from the agricultural chemist Justus von Leibig (who used it to describe how the growth rates of plants varied with the nutrient content of soils) because he found it useful in arguing, against Malthus, that the productivity of soil varies depending on historical circumstances such as methods of cultivation and techniques of land management. Thus, something that the classical economists took to be fixed in nature was shown to be a variable outcome of social relations.[55] Furthermore, as Swyngedouw points out, Marx used the term "metabolism" to describe the relationship between humanity and nature (i.e., the labour process) and held that "this socio-natural metabolism is the foundation of history."[56]

Yet, the possibility for history is located in *other kinds of relations* operating in realms that are not exclusively material, though they may always have a material aspect. The relationship between people and astronomical bodies, for example, can hardly be described as metabolic. Nevertheless, ideas spawned by the observation and contemplation of the heavens or outer space have been among the most powerful forces in history. Think, for example, of the historical consequences of the Copernican revolution and the effects on Western society of the idea of the heliocentric solar system.[57] The point is that we don't just metabolize (and represent) non-human nature; we engage (with) it in thought. All these modes of engagement, moreover, may be considered aspects of the same synthetic process – a process in which heterogeneous entities, including people, are brought into relation and by which products, representations, and ideas are formed, only to serve, in their own turn, to fuel the ongoing process.[58] The idea that everything is related in process is articulated most completely by A.N. Whitehead. Whitehead's process philosophy offers a systematic elaboration of relational thinking, and it is apposite that David Harvey draws from this philosophy for his own elaboration of relationality.[59] Whitehead, however, was far from a strict materialist, and his work provides us with a solid basis for asserting the place of ideas in the production of nature as well as in the production of history:

> In ethical ideals we find the supreme example of consciously formulated ideas acting as a driving force effecting transitions from social state to social state. Such ideas are at once gadflies irritating, and beacons luring, the victims among whom they dwell. The conscious agency of such ideas should be contrasted with senseless forces, floods ... and mechanical devices. The great transitions are due to a coincidence of forces derived from both sides of the world, its physical and its spiritual natures. Mere physical nature lets loose a flood, but it requires intelligence to provide a system of irrigation.[60]

This so-called coincidence of the physical and the intellectual and spiritual in the production of history can be taken to illustrate a kind of hybridity, whereby the things of the world emerge in a definite, but non-deterministic, fashion. Whitehead's use of irrigation to exemplify this coincidence suggests a dialectic of nature-society that hinges on internal relations and provides a contrast to Karl Wittfogel's elaboration of the external relations of water and society[61] (see pages 64-65). The example of irrigation also illustrates how the place of ideas in history may be asserted without reverting to a doctrine of strict idealism. Ideas emerge in the context of actualities – such as floods – and are realized in the marriage of the spiritual and the physical. In this sense, thought hardly occurs in a realm of its own. Thought is, rather, ingredient to things; it comes neither before nor after "reality" but is constitutive of and constituted by the process that produces the various moments of which reality is constituted. To paraphrase Whitehead, the reality is the process. It is nonsense to ask whether ideas are real – ideas are ingredient to the process of realization.[62] Ideas and concepts may therefore be seen as moments in a relational-dialectical process that cannot be considered to occur independently of their relations with the materiality of non-human nature, the material practices of people, discourse, representation, and so on. Ideas might therefore be understood as hybrid entities inasmuch as every idea forms through a process enfolding humans and the environment, or nature.

The history of water, therefore, cannot simply be traced as an extended outcome of material relations. For example (to complete an illustration begun in the previous chapter), the Cree and the Inuit of northern Quebec protested the incompatibility of their material relations of production with dams on the Great Whale River and placed an idea of water – their water – against the dominant idea and material interests of hydroelectrical producers and consumers in North America. In doing so, they contributed to a shift in the ways that the waters of their region are seen and understood by many people in North America and elsewhere. And this has, in turn, had a material effect on the rivers of northern Canada. This play of ideas was certainly bound up with material and discursive contradictions involving water that had been occurring for some time and over a much wider space than northeastern North America. But as moments in the dialectical process of northern Quebec, these ideas have helped realize the history of the region.

To summarize, my approach in this study draws from David Harvey's formulation of relational dialectics and Erik Swyngedouw's incorporation of hybridity into a dialectical-material framework. However, instead of setting on a materialist foundation the process by which nature and history are produced, I argue that it rests on no particular foundation at all. Instead, I take the process itself as foundational and avoid the need to look behind or below it to find a particular driving or determining force. This provides a useful way of analyzing both the history of water and how the idea of water articulates with its material and representative forms to produce this history.

CONCLUSION: PUTTING THINGS IN FLUID TERMS

Relational dialectics offers a compelling way of seeing things, especially in light of events and intellectual currents that reverberate and flow through contemporary society. Like all things understood dialectically, new ideas emerge in relation to social and natural processes, and these processes are affected by the emergence of new ideas. Ideas crystallize out of the socionatural flux, change society and nature, and help forge a new set of socionatural circumstances out of which new ideas are formed. There is no beginning to this process that we can know of, and no end that we can hope for. So let's jump in, considering more carefully the distinction that has already been alluded to with respect to the idea/ology that nature and society are separate, distinct entities. "This separation," as William Leiss points out, "is as old as human society itself; but only in Western society was it elevated to the status of a conscious principle for the orientation of human behaviour."[63]

This principle became methodically entrenched with the development of what is now known as scientific practice in western Europe beginning approximately in the late sixteenth century. What historians call the Scientific Revolution represents a more or less deliberate intellectual disentanglement from an older, premodern engagement with the world. In this older world, the nature of things was revealed to people through the study of ancient texts, as well as through the identification of relations, resemblances, and correspondences between things.[64] Thus, the prevalence of astrology – the sense that our very selves and life adventures were directly attuned to the cosmos such that even the movement of the stars and planets was part of our family of relations. Out of this world an intellectual revolution was effected by men who challenged the authority of ancient texts and derided the superstition of cosmological correspondences and sympathies. They invented and applied methods to pursue, probe, and torture nature so as to reveal its secrets.[65] That these men saw themselves as having to stand apart from nature in order to know it was to adopt a position that we now take as being perfectly normal. That they saw nature as a separate and distinct realm – the "other" to man and society, an innocence that could be laid bare by the application of formal methods and incisive tools – was to produce a world that has become our material and intellectual environment.

After a very successful run, the world that has been produced by the conceptual, material, and discursive tools of bisection and individuation has begun to exhibit anomalies – in a world that is simultaneously social and natural – that threaten to disintegrate it. Many observers have drawn attention to these anomalies and to the fundamental challenges they pose to modern thought as well as to the sustainability of modern society. Indications of a major shift in the modern intellectual firmament were apparent to some by the late nineteenth and early twentieth century. By 1864, for example, the diplomat and geographer G.P Marsh had identified humankind as a geomorphological agent in his classic work *Man and Nature: Or Physical Geography as Modified by Human Action.* The bedrock of an eternal external nature was further destabilized by investigations of physical reality at scales ranging from the cosmic to the subatomic. By the 1920s, A.N. Whitehead (whose ideas, as discussed above, have contributed to the argument presented in this chapter) reported the prevalence of "a muddled state of mind" arising from the revelation of a plastic nature that did not fit the static model of what nature was supposed to be. "The increased plasticity of the environment for mankind, resulting from the advances in scientific technology," he wrote, "is being construed in terms of habits of thought which find their justification in the theory of a fixed environment."[66]

Whitehead did much to unfix this mentality by, among other things, interpreting theories of relativity and quantum mechanics for a wider audience than the community of theoretical physicists within which such ideas formed. It has been in part because of the changing "habits of thought" occasioned by inuring ourselves to such a fluctuating, uncertain, and chaotic physical world that nature's plasticity is now being accepted in a conceptual as well as a material sense, and that our views of ourselves and our own nature are changing accordingly. Instead of an imperturbable realm, nature is increasingly recognized as a process of self-reflection, a kind of engagement in which are constitutionally implicated. For several decades now, the idea of this engagement has been discernible in a wide range of academic fields, as suggested by environmental historian William Cronon:

The work of literary scholars, anthropologists, cultural historians, and critical theorists over the past several decades has yielded abundant evidence that "nature" is not nearly so natural as it seems. Instead, it is a profoundly human construction. This is not to say the nonhuman world is somehow unreal or a mere figment of our imaginations – far from it. But the way we describe and understand that world is so entangled with our own values and assumptions that the two can never be fully separated. What we mean when we use the word "nature" says as much about ourselves as about the things we label with that word.[67]

The relational-dialectical approach outlined in this chapter provides a useful way of seeing that there is a great deal of ourselves to be found in water. Ever since Narcissus, we have tended to mistake water for something else – something other than a reflection of ourselves. That we can learn a great deal about ourselves from water is a corollary of water's social nature. With this in mind, in the next chapter we consider what can be learned from recent scholarly research on water and on the growing extent and depth of problems that have arisen where water and people meet.

PART 2

The History of Modern Water

3
Intimations of Modern Water

*In light of the real or perceived risks of water crises, a
review of the way in which the hydrological cycle, water
management, water politics, and water economics are
understood and theorized is long overdue.*

– ERIK SWYNGEDOUW, SOCIAL POWER AND THE
URBANIZATION OF WATER: FLOWS OF POWER

INTRODUCTION: WHAT IS WATER?

The main purpose of this chapter is to legitimize the central question
of our enquiry. "What is water?" might seem an unnecessary or even
perverse question in light of the very real and urgent water-related problems
faced by so many people today. But it is the very reality and urgency of
these problems that compels us to ask such a question. Over the past
couple of decades, researchers in a wide variety of fields have given their
attention to problems involving water. The basic argument presented
below is that, taken all together, this research suggests, or intimates, an
urgent need to rethink the nature of water – in other words, to ask plain-
ly, What is water?

One indication of the need to raise this question can be found in the
work of water experts, academics, and critics who have identified various
aspects of a fundamental shift taking place in the way water is understood:
Peter Gleick, a water expert who gained world renown for, among other
things, documenting the world's water crisis in the 1990s, has described
"the changing water paradigm" by which water problems and solutions
are becoming understood in a way that differs markedly from the predominant paradigm of the twentieth century.[1] Geographer Karen Bakker
has identified a shift from the state-hydraulic mode of water governance
to what she describes as a market-environmentalist approach, by which
regimes of water services and management have changed in different parts

of the world.[2] Meanwhile, historians and critics have been busy providing accounts of water as "both a natural and a social reality, which challenges traditional understandings in both the natural and social sciences."[3] This sense of the need to question water's identity and recognize its social reality is further evidenced in the 2008 appearance a new scholarly journal; as the editors of *Water Alternatives* point out, historical circumstances increasingly call for investigation of "the institutional, social and political dimensions of the water-society nexus."[4]

In this chapter, we examine these and other indications of change in the way water is understood in the modern, Western-influenced world. What most of the observers we examine here have in common is that *they take issue with water itself* – with the mode of conceptualizing and representing water that has dominated modern discourse. Taken all together, the work of water experts and critics alike reflects a radical disjuncture with modern water. However, for the most part, the need to acknowledge modern water, to examine its history, critique its hegemony, and identify its limitations, is only intimated in this work. To rephrase the statement by Erik Swyngedouw that opens this chapter, it seems that a review of the way that *water* is understood and theorized is long overdue. So far as the present study aims to do just that, it might thus be regarded as a kind of response to the various intimations discussed in this chapter.

"The Uncooperative Commodity"

In her study of the privatization of metropolitan water services in England and Wales beginning in 1989, Karen Bakker describes a shift from the "'state hydraulic' paradigm of water regulation, dominant in Britain (and in many countries) throughout much of the twentieth century" to what she calls the "market environmentalist" mode of regulation that swept many of these countries in the 1990s.[5] The former is characterized generally by state or public ownership and management of water and water services; an ethic of social equity in the provision of water; centralized, hierarchical regulatory apparatus; and a discursive representation of water as a resource. The market-environmentalist mode, on the other hand, involves devolution of state ownership and control, a discursive shift toward water as a commodity and economic good, and a corresponding re-fashioning of the ethics of water-service provision. The provision of water as a public good by state-owned-and-managed utilities has come under intense pressure, as it can hardly sustain (or contain) the multiple demands – financial, environmental, health, and commercial – that are now being placed on urban water and sewerage services. The state-hydraulic mode thus appears to have been superseded in many places by the market-environmentalist mode, which presumes the market to be the best vehicle for allocating water and organizing the provision of water services.

But the shift from one paradigm to another is anything but a clean one, and it is for this reason that Bakker has described water as an "uncooperative commodity."[6] Water, it seems, can no more easily be contained as a commodity than as a public good. Bakker makes it clear that we need to think of water in terms far more complex than can be represented by any single identity: that water is deemed a matter of life and death – essential for human life, health, and well-being – constitutes a moral obstacle to its pure commodification; that water and water services are of such immense and intimate importance encourages peoples' participation in water decisions, which militates against leaving such decisions to the market alone. Moreover, the physical qualities of water are hardly conducive to market purity. That water is "one of the heaviest substances mobilized by human beings in their daily search for subsistence"[7] necessitates heavy fixed infrastructure costs wherever quantities of liquid water are to be transported. In combination with the systemic nature of water distribution and sewerage collection systems, the burden of these costs produces natural monopoly conditions, whereby water supply tends to devolve to a single firm. Furthermore, that water serves as a medium for ecological functions means that its management must always entail consideration of positive and negative externalities – for example, water's role in supporting aquatic ecosystem functions, as well as destabilizing or degrading human and broader ecological health through things such as pollution. The weight of these externalities complicates the pricing of water and water services in a way that makes it extremely difficult to manage and sell them as a simple commodity.

For a variety of reasons, therefore – practical, economic, ecological, moral, and political – Bakker concludes that commercializing water is far more difficult than many had imagined. Indeed, water's refusal to stay put within the confines of the supply-demand curve has dashed the hopes of market fundamentalists, as governments have found it necessary to re-regulate the sector in the wake of initial action to deregulate and privatize. It has been necessary, in other words, for government to take a big step (back) into the water to ensure good water governance. In such circumstances, Bakker argues, the dichotomy of public versus private is not a very helpful model for understanding what is actually taking place.

Some providers of water services and community activists have pointed out that in many places around the world, what is emerging out of the initial excesses (and failures) of privatization during the 1990s is a plethora of "alternatives to both privatized water delivery and inadequate, state run utilities."[8] They describe these alternatives in terms of "people-centered public water solutions," one of the most remarkable features of which is their heterogeneity. The diversity of approaches and means to deliver water services might be recognized as a function of their embeddedness within particular communities. The picture that seems to be emerging in most places is that the state-hydraulic paradigm has given way, or is giving way, to no single clear-cut alternative, and that water's identity is very much in flux. As suggested, the very idea of a clean paradigm shift, by which one hydrological identity (water as a state-owned and managed resource) is exchanged for another (water as a commodity) is made possible by the presumption of modern water. That water can be sustained neither as purely one nor the other attests to its fundamental complexity and suggests the basic problem with our predominant, yet for the most part unconsciously held, idea of water itself.

Water may be regarded as uncooperative in various ways. For example, water has long been and remains an uncooperative resource in the sense that its multifunctionality complicates and eventually dooms every exclusive method or means by which people have attempted to manage it. As discussed in Chapter 11, the elaboration and practice of integrated water resources management is only the latest effort to reconcile the growing awareness of water's fundamental complexity with its rather one-dimensional identity as a resource. Water might even be considered an uncooperative compound, considering its proclivity for dissolving things. For our purposes, therefore, the main significance of acknowledging the "uncooperative commodity" is in recognizing water's essential non-essentialness. The more we consider how ecosystems function, how the social outcomes derived from water and water services are uneven, and how people in different places and circumstances relate differently to water, the more difficult it becomes to sustain any simple, positive identity for water, whether as commodity, resource, public good, or chemical compound. Bakker's work makes an important contribution to the general complication of water, a critique that suggests that the very foundation of our hydrological discourse – water itself – needs closer examination.

The Demo(li)tion of Dams and the Changing Water Paradigm

Although scholars such as Bakker are helping raise awareness of water's fundamental complexity (and modern water's limitations), the changing cultural status of dams and other water control structures intimates something of a large-scale, popular revulsion against its hegemony. Such structures represent and reproduce modern water – they are indeed concrete abstractions – and the extent to which they have been criticized, demoted, and in some instances decommissioned in recent years, intimates another challenge to modern water.[9]

In the 1950s, India's first prime minister, Jawaharlal Nehru, famously described dams as "temples of modernity." Few structures symbolize modernity as do large dams (dams exceeding fifteen metres in height from foundation to crest). Although not entirely a modern historical phenomenon, all but 5,000 of the world's over 45,000 large dams have been built since 1950.[10] Writer and river activist Patrick McCully has justly described large dams as "much more than simply machines to generate electricity and store water. They are concrete, rock and earth expressions of the dominant ideology of the technological age: icons of economic development and scientific progress."[11] That dams have been the subject of extensive and protracted criticism, beginning in the 1980s (around the same time that the state-hydraulic paradigm began to fall apart), suggests a breech in modern water that no Hans Brinker with a quick finger could hope to staunch.

By the late 1960s, evidence had accumulated showing that the construction and operation of modern water control structures "could have an observable effect upon soil moisture, sediment movement, water quality, or other aspects of natural water resources as well as upon social systems."[12] By the early 1980s, concerns about the disruptive ecological, cultural, and socio-economic implications of the large-scale manipulation of water were becoming impossible for water managers and state agencies, as well as the public, to ignore. In 1984, the publication of Goldsmith and Hildyard's monumental study *The Social and Environmental Effects of Large Dams* precipitated a flood of concern about the effects of large dams that continues to flow.[13] Goldsmith and Hildyard assembled an enormous volume of data "demonstrating that big dams and water projects have not only failed to achieve [their] basic objectives but are also leaving a legacy of unsurpassed cultural destruction, disease, and environmental damage."[14]

The judgment that large dams have failed to achieve their basic objectives doesn't really hold water. The water held by large dams irrigates 30 to
40 percent of the world's irrigated land and generates a fifth of the world's
electricity.[15] Dams have also improved navigation, helped control flooding,
and regulated flows for industrial and other productive purposes. If anything, it is because they *have* achieved their objectives on such a grand
scale that large dams have come under intense scrutiny. Following the
Goldsmith and Hildyard study, there was an outpouring of critical work
in the late 1980s and 1990s, ranging from global assessments of the environmental and social costs of large dams and water control techniques
to focused investigations of their social and environmental impacts in the
western United States, northern Quebec, and India, among other places.[16]
Today, it is estimated that large dams have been so effective in storing
water that somewhere between 40 and 80 million people have been dis-
placed by artificial reservoirs.[17]

Dams are instrumental to, and symbolic of, what has been described as
the modern, twentieth-century water management paradigm.[18] This (modern water) paradigm is characterized by three main features. First, it is
coincident with a period of enormous increases in water withdrawals.[19]
Over the course of the twentieth century, worldwide withdrawals of water
from lakes, rivers, and subterranean sources increased approximately seven-
fold.[20] Over the same period, the world's human population approximately tripled, making the increase in water use over twice that of
population growth. Second, the paradigm is characterized by an emphasis
on physical solutions to water resource problems, specifically by engineering increases in water supply to meet anticipated growth in demand. In
general, water resource problems have been reduced to finding the physical means to furnish users – farmers, industries, cities, households – with
additional water supplies. The conventional strategy of water management
has been to calculate the expected growth in demand for water and to
match expectations with increased water supplies: "The water management
problem then becomes an exercise in coming up with ways of bridging
this anticipated gap."[21] Bridging the gap in this fashion has necessarily
entailed the construction of physical infrastructure (dams, diversions,
canals, pipelines, and so on) to bring supplies of water to the location of
use. A similar strategy has been applied for managing apparent surfeits of
water, as in flood control and drainage projects. Here, too, physical solutions (control dams, levees, drainage networks) have been engineered to
address what have been regarded as problems that are physical in nature.[22]
Third, the traditional water paradigm is strongly associated with central

state planning, funding, and administration of water projects. The centralization and concentration of state control over water in the twentieth century was evident in all parts of the world, including regimes as diverse as the United States, post-colonial India, and the former Soviet Union.[23]

By the early 1990s, it was becoming obvious that a major shift was underway in the management of water resources. Outside of China, the rate of construction of new large dams – which peaked at 5,418 in the 1970s – began to fall in the 1980s, and this trend continued in the 1990s, when (only!) 2,069 large dams were commissioned worldwide.[24] In the United States – where the big-dam era began in the 1930s – more old dams were being decommissioned (for economic and environmental reasons) in the early years of the twenty-first century than new ones were being built.[25]

Globally, the rate of increase in water withdrawals began to fall in the 1980s, with actual reductions of total withdrawals occurring in some places – most notably in the United States – beginning around this time.[26] The reasons for these changes were several: By the mid-1980s, physical resource constraints were apparent in many parts of the world where water supplies were already being exploited to capacity. The marginal cost of increasing supplies grew enormously, and it became a truism among water management professionals that the next water project will cost twice as much as the last water project. In an era of growing fiscal restraint and reduced state investment in public works, these costs became prohibitive for most governments, and multilateral lending (which has financed most large water projects in developing countries) fell throughout the 1980s and 1990s. Second, as already noted, the ecological costs of manipulating water (which had been considered negligible) were widely recognized beginning in the 1970s and became more salient with the growing popularity of the environmental movement. Third, the economic efficiency of water use was increased, particularly in the wealthier countries, as water-saving technologies were applied in agricultural, mining, and industrial sectors and as water began to be allocated from less to more economically productive sectors. Although it is too early to declare the end of the large dam era – especially in China and India – there seems little question that the heyday of water megaprojects is a thing of the past in many parts of the world.

These developments signal what many regard as a paradigmatic change in water management. Instead of always solving water problems (perceived deficits and surfeits) by manipulating flows of water, efforts are being redirected toward "living within water's limits" by improving the efficiency with which available water is used and by adjusting human activity to better accommodate flooding.[27] Scholars with expertise in water management

have described the shift from a period of "water development" before the
1970s and "water management" in the 1970s and 1980s to an emphasis on
"sustainable water management" in the 1990s.[28] Others have stressed the
"need for a change of water paradigm" to promote "truly integrated water
resources management."[29] The new paradigm is focused mainly on rethinking the means of satisfying society's demand for water. Gleick describes
the changing water paradigm, as seen by many enlightened water experts
at the turn of the twentieth century, thus:

> Water resources management approaches around the world are changing
> dramatically. This "changing water paradigm" has many components, in-
> cluding a shift away from sole, or even primary reliance on finding new
> sources of supply to address perceived new demands, a growing emphasis
> on incorporating ecological values into water policy, a re-emphasis on meet-
> ing basic human needs for water services, and a conscious breaking of the
> ties between economic growth and water use. A reliance on physical solu-
> tions continues to dominate traditional planning approaches, but these
> solutions are facing increasing opposition. At the same time, new methods
> are being developed to meet the demands of growing populations without
> requiring major new construction or new large-scale water transfers from
> one region to another. More and more water suppliers and planning agen-
> cies are beginning to explore efficiency improvements, implement options
> for managing demand, and reallocating water among users to reduce pro-
> jected gaps and meet future needs.[30]

The new water paradigm suggests a profound and much-needed critique
of conventional water resources management. In addition to managing
water, the point is now to govern ourselves so that demands for water
resources and human impacts on aquatic ecosystems are mitigated. It could
be said that by shifting attention from water supplies to the uses that
people make of water, attention is now being directed to the *social* side of
water's hydrosocial nature. The changing water paradigm therefore constitutes a serious challenge to modern water.

CONSIDERING OTHER WATERS

An even more fundamental challenge to modern water is made by researchers who consider waters in historical, cultural, and philosophical contexts

that differ from the contexts in which most water managers, administrators, and planners work. The need for a critique of modern water is intimated most strongly in the work of those who attend directly to the "social" nature of water. The social nature of water is hardly a new idea. Anthropologists and other social scientists have long been attentive to how water takes on special meaning and is articulated in particular ways with societies other than those of the modern West. For example, research into what have been described in terms of "traditional irrigation systems," "the local subsistence mode" of irrigation, and "indigenous irrigation systems" has produced a large body of research showing how water is signified, understood, and managed in different hydrological and cultural circumstances.[31] As Groenfeldt has stated of the variety of indigenous irrigation practices in Asia, the physical properties of water are universal, a condition that imposes certain "universal management functions" such as the need to mobilize labour, to allocate water, and to institute methods of resolving disputes. Nevertheless, these "so-called" physical properties of water do not determine a homogeneous social response. "The solutions to these universal problems," he stresses, "are unique to each indigenous system, depending upon the particular social and cultural traditions, the particular physical setting, and the particular individuals concerned."[32] Such research is of interest to us because it implies not only the social nature of water but also the possibility of turning the anthropological lens upon ourselves and the particular social and cultural traditions that have produced *our* unique way of relating to water.

Among these indigenous systems, several classic examples – "classic" because of the frequency with which they appear in the water literature – have become rather familiar: the *qanats,* a groundwater supply technology supporting traditional irrigation systems thought to have originated in Iran; the *subak* system of paddy-field irrigation on the island of Bali in Indonesia; the tank (water harvesting and storage) systems of Sri Lanka and southern India; the *chinampas* (water regulation and artificial island agricultural) system of the pre-Hispanic Valley of Mexico; and the community water tribunals of the *huertas* (intensively farmed garden) regions in southern Spain.[33] As an example, consider how anthropologist Clifford Geertz describes the *subak* of Bali:

> The subak is at once a technological unit, marked out by the collectively owned dam and canal; a physical unit, an expanse of terraced land with a defined border around it; and a social unit, a corporation consisting of

people owning land in that expanse, serviced by the dam and canal. It is also ... a religious unit ... The focus of [the ritual system of water distribution in the subak] is a rice-goddess cult whose precise content [includes] special ceremonies at special times, and specific altars, gods, offerings, and prayers. These various ceremonies are symbolically linked to cultivation in a way which locks the pace of that cultivation into a firm, explicit rhythm.[34]

Subsequently, Stephen Lansing has challenged and revised Geertz's assessment of Balinese irrigation systems, putting less emphasis on the autonomy of the relatively small *subaks* and stressing instead the role of the regional water temples in integrating water allocation and social interaction:[35]

The social units that set cropping patterns and irrigation schedules are usually not individual subaks but regional water temples ... thus the practical management of irrigation is embedded within the hierarchical structure of the water temples ... Altogether it is clear that the productive process involves intricate systems of social control extending over hundreds, even thousands of hectares of irrigated fields.[36]

Despite their differences, both Geertz's and Lansing's ethnographies of water describe a society in which human-water relations define people and water in a way that renders them materially and conceptually inseparable. In these and other studies, the very presence, allocation, and disposition of water cannot be understood apart from the social relations in which it is embedded. Consider what David Mosse says of the tank (small reservoir) system in southern India:

Tank irrigation in Tamil Nadu has a long and complex history throughout which systems of "community management" have expressed and been underwritten by local and regional relations of power and patronage. Inscriptions on tank bunds (crescent-shaped earthen dams) and in temples which go back to the centuries BC ... only demonstrate the political importance of the construction and maintenance of these public resources. Water has always been a political as well as a natural resource, and the operation of tank systems regulating its flow have been influenced by changing configurations of power at both village and state level.[37]

Although the *subak* (or water temple) system and tank system are among the best known examples, researchers have described an array of indigenous hydrosocial relations that have given rise to unique but dynamic and evolving institutions, practices, and material structures that emerge at the nexus of water and people in particular places. John Bennett summarizes some of the findings of historical anthropologists up to the mid-1970s:

> The sheer variety of structures created for obtaining and better utilizing water in prehistoric and early historic societies deserves more attention than it has received from contemporary hydraulic engineers. Such societies developed suitable water control systems for every precipitation pattern in climates ranging from desert aridity to variable humidity, and the functions of many of these structures were misunderstood for generations ... As archaeological work uncovers a greater variety and intensity of water-control systems, few opportunities appear to have been missed by prehistoric peoples ... The variety of these structures, relying on delicate small-scale adjustments to local conditions, contrasts with a modern philosophy, which emphasizes large-scale projects of stereotypic design.[38]

And what is the contrast with modern water philosophy and structures that Bennett alludes to? Although most anthropological research has explored the nature of so-called traditional or indigenous water systems, the unmarked social nature of *modern water* has generally been taken for granted as that *against which all these other systems may be contrasted*. In these investigations, the nature of modern water itself is presumed to be self-evident. That modern water is an indigenous water system that deserves to be investigated in its own right is a useful conclusion we can draw from these studies.

Current interest in "other waters" reflects a growing appreciation for the functionality, sustainability, and efficiency of traditional water technologies, especially in light of the changing water paradigm described above.[39] Implicit in these assessments is a critique of modern water; indigenous water technologies are set against exogenous models and prescriptions for water and land management imposed by central agencies, development institutions, and Western-educated or inspired resource managers. This critique is expressed, for example, in terms of the opposition between ecological and industrial modes of water management, between indigenous irrigation and modern agricultural development or "externally and locally derived soil and water conservation technologies."[40] Although it is important not to idealize these indigenous systems or exaggerate their benefits, the misguided tendency of some state planners and development experts to ignore, and even to target these indigenous,

traditional systems for replacement by modern techniques has been high-
lighted along with the resiliency, despite these incursions, of traditional
water relations and systems.[41] By the mid 1990s, researchers in cultural
ecology and other fields – especially political ecology – had mounted a
powerful challenge to the imposition of exogenous water-management
techniques and to conventional thinking about problems such as desertification, soil erosion, and land management practices in arid regions.[42]

Even the term "water management" implies a particular kind of
hydrosocial relation, one characterized by deference to a kind of abstract
expertise and professionalism. It also implies a particular kind of water,
stripped of its complex social relationships such that it may be managed
by experts who are not necessarily directly involved in these relationships.
A critical examination of water management as discourse gives rise not
only to a recognition of different possible waters but to the hegemony of
"water" as an unmarked category. John Donahue and Barbara Rose Johnston highlight the need to investigate links between how water gets defined,
how it gets managed, and for whose benefit:

> What different cultural meanings does water have for the contending par-
> ties, and how do these meanings complicate mediation among the various
> interests? How are some social actors able to impose their definition of water
> on other social actors with different but equally legitimate definitions? In
> other words, how is power used in the service of one or another of the
> cultural definitions of water?[43]

The question of how power is "used in the service of one or another of
the cultural definitions of water" (and the reverse) has been considered by
anthropologists, political ecologists, and others, providing insights on how
and why some people gain access to and rights over water.[44] Attending to
the variety of meanings of water and the different hydrosocial relations
these meanings entail also has the effect of providing insights into the
hegemony of our own (modern) meaning and relations with water.

POST-COLONIALISM OF IMPERIAL WATER

Modern water's hegemony is also challenged by those who consider different waters in historical context. We consider this broad topic in the
following section. For now, I wish to focus on one strand of water history
that might be described as post-colonial waters. "Imperialism," writes

Edward Said, "after all is an act of geographical violence through which virtually every space in the world is explored, charted, and finally brought under control."[45] Of few things does this ring more true than for water, and it is against what Donald Worster calls "imperial water" that many formerly colonized people continue to wage local struggles to assert more culturally and geographically appropriate hydrosocial relations today.[46]

Bringing water under epistemological and material control in colonial settings was a major imperial undertaking and may be understood as a process that contributed heavily to the development and diffusion of modern water. The colonial state – particularly British India, British and French Egypt, and other European colonial holdings in North Africa – imposed modern water on millions of people, especially in the nineteenth and early twentieth centuries. This imposition had the ancillary effect of producing a corps of hydraulic engineers whose expertise in evacuating the (indigenous) cultural content of water could be, and was, then reapplied in the imperial centre.[47] The work of historians and others to bring these developments to light is worth considering here, as well as research on local efforts to revive or sustain traditional water relations in the face of imperial water. Resistance to imperial water continues in many places in the face of pressures from the post-colonial state working in tandem with international development agencies to promote hydrological moderniza- tion.[48] This work intimates the opportunity if not the need for a focused investigation of modern water as a discourse against which alternative and resistant discourses of water need to contend.

The specific work we consider here approaches water and hydrological management as politically contested or contestable discourses. Much of this work originates in or examines colonial and post-colonial India and is informed by a strong sense of the colonial experience as having produced the dominant hydrological and water-resource discourses.[49] The contrast between traditional and imperial hydrological relations is presented in terms of contrasting discourses – and the contemporary struggle to restore, reclaim, or update traditional water relations is seen in terms of asserting knowledges of and relations with water that directly challenge modern hydrological discourse.

Derek Gregory, who among geographers is one of its more important exponents, has described post-colonialism as a "critical politico-intellectual formation that is centrally concerned with the impact of colonialism and its contestation on the cultures of both colonizing and colonized peoples in the past, and the reproduction and transformation of colonial relations, representations and practices in the present."[50] The constitution of the present in the colonial past is an important theme of post-colonial studies and suggests how this approach may be relevant to an examination of modern water. In the encounter between European and non-European cultures in the colonial context, indigenous relations with water were often liquidated and replaced by Western hydrological discourses (including the water control structures by which they achieved material form). Because the imposition of imperial water upon such people and places had a profound impact on hydrological discourse generally, modern water is in part the legacy of imperialism.[51] Moreover, because colonial water discourses are "still abroad in the world," constituting and informing the ways water and its relations with people are conceptualized and managed, we inhabit the "colonial present," as Gregory has put it in a recent study of a different topic.[52]

Most post-colonial studies have been of the former British Empire, with a notable concentration of interest in British colonial India.[53] It is perhaps no coincidence that British India in the nineteenth century provides the most studied example of imperial water: "the wonderful history of irrigation in India," as it was regarded in the early twentieth century, featured hydraulic works of such "bold and magnificent conception" that they were legendary in their own time, often serving as a model for hydraulic engineers and water planners even in the industrialized world.[54] In the most general sense, over the course of the British Empire in India, these "bold and magnificent" hydraulic works modified or replaced indigenous hydrological relations. In the early years of the empire, there was often a tendency to adapt traditional systems of irrigation to the growing commercial and productive needs of the colonial state. But in order to expand production to supply the home market and raise sufficient state revenues, the imposition of a new hydrological discourse was deemed necessary, as described by Gregory in a passage, part of which has already been noted in Chapter 1:

> For this, water had to be "disenchanted" so that [quoting Worster] "all mystery disappears from its depths, all gods depart, all contemplation of its flow ceases." This involved not only filtering cultural residues from "water" but also replacing them with others. Thus, in the second half of the nineteenth century a new discourse of hydrology and hydraulic engineering emerged which translated "nature" into mathematical formulae. In these there would be no place for "local" knowledge and the hydraulics of irrigation channels and the mechanics of dam construction could be made the same the world over.[55]

The "filtering of cultural residues" is a significant operation and effect of modern water and suggests how an examination of the imposition of imperial hydrological relations constituted modern water's advance guard in the nineteenth century.[56] But the process by which European hydrological discourse engulfed local waters in colonial India was complex. The notion of an equilibrated, harmonious system of water relations disrupted by colonial rule sacrifices important historical nuances to a somewhat romanticized vision of hydrosocial transformation.[57] A highly nuanced history of what might be termed a traditional-imperial water nexus is offered by David Gilmartin. Gilmartin has studied the process by which the British colonial state in the late nineteenth century administered large modern water diversion and distribution canals in the Indus basin, while at the same time allowing, *and indeed relying on,* traditional, heterogeneous, locally administered water systems remaining in place in the downstream communities that received the water.[58] The British described these downstream regimes as "beyond the outlet," with the outlet of the canal representing the boundary between modern and traditional water relations:

> British canal administration was therefore marked by a strong tendency to view the canal outlet as the great theoretical divider of the irrigation system, with a system of rational environmental control operating on one side, and a world of indigenous, customary and kin-based community organization operating on the other. Even as indigenous communities were rigidly excluded from influence over the main, scientific irrigating system, their domination over the disposal of water "beyond the outlet" was largely accepted as an inevitable fact of colonial irrigation.[59]

From Gilmartin's research we learn that, for a time, these structures of local organization survived, and although they were increasingly characterized by instability and internal conflicts,[60] the bonds of local community nevertheless provided a means of water distribution and allocation that the British found useful. Ultimately, however, modern water relations overran the community-based regimes:

> [Their] increasing dependence on relations with an irrigation bureaucracy for securing the most critical of productive resources for the local environment – water – guaranteed that a meaningful definition of the environment that was purely local was impossible, as was, therefore, a structure of encapsulated, "natural," local communities operating in their outlet-defined spaces. By trying to incorporate indigenous "natural" communities into

a larger hydraulic model, the colonial state thus undercut the local environmental foundations for the very local communities that it professed
to rely on.[61]

The purpose of discussing this here is to highlight how research from a
post-colonial perspective has identified distinctions between traditional
and modern water management regimes that are still relevant in many
parts of the world. Donald Worster notes that by means of such critical
analysis, "people became more aware that imperial water was more often
than not water diverted away from traditional agrarian users: Hispanic
farmers or Native Americans in North America; village farmers in other
parts of the world."[62] But the most salient feature of imperial water is
perhaps less the enormity of the physical diversions it comprises than the
transformation of hydrosocial relations it effects. Again, Worster illustrates
how the global disciplining of humanity by means of modern water has
rippled around the world from the imperial centre by quoting Michael
Straus, US president Harry Truman's commissioner of reclamation: Straus
declared that controlling water was "a prerequisite of all development and
elevation of living standards" and vaunted that "the American concept of
comprehensive river basin development ... has seized the world imagination. Yellow, black, and white men of various religions in all manner of
garb are seeking to emulate the American pattern of development."[63]

Modern water is implied or intimated in every critique of imperial
water. To the extent that we inhabit a colonial present, an understanding
of modern water might be of value in strengthening the political relevance
of these critiques.

History and Water

Large dams can be associated with modern water, but it would be incorrect
to conflate dams with modern water. As is widely known, building and
maintaining dams have been common social practice for several millennia.[64] But from a relational-dialectical perspective, it is clear that despite
their physical similarities, dams can be vastly different from one another,
depending on the social and cultural as well as the hydrological circumstances. For example, the (sometimes very large) dams that had been built
and used in pre-colonial India sustained entirely different sets of social
relations and different waters than did the same or similar dams in the era

of British imperialism. This is an example of how a relational way of thinking can be derived from historical investigation, and of how historical investigation can show that the material qualities of a thing may be less essential than its social relations.

One of the surest signs that something is being recognized in a new light is that scholars begin to historicize it. By historicizing something, we recognize the extent to which it becomes what it is in a particular historical (social) context and how it changes as a result of changes in society. Over the past thirty years, for example, investigation of the historicity of things such as gender and race has emphasized the extent to which these are socially constructed – rather than natural – categories. And it is no coincidence that over approximately the same period, gender and race have been among the most hotly contested categories in Western society. When scholars such as Donahue and Johnston ask, "How are some social actors able to impose their definition of water on other social actors with different but equally legitimate definitions?"[65] it signals not only that water is becoming understood as a socially relative phenomenon but that even something as apparently straightforward as identifying the "nature" of water is an intensely political act.

The formation of the International Water History Association in 2001 and the appearance of the journal *Water History* in 2009 indicate that something of an exceptional nature is afoot with respect to the idea of water.[66] Such scholarly attention reflects a growing interest not just in water and water issues per se but in the sense in which water articulates in different ways with people from one social and historical context to another. Most of the hundreds of papers presented at international conferences organized by the association approach water in a way that blends the social and the natural in accounts of historical relations between people and rivers, lakes, aqueducts, networks of pipelines, drainage systems, irrigation ditches, and reservoirs.[67] In their preface to one of the published volumes of these papers, Terje Tvedt and Terje Oestigaard declare:

> Water is both a natural and a social reality, which challenges traditional understandings in both the natural and social sciences. In nature and society water is not a single phenomenon but has many manifestations and meanings. This volume aims to highlight how water has been understood, conceived and socially constructed. It investigates in what ways, how and to what extent, people have understood, conceptualized, and used different types of water as a social, cultural and religious medium.[68]

To be sure, historical investigation of the type reported here does not generally reveal water to be merely a social construction. Rather, it tends to yield "an understanding that fresh water is both an actor in its own right and that at the same time it is impossible to access this reality except through cultural and social lenses."[69]

It might be argued that this represents only a minor change from the way some historians and social scientists have investigated the complexity of water in its articulation with human society. Karl Wittfogel's 1957 book *Oriental Despotism: A Comparative Study of Total Power* is perhaps *the* classic study of the articulation of water and society.[70] Wittfogel describes a dialectical relationship between large-scale irrigation systems and centralized state power in various historical settings, including the ancient irrigation civilizations of China, Egypt, Mesopotamia, and India. In "hydraulic societies" such as these, he writes in an earlier work, "man extended his power over the arid, the semiarid, and certain humid parts of the globe through a government-directed division of labor and a mode of cooperation not practised in agrarian civilizations of the non-hydraulic type."[71] He showed how, in associating themselves with religious leaders and through developing economies with redistributive features, elites in hydraulic societies were able to entrench their powers, eventually developing despotic regimes.

Although the implications of Wittfogel's thesis have been important for social theory,[72] this does not interest us here directly. Of more immediate interest is Wittfogel's use of Marx's idea of the dialectical relation between man and nature that is discussed in Chapter 2. In Wittfogel's analysis, the transformation of the hydraulic environment produced changes in society, which brought further changes in non-human nature, and so on; "an ongoing spiral of challenge-response-challenge," as Donald Worster has put it, "where neither nature nor humanity ever achieves absolute sovereign authority, but both continue to make and remake each other."[73] Wittfogel's approach to the relationship between society and water (nature) might thus be considered something of a forerunner to more recent historical and cultural investigations into what we are calling the social nature of water. However, recalling the discussion about dialectics in the previous chapter, this approach is more along the lines of a binary dialectic of external relations than relational in the sense that water – and we might also argue, society – is not fundamentally changed as a result of the exchange. The two entities, in other words, are not understood by Wittfogel to be *internally related*.

A further step was taken by Worster, who borrowed Wittfogel's approach in describing the American West as "a modern hydraulic society, which is to say, a social order based on the intensive, large-scale manipulation of water and its products in an arid setting."[74] The idea of a *modern* hydraulic society is of obvious interest here. We have already cited Donald Worster's description of this society's (modern) water in Chapter 1. In fact, something that distinguishes Worster's dialectical approach is his attention to water itself. For his part, Wittfogel was almost exclusively interested in how the large-scale manipulation of water produces a certain *kind* of society – hydraulic society, and he has little to say about water. Worster adds to this an interest in water itself, in its particular nature or meaning in modern hydraulic society, and in the various material, or environmental, impacts this society has had on water. Although Wittfogel helped prepare the way for historical thinking about the social nature of water, Worster has perhaps done as much or more than any historian to lay the groundwork for thinking about modern water.

A considerable portion of the research that falls into the category of water history, moreover, intimates the abstract concept of what we are describing here as modern water. To take a salient example, in a recent book on the history of the Fraser River in British Columbia, Matthew Evenden describes the Fraser as a major, even unique exception to the hydraulic manipulation that characterized most North American rivers in the twentieth century.[75] Evenden's description of the norm provides an apt illustration of modern water itself:

> In the twentieth century, humans transformed the planet's rivers. On every continent, save Antarctica, they dammed, diverted, and depleted rivers ... In emerging nation-states, in empires and colonies, in capitalist and communist societies, political elites applied new technologies of power to rivers and lakes. Dreams of a hydraulic order sought to correct past ills, to raise wealth, to impose control over nature and others. Floods would be stopped, rushing, wasting water would be harnessed with hydroelectric dams, and arid lands and reservoirs would be linked with irrigation systems. A new order of previously unimaginable scope was placed on rivers almost everywhere.[76]

Evenden's book, *Fish vs. Power*, describes how, against "postwar development pressure, a fisheries conservation coalition, born of a fishing industry alliance, Canada-US cooperation in the international regulation of the salmon fishery, and the intervention of fisheries scientists critical of the effects of power development, held dams off the Fraser."[77] Following Evenden's argument, it might be suggested that the challenge of fending off the dams was compounded by the general prevalence of modern water. As Evenden points out, one of the most difficult challenges facing fisheries biologists was to convince decision makers and the public that the problem of conserving the salmon resource needed to be considered from a biological rather than a purely hydrological perspective. Especially at the height of the fish-versus-power debate in the 1950s, these scientists had to contend with a widespread belief or faith that dams and fish could be simply reconciled by engineering fishways or other hydraulic devices that would allow salmon to migrate over dams.[78] Evenden quotes from a paper by federal fisheries scientist J.R. Brett published in 1956:

> Since the problem is both multiple and complex, no delusions should be entertained concerning the possibility of some quick or simple solution. Any new mechanical contrivance expected to aid salmon at some point in their migration will create new biological problems. It is the lack of knowledge of salmon that is the great handicap. This handicap can only be surmounted by a thorough program of research directed at the fish first, from which the problems may then be resolved.[79]

The general expectation and faith held in the 1950s by practically everyone but fisheries biologists – that reconciling the fish-versus-power problem was simply a matter of engineering flows of water so as to allow for salmon migration – was bolstered by the idea that water is water, that the relation of water to salmon is little more than a medium of migration and transportation, and that engineering suitable flows of water are all that is required to ensure the conservation of salmon runs. The fisheries biologists, in other words, had to contend with modern water. The eventual success of the coalition – of which they were part – to oppose the dams was certainly a victory of fish over power. But it might also be regarded as a victory of fisheries biology over hydraulic engineering, and of the more complex, relational, view of water held by the former. This victory was, however, very rare, even unique, especially for its time. More recently, as pointed out above, the growth of the anti-dam movement – and indeed the actual decommissioning of dams for the sake of restoring fisheries – constitutes a challenge to modern water for which the Fraser might be considered to have set a precedent.

We have considered some recent approaches by scholars – in fields ranging from resource management to anthropology to history – that take issue in one way or another with the nature of water. Perhaps the most powerful indication that water is in a state of profound transition, however, is evident in the work of those who deliberately seek to destabilize its apparent naturalness. To insist or to fall back on the fact that in every particular instance water may be reduced to an essential natural kind is one of modern water's hallmarks. The denaturalization of water is therefore not merely a threat to modern water but has the effect of dismissing it entirely. We will consider a few examples from the work of researchers involved in what have been described as explorations of the "post-natural."[80] Insofar as the practice of social science has presumed a radical distinction – or dualism – between the social and the natural, this work represents something of a disjuncture with past practices.

A growing number of geographers, for example, are engaged in research that elaborates on the "socioecological nature of water."[81] Such research, if it may be so generalized, treats the very nature of water as relative to social-environmental circumstances. Water, and aquatic ecosystems, then become understood as things that combine the social and the natural as if there were never a distinction between them, much as the historian Richard White describes the Columbia River as an "organic machine."[82] The concept of hybrid freshwater ecosystems has been suggested as a point of departure for the production of healthy and desirable streams, rivers, and lakes, where "it might not be possible, and might even be detrimental, to delineate between the natural and the artificial."[83] The term "production" – rather than "restoration" – of streams is used deliberately because in more and more places subject to generations of human impact (such as the American Midwest), the baseline or so-called natural condition of streams is often beyond knowing.[84] In such places, the general condition into which streams are brought into accord is decided by local communities, who, within limits, produce the kind of nature they want.

It is perhaps in the study of urban water systems that the distinction between the social and the natural is least tenable, and that geographers have most fruitfully developed concepts of hydrosocial metabolism. The attention paid recently by geographers such as Karen Bakker to water in the urban environment is partly motivated by the need for critical reflection on the historic shift from the state-hydraulic paradigm to the market-environmentalist mode of water governance as described above. With the demise of the state-hydraulic paradigm, the tendency toward uneven flows of capital (and water) in the urban environment and beyond is a matter of utmost concern to many observers.[85]

Some scholars have begun to work with concepts that effectively dissolve modern water and thus make it unavailable for commodification. For Erik Swyngedouw, the "hybridized waters" of cities offer a politically useful means of conceptualizing how water "embodies, simultaneously and inseparably, bio-chemical and physical properties, cultural and symbolic meanings, and socio-economic characteristics."[86] The way humans and water collaborate in "the making of metropolitan nature" is also politicized by Matthew Gandy, who offers a way of thinking that allows us "to 'rematerialize' the city" so as to promote environmental justice and uphold "the continuing political salience of the public realm."[87] And increasingly, it is the term "hydrosocial cycle" that describes the process by which flows of water reflect human affairs and human affairs are enlivened by water.[88] The task, already begun, is to put the hydrosocial cycle to work in helping promote social equity and environmental sustainability, not just in cities but wherever human intervention in the hydrologic cycle has produced inequitable or uneven access to water and water services.[89]

Modern water can be recognized in this work by its implication in the conquest and domestication of water in the (mainly nineteenth- and twentieth-century) city.[90] "Indeed," writes Swyngedouw, "the very homogenization and standardization of 'potable' urban water propelled the diverse natural flows and characteristics of nature's water into the realm of commodity and money circulation with its abstract qualities and concrete social power relations."[91] And as Gandy has observed of the flow of water in cities and beyond, "recent insights from geographers, historians and others ... emphasize how abstract flows of capital have driven the material flow of water through space."[92] In one sense, this insight is not all that recent. Consider, for example, the decades-old saying in the American West, "Water flows uphill towards money." What is more recent, and potentially transformative, is a rejection of the necessity of such flows by means of theorizing water as a hydrosocial hybrid.

Manufactured Scarcities

Just as we say that "water flows uphill towards money," it could be said that drought is attracted by poverty. Our normal understanding of water's nature makes these notions seem absurd. But as more people explore the post-natural status of water, it is our normal understanding of water that seems increasingly absurd. Although resources such as soils, forests, and oil have been the main focus of research in the field that identifies itself as political ecology,[93] questions involving various aspects of access to and control of water are of growing interest. Much of this work takes up the same issues addressed in the history and cultural ecology of water: the cultural embeddedness of local or traditional approaches to water management and water conservation, the social and environmental sustainability of these practices, their adaptability to changing environmental and economic circumstances, etc. However, research in the political ecology of water foregrounds a number of critical questions relating to the political, economic, and discursive circumstances impinging on hydrosocial relations. Questions involving whose understanding of water – and whose mode of relating and gaining access to it – attains dominance is what we want to highlight here. As Vandana Shiva has argued, the way water is conceptualized and represented is instrumental in determining who gains access to it and on what terms. This necessarily produces conflict over the meaning and definition of water, a kind of conflict that she describes as "water wars" – "paradigm wars – conflicts over how we perceive and experience water."[94]

One of the key tasks of political ecology is to analyze and politicize the ecological categories and suppositions that are often taken for granted in expert environmental discourse so as to reveal the social location of the interests and privileges, advantages and disadvantages they entail.[95] Discourses of water resources management, water conservation, urban water services, drought, and desertification are good candidates for critical analysis in the political ecology tradition because they often tend to internalize the interests and advantages of privileged groups while dismissing or rendering mute the hydrosocial needs of the disadvantaged.[96] "The water problem," as Swyngedouw has shown, "is not merely a question of management and technology, but rather, and perhaps in the first instance, a question of social power."[97]

Political ecology makes an especially important contribution to understanding the social production of apparently natural processes such as drought and water scarcity.[98] In contrast with conventional resource management, which seeks to identify "the essence of the problem and point toward essential solutions," political ecology focuses on "analysis of discourse and ideology, seeking to define new ways of collapsing the nature/ society duality and presenting a coherent view of hybrid environmental problems such as drought."[99] In other words, the notion of drought as a natural disaster or natural hazard is increasingly challenged by awareness of the political, economic, and discursive contexts in which it is always embedded. Drought, as Piers Blaikie and his colleagues have put it, is "the product of the social, political, and economic environment."[100] The notion of drought as a natural disaster depends on an acceptance of modern water, which allows something like drought to be perceived simply as a consequence of its absence. The socio-economic circumstances, technological changes, income disparities, geographies of power, commercial activities, and demographic factors that convert a shortfall of water into "drought" – the social causes of drought – can be largely ignored when water is seen in terms of volumes of precipitation and streamflow rather than in relation to the social circumstances that make it what it is. By challenging the apparent naturalness of drought and water scarcity, research in political ecology is in effect challenging modern water.

Researchers have shown how the material production and discursive construction of water scarcity have been deployed to promote the commodification and privatization of water services.[101] This kind of analysis follows a line of thought set out by Neil Smith over twenty years ago, describing the symbiotic relation between the capitalist production process and what he called "produced scarcity in nature."[102] Applying this idea of produced scarcity to water, Erik Swyngedouw argues:

> A climate of actual, pending, or imagined water crisis serves not only to instigate further investment in the expansion of the water-supply side ... but also fuels and underpins drives towards commodification. As the price signal is hailed as a prime mechanism to manage "scarcity," the discursive construction of water as a "scarce" good becomes an important part of a strategy of commodification, if not privatization.[103]

Representing water as something devoid of social content – that is, as a part of nature, a natural resource, or commodity – allows nature to be used as the explanation for water scarcity instead of, for example, the lopsided distribution of water services in cities (with poor sections being disproportionately underserviced). Given the antisocial nature of modern water, nature always takes the blame as "the principal 'cause' of water scarcity rather than the particular political economic configurations through which water becomes urbanized [or otherwise made available to people]."[104] Recognition of water's social nature, by contrast, draws critical attention to the socio-economic circumstances that occasion every instance of water scarcity.

The specific historical contexts in which "drought" has been critically analyzed in political ecology have varied, from the Yorkshire drought of the mid-1990s in England to "the social construction of drought as a disaster" in rural Queensland and New South Wales, Australia; the California drought of the late 1980s; and the drought affecting the city of Athens between 1989 and 1991.[105] These analyses treat drought as a partially discursive construction that is shown to serve specific social, political, and economic interests. They show, respectively, how the construction of drought bears on the privatization of water services, how the meaning and effects of drought vary between men and women in affected agricultural regions, how discourses of drought serve to authorize large-scale state water projects, and how drought helps produce a climate of crisis that then helps promote the commodification of water.

"Water scarcity" has also been considered as a social artifact brought on by changes in demand for and use of water associated with industrialization, commercialization, and technological change. It has been argued, for example, that private ownership of groundwater rights and growing reliance on commercial crops have been instrumental in producing socially constructed scarcity in the Canary Islands, and that the shift from traditional rainfall capture techniques to widespread dependence on desalination technologies in the Virgin Islands has been a cause of water scarcity.[106] Elsewhere, it has been shown that the displacement of traditional irrigation schedules brought on by Green Revolution techniques of continuous cropping of rice in parts of Asia have had the effect of producing general water scarcities where none existed before the introduction of these technologies.[107] The growing incidence of water scarcities induced by the proliferation of golf courses in Southeast Asia (indeed, in all places easily accessible to the world's Westernized urban class) might also be mentioned in this context.

In an article introducing "the political ecology of water," Barbara Rose Johnston gives numerous illustrations of the manufacture of water scarcities in different parts of the world (including industrialized countries such as the United States).[108] These show the extent to which water scarcity has been a result of social factors: the increased water demands entailed in Western-style industrial development and economic growth; the ecological and social effects of large dams and other hydraulic structures; surface and groundwater pollution; and the effective enclosure of formerly successfully managed common water resources to fit the demands of industrial capitalism and, in the case of many developing countries, structural adjustment requirements imposed by international financial institutions.[109] The latter demands have included the privatization of public water services, which, as Johnston points out with the case of the privatization and commercialization of water services in South Africa, resulted in over 2 million South Africans being evicted from their homes for failing to pay water bills.[110] The point of these various critiques is certainly not to question the reality of water or the real human suffering that arises when people do not have adequate access to it. As with work revealing the social construction of drought, their purpose is to highlight the extent to which any instance of water scarcity needs to be considered as a socio-natural phenomenon.

CONCLUSION

In this chapter, we have considered research in various fields that suggests the possibility, if not the need, to pay critical attention to modern water – the hegemonic yet unmarked nature of water in the Western, industrialized world: research proving water to be an "uncooperative commodity" suggests water's fundamental complexity, in contrast with the simplicity with which we have hitherto allowed ourselves to assign it identity and function; recognition of the growing imbalance between water supplies and water needs is giving rise to a changing water paradigm in which many of our presumptions about managing water have been thrown into doubt; historical and anthropological research reveals a wide variety of waters and social relations with water in different historical and cultural circumstances, suggesting the need to cast the spotlight on our own hydrosocial relations; scholarly involvement in attempts to liberate people from the hydrological colonial present suggest the possibility of liberating everyone subject to modern water from the same predicament; and the exploration of post-natural, socio-natural, hybrid waters suggests that this might already be taking place.

Although the (social) nature of water in dominant, Western discourse is often implied or intimated in these investigations, it is seldom considered in its own right. By attempting to bring the social nature of modern water to light, the remainder of Part 2 is intended as just such a consideration, beginning in the next chapter by showing that the modern, Western way of seeing and relating to water has a definite history.

4

From Premodern Waters
to Modern Water

*For not only does the way an epoch treats water and
space have a history: the very substances that are shaped by
the imagination – and thereby given explicit meanings –
are themselves social creations to some degree.*

– Ivan Illich, H₂O and the Waters of Forgetfulness

*The obscurity of the olden days, which illegitimately
blended together social needs and natural reality, meanings
and mechanisms, signs and things, gave way to a luminous
dawn that cleanly separated material causality from
human fantasy. The natural sciences at last defined what
Nature was, and each new emerging scientific discipline
was experienced as a total revolution by means of which it
was finally liberated from its prescientific past, from its
Old Regime.*

– Bruno Latour, We Have Never Been Modern

Introduction

Like the water systems of indigenous and traditional cultures, modern water takes shape within the context of the particular ensemble of social relations that makes it what it is. Modern water, in other words, is as deeply embedded in the matrix of modern, Western culture and history as the *subak* or water temple system is in Balinese history, or the tank system is in the history of southern India. However – and this also seems to be the case for the people of Bali and Tamil Nadu – we do not regard it this way *ourselves*. To us, water exists in a way that has nothing at all to do with our culture; it is something that we simply take for granted as

73

having always been there in the same way that it discloses itself to us today. But modern water is *deliberately* non-social and non-historical in a way that the waters of other places and times are not. Its distinguishing yet unmarked feature is that it is disclosed to us primarily through modern scientific practice – we have learned to know water in the way that we do through scientific discourse and modes of representation. As the dominant epistemological mode of Western culture, scientific practice has produced a distinctive way of understanding and representing water that makes it appear timeless, natural, and unaffected by the contingencies of human history. A relational-dialectical approach would hold that any such entity must have human relations existing in an historical dimension. It is to these relations and this dimension to which we turn our attention in this chapter.

The approach that will be used here is historical and comparative. To gain insight into modern water, we will consider ways that people in the West have represented and related to water in the more distant (pre-scientific) past, and contrast it with ways that water became known through the emerging practices of modern science, particularly in the fields now known as chemistry and hydrology.

WATERS IN HISTORICAL CONTEXT

*Following dream waters upstream, the historian will learn
to distinguish the vast register of their voices.*

– IVAN ILLICH, H₂O AND THE WATERS OF FORGETFULNESS

As discussed in Chapter 3, the history of water is of growing interest to many scholars, indicating a new-found historical appreciation for the extent to which the water of our own time and culture is unique in itself. Among the works that may be included in this category are several that point to a fundamental change in the way water was perceived and understood in the West, occurring approximately from the seventeenth to the late nineteenth century. This change has been described in various ways, such as "the paradigm change from waters to water," "the conquest of water," the change by which "the waters of forgetfulness" were transformed to "H_2O," and the transition "from tangible water to H_2O."[1] These works clearly have something to say about modern water and how it came about.

Written by a social critic who loved inciting new ways of seeing things, Ivan Illich's monograph, *H₂O and the Waters of Forgetfulness,* can be read as an application of historical method to incite alternative ways of knowing and relating to water.[2] According to Illich, water is known to us moderns as a scientific abstraction, a way of knowing that entails the loss of a capacity to experience water as "stuff." This "ineffable stuff called water," he argues, is constitutionally ambiguous and able to disclose itself in practically infinite ways. In treating the historicity of water, Illich laments the loss of our capacity to know the waters "of the deep imagination ... that stuff which can gurgle, and chant and sparkle and flow and rise in a fountain and come down as rain ... but also comes down [among the Lacandon people in the south of Mexico] as the souls of women who have died and who seek reincarnation."[3] The very idea of studying the history of water, he points out, may seem like a rather strange notion, simply because we have grown so accustomed to thinking that water itself has no history. In this brief and apparently unfinished book, Illich hints at the histories of far richer, culturally impregnated waters of places and times that differed from our own modern, scientific environment. Even our tendency to naturalize water's beauty – "the natural beauty of water ... the beauty intrinsic to H_2O" is subjected to Illich's rather hard-eyed scrutiny, whereby it is shown how less abstract modes of apprehending have revealed various waters in vastly different, and not always beautiful, ways.[4]

Notwithstanding Illich and his seminal work, most who have studied the history of the concept of water in such a broad historical fashion have approached the subject as a facet of the history of science, particularly the history of chemistry and of public health. In the following sections I will draw critically from an article by historian of science Christopher Hamlin, which offers a particularly useful place to begin our consideration of the history of modern water.[5] This article considers how, at earlier stages in Western history, water was represented in four literary traditions: classical natural philosophy, classical natural history, folklore and religion, and the medical literature promoting mineral springs. Examining these traditions, Hamlin has identified a transition from an empirical emphasis on diverse "premodern waters," which were regarded as heterogeneous entities exhibiting different properties and qualities, to a modern "essentialist conception of water itself"[6] – a transition he argues had taken place throughout the industrialized world by the end of the nineteenth century. The main contribution that I will add to Hamlin's thesis is to elaborate on an idea that is found in nascent form in his article: that an "essentialist conception of water itself" not only destroyed the irreducible variety of waters encountered in premodern times but had the effect of disembedding these waters from the myriad social contexts and relations that had constituted them in the first place.

According to Hamlin and others, the pivotal moment of this change from waters as heterogeneous entities to water as an essential substance was its identification as a compound of hydrogen and oxygen by proto-chemists in the late eighteenth century. Illich, in one particularly insightful passage, likens "the H_2O ... which industrial society creates" to "a discordant sound that is foreign to waters ... a fluid with which archetypal waters cannot be mixed."[7] The naming of water as a chemical compound overturned a tradition of belief – generally upheld for some two thousand years – by which water was understood as an element. Here, a brief diversion into classical natural philosophy is in order, as the classical idea of an element differs from its modern namesake, especially as applied to water.

Thales of Miletus (ca. seventh or sixth century BC), who is usually considered the first philosopher in the Western tradition, held that all of material reality was derived from a single elemental substance and process: water.[8] Thales propounded the view that water – perhaps owing to its mutability from solid to liquid to gas, its formlessness and its association with life – was nothing less than the fundamental basis of everything encountered in the universe. Both substance and process, "the watery principle," as Hamlin puts it, was considered by Thales to be the fundamental element of which the world was made.[9] Despite subsequent philosophical developments, the persistence of this principle was such that Pliny the Elder (AD 23-79), from whose writings we learn a great deal of how nature was understood in classical times, described water as the "lord of all" the elements.[10] Subsequent Milesian thinkers argued for air rather than water as the foremost element. However, the monistic elementary theories were effectively overturned by Empedocles (ca. fifth century BC), who was the first to put forward the classic doctrine of four elements: earth, water, air, and fire. These elements were understood by Empedocles in terms of "four primal divinities, of which are made all structures in the world."[11]

Aristotle (384-22 BC) refined Empedocles' doctrine by, among other things, assigning each element to a place in the cosmos, with earth at the centre (or bottom), surrounded (or overlain) by water, then air and fire. Furthermore, he postulated that each element was produced by a particular combination of four primary qualities: heat, coldness, dryness, and moistness. Accordingly, earth was formed of coldness and dryness; water, of coldness and moistness; air, of heat and wetness; and fire, of dryness and heat. In Aristotle's view, the contiguous elements underwent transubstantiation – they were, in his words, "transformed into one another" – by means of undergoing changes in quality.[12] Water, for example, was transformed into air by replacing coldness with heat (i.e., by heating it so as to produce what we now know as water vapour) and was condensed into earth by replacing wetness with dryness (i.e., by producing ice – which was regarded as a kind of earth). Similarly, air became water by cooling (condensing) it, and earth became water by wetting (rarefying) it. According to the Aristotelian scheme, all matter consisting of wetness and coldness was understood as the water element, an element that was nevertheless subject to change in relation to differing qualitative circumstances. This elementary concept was, as Hamlin points out, "very much an antiessentialist view," one that differed substantially from the modern concept of an element as an irreducible and intransigent substance. Hamlin also contends that the vast differences between empirical waters found in antiquity were usually understood not in terms of mixtures of various substances with elemental water but as qualities that inhered uniquely in each separate instance of water: "Though there was disagreement on the subject," these qualities "were not necessarily conceived as adventitious substances or imposed states, but rather as properties unique to a particular water, often unanalyzable, inimitable, and utterly non-reductive."[13]

While enduring increasing competition, the basic Aristotelian fourfold classification of the elements remained the canonical understanding of the nature of matter throughout Roman, medieval, and Renaissance times and was not definitively overthrown until the late eighteenth century.[14] Even Robert Boyle (1627-91), one of the leading figures in what is widely known today as the seventeenth-century Scientific Revolution, considered water a mutable element, observing: "It seems evident that water may be transmuted into all the other elements."[15] As late as the 1770s, water was still commonly held to be an element in the classical sense. Thus, for some two thousand years, at least in discourses of natural history and philosophy, water retained an identity that was at once elementary, unstable (in its transmutation into the other two proximate elements, air and earth), and heterogeneous (in its unique manifestation in different places and circumstances). Perhaps it was the very ambiguity of premodern water(s) that allowed it such longevity. In any case, such an animal as this was doomed to extinction in the face of the radical change of intellectual climate associated with the rise of modern scientific practice.

Antoine Lavoisier (1743-94), perhaps the most important figure in the foundation of modern chemistry, is usually credited with being the first

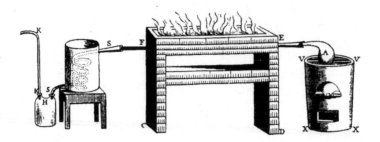

FIGURE 4.1 Water dethroned in Lavoisier's laboratory
This diagram, from Lavoisier's *Traité élémentaire* (1789), illustrates the equipment used to
analyze water in his lab. Water is put to the boil in a flask (A) and passed through a hot
glass tube (E-F) that contains small bits of iron. The oxygen in the water vapour combines
chemically with the iron, while hydrogen gas is collected and weighed at left (H). Careful
weighing shows the equivalence of water and the products (hydrogen and oxygen) into
which it is broken down. | P. Dear, *The Intelligibility of Nature: How Science Makes Sense of the
World* (Chicago: Chicago University Press, 2006), 79. © 2006 University of Chicago. Reproduced with
permission.

to describe water in the fashion that has found its way into *The Concise
Oxford English Dictionary*, that is, as a "colourless transparent tasteless
scentless compound of oxygen and hydrogen."[16] Lavoisier showed that
such a compound could be synthesized – "made artificially and from
scratch," as he put it – by burning an element that he subsequently named
"hydrogen" (and thus combining it with another element, which he subsequently named "oxygen"). He also showed that the compound could be
broken down by the simple means of rusting (oxidizing) iron filings in a
beaker of water. "Thus," Lavoisier declared in 1783, "one is led still more
nearly inevitably to conclude that water is not a simple substance at all,
not properly called an element, as had always been thought."[17] It was
another half century before the water molecule became represented as
"H_2O," a compound comprising a single atom of oxygen linked with two
atoms of hydrogen. However, it seems the signal move in the dethrone-
ment of water from its position as "lord over all" the other elements, as
Pliny had described it in the first century, took place in Lavoisier's laboratory (see Figure 4.1). This exercise might thus be regarded as a singularly
powerful instance of the production of local knowledge that was, eventually, made to engulf the entire hydrosphere.[18]

   John Dalton (1766-1844), another key figure in the founding of modern
chemistry, effectively took up from where Lavoisier left off by constructing
an imaginative visual representation of the way oxygen and hydrogen

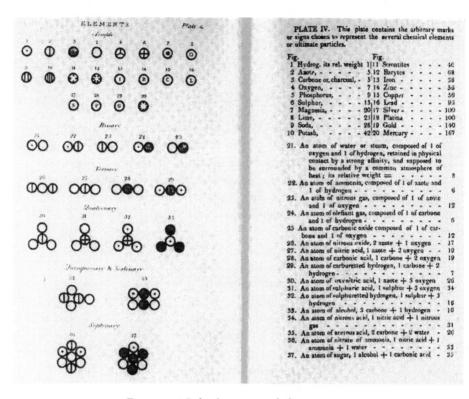

FIGURE 4.2 Dalton's atomic symbols
Water is depicted fourth row from top at left, annotated by Dalton as: "An atom of water or steam, composed of 1 of oxygen and 1 of hydrogen, retained in physical contact by a strong affinity, and supposed to be surrounded by a common atmosphere of heat; its relative weight = 8." | J. Dalton, *A New System of Chemical Philosophy*, vol. 1 (Manchester: S. Russell for R. Bickerstaff, 1808-11), 218-19.

combine to form water. By drawing atoms and molecules, Dalton made an important contribution to the advancement (and the propagation) of scientific chemistry; he also rendered the first visual representation of the water molecule (see Figure 4.2), enabling us to *see* the essential nature of water. Dalton's depiction of the water molecule underwent several changes before it was stabilized in scientific discourse as H_2O in the mid-nineteenth century. In the early nineteenth century, Swedish chemist Jöns Jakob Berzelius (1779-1848) proved that the ratio of hydrogen to oxygen atoms in the water molecule is 2:1.

Following Berzelius, the water molecule was represented as two upper case Hs and a dot, signifying two atoms of hydrogen and one of oxygen.

It fell to a French chemist, Louis Jacques Thénard (1777-1857) to render
the now familiar H_2O in the 1836 edition of his *Traité de chimie* (here
describing the chemical reaction of water, sulphuric acid, and zinc)[19] as

$$H_2O + SO^3 + Zn = H^2 + (Zn\ O, S\ O^3).$$

As described in more detail in Chapter 5, Dalton proved by experimentation that water retains this fundamental nature when it vaporizes,
rather than being transformed into air, as had hitherto been the most
popular explanation of this phenomenon. Hamlin offers a useful assessment of the effect of the (new) way of understanding and representing
water that became increasingly common from the late eighteenth century
onward:

> A richer and deeper range of conceptions of water and its effects on the
> body existed before the achievements of Lavoisier et al. A paradigm shift in
> the concept of water occurred ... in which water went from an [sic] class of
> infinitely varied substances to a monolithic substance containing a greater
> or lesser concentration of adventitious ingredients, known as "impurities."
> A vocabulary stressing qualitative and geographic uniqueness gave way to
> a dichotomous determination in which water was pure or impure ...
>
>   That premodern paradigm took as its normative starting point the apol-
> ogy we now make that water is never found pure in nature. It did not neces-
> sarily reject an understanding of water as a basic substance, known as
> simple, common, or sweet water, which might be more or less "good" ac-
> cording to circumstances in which it was found. But it was more interested
> in how waters were different than in the ways they were the same ... Waters
> were aspects of the histories of places. Although there were general types,
> there was also a sense of infinite variation; waters had qualities or properties
> that went far beyond taste and salubrity and even beyond the poorly marked
> borders of the natural.[20]

That premodern waters were "aspects of the histories of places" and
that they had qualities that transcended "the poorly marked borders of
the natural" is of central importance to my argument. In scientific chemistry, the nature of water was defined as some *thing* to which all waters
could be reduced – H_2O – a basic chemical compound that drove out all
the socially specific qualities of different waters. The shift from premodern waters to modern water was the product of an epistemological revolution; it occurred as a consequence of a particular way of knowing and

representing water in modern scientific practice – a practice that presumed a fundamental separation between the natural and the social. By contrast, the "premodern waters" that disclosed themselves to people in the West prior to the chemical revolution were of an entirely different nature, owing to the different epistemological contexts in which they were sustained. Hamlin offers a sketchy summary of the way premodern waters were sustained in the contexts of natural philosophy, natural history, folklore and religion, and early medical discourse. Although it is not necessary to produce an exhaustive study of what Hamlin describes as the "premodern paradigm," we need to examine each of these four contexts more carefully and give critical thought to the overall thesis that "there was also a sense of infinite variation" in the waters encountered and described before the eighteenth century. In addition to helping clarify the meaning of modern water, this will set the stage for considering some of the other factors that have contributed to the pivotal shift that Hamlin identifies.

"Premodern Waters" in Classical Natural History and Philosophy

An early twentieth-century painting by Zeno Diemer (see Figure 4.3) depicts the convergence of five Roman aqueducts as they would have appeared toward the end of the first century.[21] The ancient Roman aqueducts are commonly invoked as an engineering phenomenon that anticipated modern water infrastructure by almost two thousand years. However, if we inspect Diemer's painting carefully, we find a sign of something that is distinctly *not* modern: In the section under repair, we can see that what might appear to us as a single aqueduct is in fact three separate conduits. Why three instead of one large conduit? From a modern perspective, the inefficiencies involved in maintaining three channels instead of one might appear striking. But our modern perspective doesn't register the need to respect the integrity of different waters by maintaining separate conduits for each. In our view, the idea of efficient transportation of water completely overrides the idea of maintaining a distinction between the water(s) that are to be transported. Diemer's painting suggests that the Romans were more attentive to the differences between *waters* than to disposing of *water* in the most efficient way.

"Why so many aqueducts?" asks the classicist Trevor Hodge; "given that the conduit was often only partially filled, why not simply divert into it more water from a new source instead of building a whole new aqueduct? The Roman practice of keeping their aqueducts separate no doubt sprang from a recognition of the different quality of water from different

FIGURE 4.3 The Roman aqueducts by Zeno Diemer, ca. 1920
Five aqueducts are depicted in this scene: the Anio Novus superimposed on the Claudia
in left foreground; the Marcia, Tepula, and Julia are likewise borne on a single arcade in
the right foreground. The road depicted is the Via Latina, and Rome can be seen in the
distant background. | Image © Deutches Museum. Reproduced with permission of Deutsches
Museum.

sources."[22] According to Sextus Julius Frontinus (ca. 40-103 AD), Rome's
water commissioner at the end of the first century, "It was determined ...
to separate them all and then to allot their separate functions so that first
of all Marcia should serve wholly for drinking purposes, and then that the
others should each be assigned to suitable purposes according to their
special qualities."[23] Recognition of these so-called special qualities appears
to have included a respectful attitude toward the *genius loci* of particular
waters, particularly those of certain springs. For example, at least three
classical authors are known to have commented on the Aqua Virgo (which
was eventually to supply the Trevi Fountain in Rome) and its purity, with
Pliny the Elder reporting that "it refused to mingle with the waters of a

nearby stream sacred to Hercules, and therefore was named Virgin." Cassiodorus, a Roman Senator who lived from about 490 to about 585, writes: "Purest and most delightful of all streams glides along the Aqua Virgo, so named because no defilement ever stains it."[24] The classicist R.G. Tanner finds evidence of a general "hostility to blended waters" in classical literature from the time of Hippocrates (ca. 460-377 BC). Such hostility found material expression in the very structure of the Roman aqueducts: "Pity towards the deities of individual springs and medical strictures against blended waters would have obliged Roman engineers to conduct supplies through separate channels to the city wherever possible."[25]

That the Romans regarded their aqueducts in a manner not unlike the way we moderns do – as marvellous feats of human engineering – is readily apparent. Frontinus' famous boast is cited in practically every historical treatment of the aqueducts and ancient water management: "With such an array of indispensable structures carrying so many waters, compare, if you will, the idle Pyramids or the useless, though famous, works of the Greeks!"[26] However, it is the differences in the "many waters" carried by these structures that interests us here. The waters of the aqueducts flowed freely from the fountains and baths of Rome, through the streets of the city and into the Tiber River. There were no taps, no technical means of stopping these waters. Nor was this merely a matter of failure to invent the valve; respect for water demanded that allowing it to flow remained "a necessary condition of its proper use."[27] Containing water was proscribed in "the old Roman law 'Aqua currit et debet currere ut currere solebat' – water may be used as it flows and only as it flows."[28]

Also obscured by our tendency to see the aqueducts in a modern light is that they seem to have been regarded by contemporary Romans in a way that complicates our modern distinction between the natural and the artificial. The Romans usually referred to their aqueducts in terms of the different waters they carried, not in terms of the structures that carried them. Thus, according to Frontinus, the first aqueduct built was *aqua Appia,* named after Appius Claudius Crassus, the censor who brought this particular water, drawn from springs some ten kilometres to the west, into the city in the fourth century BC. Other aqueducts carrying spring waters were similarly referred to by the waters they carried; some were named after the person responsible for bringing them into the city (e.g., *aqua Marcia,* apparently after the Roman censor Gaius Marcius Rutilus), others after the attributes of their waters (e.g., *aqua Tepula,* or "tepid water"), and still others in accordance with stories surrounding the location of the spring from which they were supplied. Two of the eleven Roman aqueducts carried water drawn from the Anio River. When the first was built in the third century BC, it was simply called Anio, the name of the river itself. When a second aqueduct carrying water from the Anio was constructed some three hundred years later, it was given the name *Anio Novus,* or New Anio. The social nature of Rome's water is suggested by the fact that there was no common distinction made between the aqueducts and the "waters" or "rivers" they carried. Moreover, these anthropogenic waters or rivers were distinct in ways that, whether rooted in local legend or practical application, can be appreciated only in relation to their historical and cultural contexts.

Attention to different waters is still found in modern, Western society, especially where this is recognized as having commercial advantages, as in bottled spring waters. But the commodification of specific waters remains marginal to the homogeneous tendency of modern water. In contrast, one cannot help but be struck by the extent to which "the different kinds of water were a subject of close interest" not only to the Romans, but to all societies of antiquity.[29] This attention to difference is particularly apparent in the category of writing described by Hamlin as classical natural history.[30] Hamlin relies for the most part on two Roman sources: Book 8 of Vitruvius' *Ten Books of Architecture,* published around 30 BC, and Pliny the Elder's *Natural History,* published a century later. Both these works draw from Greek as well as contemporary writers (and Pliny draws from Vitruvius) to offer encyclopaedic compendiums of virtually everything known – including everything that had ever been written or was known to them by word of mouth – of the variety of waters in the classical world. These authors pay a great deal of attention to the (often astonishing) variety of waters, particularly those occurring in the form of springs, drawn from wells, encountered in rivers, and, occasionally, in the diversity of hydrometeorological phenomena such as rain. The list of the contents of Book 31 of Pliny's *Natural History* provides a useful (and to modern readers, a rather amusing) illustration:[31]

> Remarkable facts as to waters ... Differences in waters ... Medicinal properties: what sorts of waters are good for the eyes, what sorts produce fertility, what sorts cure insanity, what sorts gall-stone, what sorts wounds, what sorts protect the embryo, what sorts remove tetter, which make dye for wools, which for human beings, which produce memory, which forgetfulness, which keenness of sense, which slowness, which a musical voice, which dislike of wine, which intoxication ... Remarkable waters: waters in which

all objects sink, in which no objects sink; waters that kill ... waters that turn into stone, or produce stones ... Health-giving property of waters ... differences of waters according to kinds of earth; variation of springs with the seasons.[32]

Pliny makes further observations on "the marvels of many waters," including those specific waters in which everything sinks, or floats; waters that offer prophesies, harbour poisonous fish, kill everything that touches them, corrode bronze and iron, nurture unique flora, calcify the land they irrigate, grow rocks, turn into stone whatever is thrown into them, and so on.[33] And should anyone think "that some of these statements are incredible," avows Pliny, "he has to learn that in no sphere does Nature show greater marvels." For our purposes, let's consider the expressly social nature of these marvellous waters. Like the waters of the Roman aqueducts, the stuff of springs and rivers is understood in (and was constituted by) the social context and through the relations by which it became known to people. The hydrological marvels described by Pliny as well as the more mundane attributes of the various waters he describes are always in relation to people, their productive activities (as in waters that serve particular industrial purposes), and principally, their health. The effects of these various waters on the bodies of the people who internalize them constitute the most important means of knowing and ascribing significance to them: "The human system (and not just its senses)," as Hamlin notes, "was seen as the most sophisticated instrument in a qualitative assessment of waters."[34] The distinction between subjective and objective modes of knowing water is well illustrated in the contrast between these premodern waters and H_2O.

However, it must be acknowledged that when he moves to a discussion of what *we* would consider more general *hydrological* phenomena, Pliny shifts from the variety of "waters" to "water" in general. In the case cited below, he elaborates on the then commonly held notion of the subterranean hydrologic cycle (discussed in Chapter 5) in the course of giving an answer to the perennial question of why the sea remains at the same level despite the constant inflow from rivers:

Water permeates the earth everywhere, inside, outside, above, along connecting veins running in all directions, and breaks through to the highest mountain summits – there it gushes as in siphons, driven by pneuma (spiritus) and forced out by the weight of the earth; it would seem that the water is never in danger of falling; on the contrary, it bursts through to high places and summits. Hence, it is clear why the seas never grow from the daily influx of river water.[35]

Pliny reverts to a discussion of "waters" when he turns to describing the qualities of various rivers.[36] It may be said that depending on the specific sort of question investigated in Pliny's *Natural History*, the variety of "waters" – manifested in relation to different people and places – gives way to the nature of "water" – which manifests as a process of nature. A similar observation can be made of Vitruvius. Like Pliny, and owing to his ultimate concern with the water supply of towns, Vitruvius is greatly interested in "the characteristics of waters" associated with different springs. Of these, "it is hardly to be wondered at that in the great immensity of the earth itself there are indeed countless varieties of sap to be found, and that when water's power flows through the veins of these earths it arrives at the heads of springs adulterated, and thus for this reason fountains are made diverse and various according to their types, because of the discrepancy of their localities and the dissimilar characteristics of their regions and soils."[37] But when he turns to discussing phenomena such as rainfall and streamflow, Vitruvius speaks of water as a process, and he describes it in what seem almost like modern, hydrological terms.[38]

Evidently, both Pliny and Vitruvius address questions relating to the timeless and placeless nature and behaviour of water as well as the culturally and geographically contextualized nature of particular waters. This difference might be explained in terms of the distinction between natural history and natural philosophy. Natural history has been defined as "a former branch of knowledge embracing the study, description and classification of natural objects" and natural philosophy as "the study of nature in general."[39] As natural historians, classical writers attended to the geographical and cultural diversity of waters as particular and distinct phenomena, while, as natural philosophers, they attended to the nature of the water process in general. Moreover, despite Hamlin's suggestions to the contrary, it has to be admitted that Aristotle and other authors in the category of "classical natural philosophy" were equally or more attentive to the nature and behaviour of water in general (hydrological) terms than to the heterogeneity of specific waters.

The instances of specific waters found in the early natural histories, it may be said, occurred where the water process described in classical natural philosophy met with particular social processes (such as the functioning of human bodies, engagement of people in various industrial and agricultural activities, and so on) that gave distinction to the specific waters described in classical natural history. The early attention given to water in general – which, as we know from Thales, can be traced to the earliest stages of Western philosophy – usually involved questions about such things as the origin of springs and rivers, the cause of floods, the cause of rainfall, the reason that sea levels remain constant despite the inflow of rivers, the process of what is now called evaporation, the source and disposition of water flowing in the ground, and why the Nile floods later in the year than other (European) rivers known at the time. Today, we describe these as *hydrological* questions in deference to the modern branch of science – hydrology – that addresses them. "The importance of hydrology to many of mankind's most basic activities," states *Encyclopaedia Britannica,* "made such studies of interest to the earliest natural philosophers."[40] And, we might add, to the extent that water was considered in relation to such questions, it was water in general – or the general process of water – to which natural philosophy addressed itself.

Hamlin argues that "the natural historian's view of waters as many, however is fully compatible with the philosopher's."[41] Although this is undoubtedly true, the inclusion of water in general – as well as an empirical attention to specific waters – in the premodern paradigm suggests that rather than a simple temporal shift from premodern waters to an essentialist idea of water itself, the more salient move has been from natural history as a legitimate approach to the study of water(s) to an exclusive reliance on natural philosophy – and its heirs in modern scientific disciplines such as chemistry and hydrology – as the accepted mode of knowing and representing water. In other words, ways of knowing waters that are reflected in the category of classical natural history are no longer valid – they strike us as odd, or even ridiculous, just as does the notion that engineers would go to great lengths to keep the flow of different streams in separate aqueducts despite the obvious efficiencies to be gained by combining their flows.

The demise of natural history was a long and complex affair associated with the reciprocal process by which modern scientific practice came to be accepted as the authoritative approach for gaining and representing natural knowledge.[42] Works such as Pliny's *Natural History* were popular throughout the Renaissance, when interest in restoring the glories of classical antiquity was reflected in a reverence for classical texts. But by the turn of the sixteenth century – at the beginning of the era now described as the Scientific Revolution – natural philosophers were beginning to show a disdain for such works. William Gilbert (1544-1603), for example, dismissed the testimony of traditional natural historians such as Pliny as "the maunderings of a babbling hag."[43] Such dismissal is suggestive of how, eventually, modern science managed to render water in a way that has driven out its heterogeneous social nature.[44] The social nature of premodern waters, it may be surmised, was able to coexist alongside the more abstract notion of the natural water process in natural philosophy because these waters internalized both social and natural water processes. The modern separation of nature and society at the hands of the early practitioners of scientific method made such an intermingling of society and nature impossible, leaving an asocial hydrological discourse the only valid way of knowing and representing water.

There is an association between the disenchantment of waters and the demise of natural history as a legitimate means of knowing and representing the natural world. Knowledge of other natural phenomena, we might add, underwent a similar transition. By the early eighteenth century, for example, naturalists began to dispute the plurality of symbolic, emblematic, or portentous (i.e., social) meanings that had traditionally been attached to things such as hurricanes and floods:

> They claimed that meteorological phenomena did not possess other than "natural" significance. Meteors were to be divested from their spectacularity and treated as ordinary products of the material world ... The general trend was ... toward an epistemic recontextualization of prodigious meteors, i.e., toward a complete evacuation of their "non-scientific" meaning.[45]

The evacuation of non-scientific meaning – or the "disenchantment of the world," as Max Weber famously described it – is a major theme of the intellectual history of the seventeenth and eighteenth centuries, when modern scientific practice began to distinguish itself from classical natural philosophy.[46] Like hurricanes and floods, as water was increasingly fed through scientific discourse, its cultural content was filtered out.

Premodern Waters in Folklore and Religion

Folklore and religion are extremely broad, overlapping, and rather arbitrary categories. Particularly for the modern reader, the special attributes and qualities of spring waters discussed by the likes of Pliny and Vitruvius might just as well fall into the category of folklore as natural history. Such categories nevertheless provide a useful heuristic approach for understanding different modes of relating to water. Hamlin asserts that "the more we know of the subject ... the more it becomes clear that the investment of local waters with some kind of sacred status was important, widespread, and varied in nature."[47] Approaching this question from the perspective of a comparative study of religions, Eliade makes a generalization that is reminiscent of Thales' cosmogeny: "To state the case in brief, water symbolizes the whole of potentiality; it is *fons et origio,* the source of all possible existence."[48] The sacred status and the symbolic as well as the actual potentiality of water has been memorialized in myriad ways as the water process has got mixed with different cultural practices and processes.

Hydrolatry

The pervasiveness of spirit(s) in local waters of other times is particularly striking. Along with stripping it of its sacred qualities, modern water, it may be anticipated, has almost entirely destroyed the geography of the sanctity of water.[49] Among the better-known examples in classical literature is that of the Greek poet Hesiod, who, around the seventh century BC, made reference to some ten thousand nymphs (the resident divinities of specific springs) in the ancient world.[50] Throughout antiquity, the waters of local springs and rivers were associated with deities and honoured with rites of worship, propitiation, and offerings, of which the tossing of coins into fountains remains a common vestige. The powers and qualities ascribed to the specific waters of ancient springs varied with the site – from sources of spiritual danger, healing, or purification to the more profane influences on the body, as already noted from the accounts given by Pliny.

The cult of water, or "hydrolatry," as it has been called by James Rattue, may be considered "a single religious motif spanning the whole Eurasian continent and beyond," from Greek times to the modern age, and until very recently in some places.[51] Although a single motif, it was manifested in ways that were far from uniform and had everything to do with the specific historical and cultural relations in which they occurred. Although some general principle of "the sacredness of water as such" appears to have prevailed widely, the myriad water cults were founded on what Eliade describes as "the local epiphany, or the manifestation of a sacred presence in some particular watercourse or spring."[52] The significance of this "sacred presence" has to be understood in every instance from an historical-geographical perspective. An historian of holy wells, Rattue points out that a peculiar aspect of the Greek worship of springs, not repeated in later cultures, was in "their power to inspire prophecy." Unlike the sacred springs among people of many other cultures, the Greeks did not appear to make offerings to their wells. The Romans adopted many formerly Greek well sites but "reinterpreted the prophetic powers of the wells and turned them into inspirers of poetry."[53] In the later Roman Empire, Rattue suggests, many of the same springs and wells were incorporated into the mystery religions, secret cults that afforded initiates a mode of religious experience not sanctioned by the state.

The Christianization of waters that had been held sacred to pre-Christian cultures is a common theme in the study of hydrolatry and offers a good example of how individual waters form at the confluence of the water process and social processes. Persecution of pagan water cults was common in the Middle Ages, but these managed to persist despite the strenuous efforts of the Church. In England, during the reign of the Saxon King Edgar (943-75), a cannon was issued exhorting "every priest industriously [to] advance Christianity, and extinguish heathenism, and forbid the worship of fountains."[54] Similar edicts continued to be issued into the twelfth century, but in the face of its continued popularity, hydrolatry was eventually accepted by the Church and incorporated into Christian observances. Eventually, many pagan wells and springs were Christianized – reinvested with new layers of religious significance. A variety of methods was used to effect this reinvestment. For example, water from holy wells was used for baptism, with baptisteries built at ancient well sites; wells were dedicated to Christian saints, thenceforth bearing their names; and in some cases, churches were built nearby or actually constructed overtop sacred wells. Holy springs were thus reinvested with meaning and became "an accepted, established part of the Church's spiritual weaponry."[55]

But incorporating sacred waters into the catechism, if not the body, of the Church did not preclude them from serving more prosaic and profane purposes. The spiritual uses of holy waters (e.g., for baptism) blended rather seamlessly with their continued use for medicinal purposes and did not even exclude them from such mundane applications as furnishing supplies of water for growing towns of the Middle Ages. The premodern waters of springs thus dissolved certain categories and distinctions – such as sacred and profane – that seem only natural to us today. Moreover, in their infusion with medieval society, they simultaneously held together and were constituted by contemporary social relations. Rattue emphasizes the social embeddedness of hydrolatry in medieval England: "Wells," he notes, "were part of [a] battery of social control and protection mechanisms.

Some were haunted, others bottomless, all were awesome ... and a little fearsome. The rituals in which they played a prominent part helped reinforce communal and group unity, and the position of the Church as the prime medium of cultural expression."[56]

Depending on the particular place, sacred waters withstood the onslaught of the Reformed Church, the relative depopulation of the countryside, and the incursions of modern science for a surprisingly long time. As suggested above, Lavoisier's reduction of water to a chemical compound is often thought to have been decisive in stripping waters of their mythology.[57] But the process of purifying water of its sacredness occurred over centuries. "Water retained its sacred, purifying role for a long time," writes the historian Ladourie, "whether this was reflected in the mineral springs ... or quite simply in the holy water at the entrance to our churches."[58] Indeed, it is because of their chorological and cultural idiosyncrasies, Hamlin suggests, that sacred waters had the effect of "offsetting the universalizing tendencies of the incipient science of chemistry."[59]

Nevertheless, it appears to have been the collapse of the social relations that had held them in place that caused the demise of sacred waters in most locales by the late nineteenth century. Again, Rattue's study of holy wells in England provides an illustration of the essentially social nature of these waters:

> The social functions of holy wells can hardly have altered since the Middle Ages; yet these functions were largely being provided by a whole new series of social mechanisms. Local community solidarity was the casualty of the "Great Change," and was nationalized into national politics and, later, the Welfare State, after the 1867 and 1885 Reform Acts. Church-organized youth groups, schools and scouts, trade unions and political parties provided new class- or age-based forms of social solidarity.[60]

With the growth of scientific medicine, another vital social function of the springs fell into disuse, as reflected upon by a writer living in the vicinity of Brampton, England, in 1841:

> The water is still of the most pellucid clearness, sweet to the taste, though much neglected, full of fallen leaves and haunted by vermin ... The present generation, however, have ceased to avail themselves of the medicinal properties of the waters, which have lost their virtue, or are eclipsed by the superior abilities of the Medical Practitioners to whose charge the health of Brampton is consigned.[61]

FIGURE 4.4 Rivers as social entities
The Nile, a second-century Roman copy of a Hellenistic statue, probably Alexandrian in origin. This statue offers a good illustration of the inseparability of Egyptian society and the river. Sixteen children tending to and reveling in/upon the Nile is an allusion to the sixteen cubits by which the river rose to its most socially productive flood level. | Scala Art Resource, New York, ART57764.

So far, we have been considering the investment of local spring waters with sacred and spiritual qualities. Premodern waters also included rivers. Like the anthropogenic rivers of the Roman aqueducts, river waters of antiquity were often regarded as unique, with some even refusing to mingle with the waters of lakes and other rivers.[62] Not unlike the way rivers, especially the Ganges, are still regarded by many in India, spiritual and emotional as well as cultural attachments to specific rivers were among the most pronounced characteristics of ancient hydrosocial relations. Each of the major and many (perhaps all) of the minor rivers known to the ancient Greeks and Romans were considered deities in themselves or were associated with a particular god or gods, and each was imbued with its own myths as well as cultural, economic, and religious traditions and sacraments. As a rule, rivers were not – could not – be known as abstract hydrological phenomena but were known, rather, in relation to the social context in which they became manifest. Rivers were more than just influences or determinants of society; they were *part of* society, figuring as full actors in the founding myths, the legends, and the rituals out of which social relations were spun (see Figure 4.4). Many of these attachments survived the Middle Ages or were adapted to changing cultural circumstances. The historian Simon Schama reports on a pair of Renaissance geographers known as the Champier brothers, "who in the sixteenth century produced a comprehensive anthology of the myths and legends associated with the entirety of known streams: the crystalline waters of the Auvergne that could wash away cataracts of the eye; those which could naturally polish pebbles so that they sparkled like true brilliants."[63]

According to Tanner, because rivers and springs were considered sacred beings, people often expressed concern about angering them by channelling or damming their waters.[64] Even the acts of bridging rivers and (the very common practice of) adorning springs were considered on certain occasions to be contrary to the wishes of the waters. Of course, given that a great deal of damming, bridging, draining, channelling, and adornment did take place throughout antiquity, such a statement needs to be read with care. One might plausibly surmise that there was at least as much political interest involved in such contrariness as there was a respect for water. Nevertheless, it does seem that the urge to manipulate water in the modern sense of exploiting it as an abstract resource was at least tempered by recognition of its sacred dimension and an appreciation of the social context in which various waters were sustained. The Roman historian Tacitus (AD 56-ca. 117), for example, called attention to "the effect of religious belief in preserving the natural order, mentioning popular objections to diversions of rivers on the ground that nature herself had made the best provision for their sources, their courses, and their mouths."[65] In a discussion in the Roman Senate about whether to divert rivers tributary to the Tiber for purposes of flood control, Tacitus reports the protests of people from the affected region: "Nature," they said,

> had made the best provision for the interests of humanity, when she assigned to rivers their proper mouths – their proper courses – their limits as well as their origins. Consideration, too, should be paid to the faith of their fathers, who had hallowed rituals and groves and altars to their country streams. Besides, they were reluctant that Tiber himself, bereft of his tributary streams, should flow with diminished majesty.[66]

That the premodern waters of springs and rivers articulated themselves differently with people in different places prompts Hamlin to suggest the practice of a contemporary "chorographic science" of waters: "In a chorographic science, focused on the uniqueness of a locality rather than on the general characteristics of a group of objects, the lore of sacred waters [would be] a valuable indication of the uniqueness – here the divinely generated uniqueness – of a particular place."[67] In relational-dialectical terms, such a practice might be regarded as the study of how and why water becomes what it is in different historical and geographical circumstances. In any case, the thought of such a science is difficult to reconcile with the science of modern water, rendering as it does the undifferentiated stuff (H_2O) occurring in a universal form (the hydrologic cycle) that leaves behind all chorographic and cultural uniqueness.

Premodern Waters of the Mineral Spring

A fourth major field of discourse in which premodern waters were sustained is found in the large body of literature dealing with mineral waters that flourished from the sixteenth to the mid-nineteenth century.[68] The literature of mineral waters, as identified by Hamlin, is mainly associated with the testimony and declamations of medical practitioners interested in the effects of various waters on human health. As already suggested, we could certainly reach further back in time for illustrations of the diversity encountered in various waters and their effects on the human body. Hippocrates' *Airs, Waters, and Places* – often taken as a foundational text in the medical sciences – treats of "waters" – water in its plurality – stressing their various effects on bodies and their influences on the cultures of peoples in different parts of the ancient world.[69] The commentaries on waters found in the works of Vitruvius, Pliny, and others reflect how strongly these concerns prevailed in Roman times. Furthermore, testimonials of cures and descriptions of the physiological effects of different waters are found in a wide variety of medieval and Renaissance sources.[70]

The outpouring of mineral water analyses identified by Hamlin as having occurred from the sixteenth century onward appears to have been the result of the growing need to render personal testimonials and unsubstantiated explanations into proofs that accorded more with the standards being developed in the nascent chemical and medical sciences. The period beginning in the sixteenth century is associated with a general shift in discourse from holy wells to mineral waters, a shift that can be associated with the emergence of competition for paying clients between the owners of various mineral springs. Scientific medical men appear to have found a niche in developing various means of comparing different waters so that their promoters could make solid-sounding claims about their respective benefits. Hence, the concept of systematic water analysis: "Ideally, a water analysis simply translated [the putative qualities of waters] into the universal language of science, the better to allow the promoter [of a particular mineral spring] to highlight the uniqueness of the spring and distinguish its water from anything available elsewhere."[71]

Elsewhere, Hamlin has written a monograph on the struggle between different communities of proto-scientists over which could legitimately claim authority to analyze the differences between various mineral waters – or to use Hamlin's felicitous expression, over who could legitimately "speak for water."[72] Of particular interest to us here is his assessment that the eventual victory of chemists over physicians in this matter was an important factor in the conversion of myriad, unique waters to universal, modern water. Hamlin quotes the physician John Barker from his *Treatise on Cheltenham Water* (published in 1786), in which he insisted that "there are specific properties in almost every mineral water, wherein it differs from every other of the same class. Nay, there are qualities in the water, and even in the spirit of every common spring, whereby it is peculiarly different, in many respects, from all others." Barker, however, cautioned that the chemists could never "elucidate things of so high a nature."[73] The eventual dominance of chemistry, Hamlin argues, had the effect of driving out these things of "so high a nature" from water. "Oddly," he remarks of the substance to which water was reduced by the chemists, "the new compound was more elemental than the old element."[74] The differences between what had formerly been regarded as "waters" could now be explained in terms of H_2O plus whatever adventitious substances were found mixed in it. Although certain minerals continued to be seen as desirable components of the admixture, chemistry, and later bacteriology, shifted an entire discourse from an effort to locate the intrinsic properties of different waters to "the quest for pure water."[75]

Although the influence of scientific chemistry was clearly a pivotal factor in the reduction of premodern waters to essential "water," this is hardly the only explanation. Hamlin points out that even the chemists, though they may not have seen the same things in the water that were seen by a physician such as John Barker, nevertheless upheld the notion of distinctive "waters" long after the advent of H_2O. Indeed, up to at least the mid-nineteenth century in Britain, and during the next fifty years elsewhere in the industrialized world, a variety of "waters" persisted in the environment and the language of many people: "We do not know fully," admits Hamlin, "why waters ceased to be many and became one (albeit, one with contaminants); and equally why attention shifted from variability to uniformity."[76]

The need for further explanation suggests that we consider the advent of modern water in discourses outside the history of medicine and chemistry. It has already been noted that in certain contexts, such as natural philosophy, the variety of waters gives way to considerations of the nature of water itself in writing that predates the advent of chemistry. Furthermore, it must be admitted that the nineteenth-century demise of "many waters" is not as clear-cut as Hamlin suggests. In other contexts, we find that attention to different waters remained very much alive well into the twentieth century.[77] For example, the treatment of disease by internal and external application of "natural mineralized waters of reputed value" remained a subject of scientific interest until at least 1930 in Great Britain.[78] This field, referred to as "hydrology," was centrally concerned with the peculiar, local qualities of various "waters" and dealt with the question of how their medicinal effects "are due to water, pure and simple, and to what extent they are modified by the peculiar properties of the various 'waters' found at the different spas."[79] As long as the primary identity of water could legitimately be detected by an instrument as complex and culturally embedded as the human body, a variety of waters was entirely natural. But as water became more commonly understood as an abstract substance – something that could in every instance be accounted for in quantitative terms – the heterogeneity of waters was eventually banished, leaving only the essential substance.

In Chapters 5 and 6, we consider how modern water has been constituted, in part, by the expulsion of practices such as those described in the paragraph above from the field of "scientific hydrology" and by the reservation of the title "hydrology" to scientific practices that deal with quantities of water, "pure and simple." In the process of becoming identified as the one and only legitimate brand of hydrology, it was necessary to get rid of "the peculiar properties of ... various 'waters'" that had been the subject of less scientific hydrological practices since ancient times. The most expedient way of accomplishing this was to consider water exclusively in its quantitative dimension. But now we are getting slightly ahead of the argument. The salient point here is to stress the need to look beyond Hamlin's focus on water in the history of medicine, chemistry, and public health in order to reveal a more complete, albeit more complex, picture of the emergence of modern water.

"All This Is No Concern of Mine" – Scientific Hydrology and Modern Water

At the outset, identifying the broader historical context in which it occurred will help explain the apparent shift from "many waters" to "one" that Hamlin has identified as having taken place in Britain by 1850 and elsewhere in the industrialized world over the next half century. This period corresponds to the advanced stages of the Industrial Revolution and the emergence of industrial capitalism in these very places. These momentous changes in political economy brought new demands on water – particularly as a source of power and a factor of production – and on ways of knowing and accounting for water that would allow it to serve these demands. Also, the migration of people from rural to urban areas associated with the rise of industrialism inevitably had a profound effect on the way water was regarded. Social relations with water are inevitably transformed as people move from one place to another. In the mass movement of people from the countryside to the city – especially when served by piped water through urban distribution systems – those relations are attenuated in a pronounced way. In other words, an obvious explanation for the demise of heterogeneous waters in the nineteenth and early twentieth centuries is that for a growing number of people in the industrializing world, the most significant (and intimate) hydrological experience was occurring in the city, specifically at the mouth of a tap or in relation to a toilet. As Matthew Gandy has written in relation to the nineteenth-century transformation of Paris' sewer system, "the changing place of water emerges as a central element in the shifting boundary between premodern and modern conceptions of nature."[80]

In his study of the urbanization of water in the nineteenth and twentieth centuries, Jean-Pierre Goubert describes the process by which rivers and streams were made to flow through subterranean urban space as "the conquest of water."[81] Recently, as noted in Chapter 3, geographers and others have taken a critical interest in this "conquest," showing more forcefully how it occurred in the context of Western capitalism. But the fundamental *intellectual* abstraction of water from its cultural embeddedness is perhaps the most significant operation at play here. In a recent study of the history of sewage, Jamie Benedickson describes how, when combined with eighteenth-century legal and economic thought, the rendering of water as a mere chemical compound produced the secularized substance that helped pave the way for what he describes as "the culture of flushing" in the nineteenth century.[82] Returning to Goubert, he points out that in order to be conquered, water was first "besieged by science and technology."[83] On this view, it was primarily through modern science and technology that a new way of knowing and relating to water emerged. And it is to this new way of knowing and relating to water to which we turn in the remainder of this chapter.

To provide a more sufficient explanation of the conceptual shift from premodern waters to abstract modern water therefore requires that we consider the history of scientific traditions other than chemistry and public health. The conquest of water must also be considered in relation to the development of the hydrological sciences. This suggests a different periodology from that presented by Hamlin; it locates the historical origins of modern water somewhat earlier, namely in the mid-seventeenth century. Moreover, it suggests that the ascendancy of modern water grew incrementally as the intellectual and material consequences of seeing and relating to water *hydrologically* took root and spread, reaching a kind of apogee only late in the twentieth century with the advent of what I call "global water" (see Chapter 8). Finally, a point that bears repeating: Considering modern water from the perspective of the history of the hydrologic sciences suggests that the paradigm change was more than simply a shift from waters to water; it entailed the effective removal of water – its extrication, or abstraction – from the social relations that had given water(s) a variety of meanings and manifestations in different places and times.

The history of hydrology is treated in greater depth in later chapters and here I will only introduce it and suggest its contributions to modern water in the very broadest of terms. There are many definitions of hydrology, but as the eminent hydrologist James Dooge has pointed out, they all centre on the scientific study of "the occurrence and the movement of water on our planet."[84] Scientific hydrology deals with many of the same questions and arguably builds on the general approach to *water* found in classical natural philosophy, in contrast to the attention to specific *waters* encountered in classical natural history. Indeed, most historians of the science have located hydrology's origins in writings that deal with the occurrence and movement of water (the nature of water) found in the works of Aristotle and other classical natural philosophers. Modern, scientific hydrology, however, is distinguished by its emphasis on quantitative methods; the *sina qua non* of modern hydrology is the abstract quantification of water.

Perhaps the most important contribution of scientific hydrology to modern water is suggested by the fact that the very act of measuring water abstracts it from its qualitative dimensions. "The creation of a 'measure' is a complex mental act," writes Witold Kula in his social history of standards of measurement; "it demands that we abstract from a great many qualitatively different objects a single property common to them all, such as length or weight, and compare them with one another in that respect."[85]

Measuring water, of course, is hardly a uniquely modern practice. As Kula and others have shown, measurement is a social activity common to virtually every society of the past as well as the present. And as we know from the standard histories of hydrology, measuring rainfall is practically as old as civilization itself.[86] But unlike modern measures, older standards of measurement were inextricably related to their social and historical contexts. The famous nilometres could be cited as an example – the oldest existing hydrometric installations on earth, they are commonly featured in general histories of hydrology and fluvial hydraulics. But what did these instruments measure? To be sure, they measured the height of water, or the river stage, at various locations along the lower Nile. But rather than being measured against an abstract mathematical index, at least some of the ancient nilometres were gauged according to the anticipated effects of different water levels on contemporary Egyptian society – thus, what Dooge describes as the "social calibration" of nilometre readings:

> It is interesting to note that the first hydrological measurements were made to serve social purposes rather than to assist in the design of hydraulic works or the understanding of hydrological phenomena. Thus nilometers were installed four thousand years ago because the level of the Nile was used as a basis for taxation. In this connection it is interesting to recall that the first calibration of nilometer readings was not in terms of flow but in terms of economic and social effects ... Similarly the first recorded reference to a rain gauge which dates back over a thousand years refers to it as the basis of the land tax in India.[87]

The ancient nilometres could be described as socio-hydrometric instruments in the sense that they measured water levels as social phenomena. Thus, on the nilometre described by Dooge, a height of approximately 20 metres was indicated in terms of "Disaster," a height of 16.5 metres as "Security," and a height of 13 metres, as "Hunger."[88] This approach to measuring water differed from that employed by the administrators of the water supply for ancient Rome (discussed below) and again from that used to measure and control the distribution of water in traditional Balinese irrigation systems as described in the previous chapter. Such idiosyncratic ways of measuring the different waters of different places (as for other things, such as grain and salt) were both constituted by and constitutive of the hydrosocial relations through which they were practised.[89]

The notion that things can be measured in the abstract – that is, by abstract standards that bear no reference to, or have lost their association with, concrete cultural references – represents a modern departure. "The West's distinctive intellectual accomplishment," argues Alfred W. Crosby in a memorable phrase, "was to bring mathematics and measurement together and to hold them to the task of making sense of a sensorially perceivable reality, which Westerners, in a flying leap of faith, assumed was temporally and spatially uniform and therefore susceptible to such examination."[90] Abstract, mathematical stock-taking of the world thus became the "measure of reality," as Crosby puts it. Such an accomplishment was made possible by the fusion of mathematics with empirical natural philosophy, an intellectual feat associated with Kepler (1571-1630), Galileo (1564-1642), Descartes (1596-1650), and others early in the seventeenth century.[91] Historians of science have referred to this innovation in terms of "mixing mathematical practice and natural philosophy," "the geometrization of nature," and the rendering of "the differences between things ... to simple quantifiable proportions."[92]

Regardless of how we describe it, this intellectual advance made the social nature of things – the sense in which things are fundamentally bound up with the human societies in which they become present – illegitimate. This illegitimacy was a corollary of the contemporary impoverishment of reality. Historians and philosophers of science refer to the seventeenth century – Cartesian and Galilean philosophy in particular – as having reduced the reality of things to their "primary qualities," or to their *objective* and *measurable* features (e.g., physical dimension, shape, and mass), while banishing their "secondary qualities" (e.g., smell, taste, and appearance) to the realm of merely subjective phenomena.[93] From a world of qualitative, often incommensurable differences, the new mathematical natural philosophy effected "the restriction of natural reality to a complex of quantities" of which "nothing is scientifically knowable except what is measurable."[94] The "vast book" of nature, as Galileo famously proclaimed, "is written in mathematical language."[95]

This innovation of combining mathematics and natural philosophy made an important contribution to the way space has been experienced, conceptualized, and materialized in the West. Space, as emphasized by Henri Lefebvre, is always a product of history – an historical artifact of sorts. Lefebvre distinguishes modern spatial experience and practices in terms of what he calls "abstract space," the product of a particular set of material and intellectual (historical) circumstances with ancient roots but coming to flower in industrial capitalism. Abstract space is the particular kind of space that "tends towards homogeneity, towards the elimination of existing differences or peculiarities."[96] For Lefebvre, abstract space is isometric space, defined by its homogeneity, "a property which guarantees its social and political utility."[97] Lefebvre's critical analysis of (the history of) space is worth mentioning because of what seem important parallels between modern water and this abstract space. "The water that we have set out to examine," declares Illich, "is just as difficult to grasp as is space."[98] But in the sense of its isomorphism, its indebtedness to mathematical natural philosophy, and its social and political utility, modern water might be considered a liquid form of modern, abstract space.

In any case, the mixing of mathematical practice with natural philosophy produced the circumstances in which the modern nature of water could be revealed. More than sixty years ago, philosopher of science R.G. Collingwood pointed out that "when chemistry correlates the qualitative particularities of water with the formula 'H_2O' [it implies that] a thing's 'nature' ... is not what it is made of but its structure, as that structure can be described in mathematical terms."[99] Like modern chemistry, modern hydrology also treats of water in essentially mathematical terms, though at a much larger scale than that of the water molecule.[100]

The history of scientific hydrology is strongly associated with the concept of the hydrologic cycle, as discussed in the following chapters. For now, I will sketch in the briefest of terms the West's distinctive intellectual contribution of rendering water as something that may be measured in the abstract. To begin, a contrast with the ancient Romans is in order: When considering the discharge of water from pipes, we learn from Frontinus that the Romans applied a standard measure of cross-sectional area of flow known as the *quinaria*. However, they found it unnecessary to measure the discharge of water *itself*, a measure that would have included consideration of the velocity of flow.[101] Instead, their concept of the water "rate" was deemed adequate and appropriate for purposes of ensuring a just distribution of water to the citizens of Rome.[102] Throughout Europe, this approach to measuring the flow of liquid water was generally applied to open channels as well as to pipes until the seventeenth century.[103]

It was not until the first half of the 1600s that water began to be studied systematically as a moving fluid – by Galileo's students at the

Academia del Cimento.[104] One of these students, Benedetto Castelli (1578-1643), formalized the abstraction of the flow of water in what is now known as the "law of continuity" in his *Della misura dell'acque correnti* (The Measurement of Currents of Water), published in 1628. In essence, this principle applies the conservation of mass to the flow of fluids and can be expressed as $Q = A \times V$, where Q represents discharge, A is the cross-sectional area, and V is velocity. In providing a convincing proof of the law of continuity and a scientific basis for the measurement of streamflow, Castelli's work made a contribution to the development of the hydrological sciences. In a particularly illuminating passage, historian of science Cesare S. Maffioli describes Castelli's contribution in terms of "annexing the subject of waters to the territory of the Galilean mathematical sciences."[105]

Although Castelli's law of continuity – drawn from his experimentation with the rivers around Florence – made an important contribution to the development of hydrology, it eventually followed a channel of its own and fed the related modern science of hydraulics.[106] Meanwhile, the spirit of the mathematical philosophy spread to annex the waters of other parts of Europe. It was in the mid- to late seventeenth century that water was first subjected to scientific, systematic measurement by natural philosophers whose work comprised the core of what modern hydrologists consider their specific field. The famous rain gauges of Christopher Wren (1632-1723), some of which were tested on the grounds of Gresham College in London in the 1660s, are cited by David Livingstone as an example of the "mathematical baptism" by which "several ... areas traditionally coming under the rubric of geography were beginning to experience the first flush of quantification."[107] The specific area of which Livingstone speaks in this instance is the field of hydrology. It is appropriate, therefore, that Wren, along with a coterie of seventeenth- and eighteenth-century measurers and calculators of rainfall, evaporation, and streamflow are cast as protagonists in the conventional narrative of the development of the hydrological sciences.[108]

We have already noted that questions about the nature and behaviour of water in general had been entertained in the context of classical natural philosophy. Problems such as the cause of rainfall, the link between evaporation from the seas and precipitation, the relation between rain and the flow of rivers, the behaviour of underground water, and the origin of springs had been of interest to natural philosophers from the earliest times. It is when the mathematical approach to water gets grafted onto such classic "hydrological" questions – when water became systematically swept up in the metrological fervour engulfing Western thought – that modern water began to take form.

Conclusion: A Manifesto for Modern Water

Around the time that Christopher Wren was busy measuring rainfall at Gresham College, researchers in France were conducting embryonic studies on the mathematical relationship between evaporation, rainfall, and streamflow in basins tributary to the Seine River. The person considered by most modern hydrologists to be the first of their line (by virtue of his pioneering attempts to quantify flows of water in different stages of the hydrologic cycle in the 1760s) was one Pierre Perrault (1611-80). Until his reputation was resuscitated in the writing of hydrological history, Perrault was practically unknown, except perhaps for having lost the post of receiver-general for Paris when he was discovered helping himself to funds from the public treasury in 1664.[109] Nevertheless, it was Perrault's quantification of water rather than public funds that has stood the greater test of time. His measurements of precipitation, evaporation, and streamflow in a basin at the source of the Seine were made with the ostensive aim of proving that the water of springs was derived from rainfall (precipitation) rather than from the oceans and seas via subterranean channels. More will be said of Perrault in the following two chapters, but for now let's consider a quote from the concluding paragraph of his famous (to historians of hydrology at least) monograph, *On the Origin of Springs*. In it Perrault delivers a statement that might be taken as a kind of manifesto for modern water:

> For me, who have undertaken to speak only about the Origin of Springs, it is sufficient to have done so, and by this means to have given them birth. Their fate is to run upon the Earth and throughout the World, I shall let them do so without taking any interest in what may happen to them, good or bad; if the ones become famous through the various good or bad qualities they have contracted in their travels, according to the lucky or unlucky meeting they may have made with favourable or unfortunate soils; if others attract the admiration and amazement of curious people by their flow and by their surprising effects, if others remain by nature mild and peaceful, as they were at birth. All this is no concern of mine, it is enough that they should be simply springs, their quality being only an accident which can happen or not happen to them without changing their essence.[110]

This quotation suggests the main contribution of science to modern water: the disentanglement of the waters of the earth from the chorological and cultural contexts that otherwise give them meaning for people. Here we have a founding statement – at the very doorstep of the modern era and the very moment of the birth of scientific hydrology – signalling the intention to abstract water from these contexts, to establish its essence in terms of what can be measured, and relegate its qualities and meanings to the status of mere accident. Here, in other words, we have a prescription for the death of water's social nature.

5

The Hydrologic Cycle(s):
Scientific and Sacred

In the previous chapter, I discussed how the scientific approach to hydrology has made a seminal contribution to the way we understand water in the modern world. Essentially, this contribution consists in performing the necessary intellectual move that allows us to quantify water and to abstract it from the cultural contexts that otherwise define its social nature(s). Let's begin putting this idea to work, specifically by examining the concept of the hydrologic cycle. Along with the construction of H_2O, the development and dissemination of the concept of the hydrologic cycle represents an important contribution to the idea of abstract, modern water. The hydrologic cycle is the central concept or model of modern hydrology, but it has spilled beyond the confines of scientific discourse and inundated the popular imagination. The hydrologic cycle's popularity is suggested by its ubiquity – especially in diagrammatic form – in academic and popular visual discourse. There can scarcely be found a standard earth sciences or geography text that does not feature a description and diagram of the hydrologic cycle. A quick search on the Internet for images of the "hydrologic," "hydrological," or "water" cycle yields hundreds of illustrations.[1]

Like H_2O, the hydrologic cycle relates to modern water internally, that is, as both cause and effect. Both these ideas are made possible by considering water in the abstract – as a phenomenon that exists independently of people – and both have the effect of making it appear natural that this should be the way of understanding water. The main difference is that "H$_2$O" presents water in its essentially static form (as physical substance), whereas the hydrologic cycle represents its essential dynamic character (the natural circulation of this substance). As argued in Chapter 9, by abstracting the circulation of water from its social context, the hydrologic cycle contributes to difficulties in the conceptualization of the relationship between people and water. Because it is understood as the natural circulation of water on earth, the only possible way that people can involve themselves in the hydrologic cycle is to *alter* it, thus inevitably producing an antagonistic kind of relationship. Instead of allowing for the increasingly hybrid (socio-hydrological) nature of the circulation of water, the hydrologic cycle conditions an understanding that keeps water and people in separate, externally related spheres.

The hydrologic cycle represents an important scientific achievement in understanding the behaviour of water in the hydrosphere. The theoretical – and especially the empirical – work out of which it came to be known is associated with the correction of deeply held beliefs and conjectures that today can be described only as ludicrous. (See the discussion of the origin of springs below.) There is, moreover, no question that the hydrologic cycle "works" in the sense that it describes the potentiality and complexity of water within a range of possibility defined by the physical properties of H$_2$O on earth. The aim in this and the next chapter is to consider the more fixed idea of water that is conveyed or effected by standard, conventional representations of the hydrologic cycle. In this respect, the significance of the hydrologic cycle is its fixing a particular idea – or picture – of this circulation (see Figure 1.4). In fact, the proportion of H$_2$O molecules that follows the complete path depicted in most diagrams of the hydrologic cycle, at least without diversion through countless subcycles, is very small indeed. Moreover, the standard image of the hydrologic cycle, by featuring the presence of copious supplies of liquid, flowing water, betrays the northern temperate bias of its construction.[2] By considering the history of the hydrologic cycle, I do not mean to question the reality of the water process but, rather, wish to point out the mistake of confusing the representation for reality.[3]

Like (modern) water itself, the hydrologic cycle seems so obvious that it is normal to regard it as "one of nature's grand plans" or "a great natural system."[4] But as natural(ized) as it has become, the hydrologic cycle nevertheless internalizes a human story. To the extent that a grand plan is involved, much of the planning has been on the side of the people who, quite literally, have *drawn* the hydrologic cycle from water's ineffable flow. In this and the following chapter, I present the hydrologic cycle as something that takes shape where the water process meets with the practice of scientific hydrology. I therefore treat the hydrologic cycle as a hydrosocial phenomenon, albeit one that represents water in a way that erases its own social content. In presenting the hydrologic cycle in this light, my overall aim is to highlight the important part it has played in disseminating modern water. This is an argument that will be taken up once again in Chapter 8, when we push modern water to the ends of the earth with the concept of global water.

For now the aim is to consider how and when the hydrologic cycle concept came into being, an approach that is best described as historical in the same sense that we considered the general history of modern water in Chapter 4. Like everything that we proclaim to be natural, the hydrologic cycle has a history – or rather, it has histories, for there are at least three ways of telling the story of how it came about. In keeping with our relational-dialectical approach to things, each of the three histories presented here tells a story about something that is slightly different from the others. In each case the observers, storytellers, or historians in question see and describe a slightly different hydrologic cycle, depending on the particular practices in which they are involved.

The first story is the one usually told by modern hydrologists about how the scientific concept of the hydrologic cycle was discovered. This is the most common story of its origin. In fact, with only one major exception (discussed below), it is the only kind of hydrologic-cycle history that has been published, and so we may refer to it here as the conventional history. Because it is told mostly by scientists for scientists (or for students of science), this conventional history tells the story of what I call the "scientific" hydrologic cycle – it presents the hydrologic cycle not as a hydrosocial construction but as something that has always existed in nature but which was imperfectly or poorly understood until relatively recently. Although the hydrologic cycle was the subject of much speculation in ancient and medieval times, the conventional history relates the story of how the correct understanding of it got revealed. (Pierre Perrault, the proto-hydrologist introduced at the end of Chapter 4, figures prominently in this history.) This story is commonly found in textbooks and reference books on hydrology and other earth sciences, environmental sciences, and geography. It provides the main theme in most accounts of the history of hydrology.

The second story is the exception mentioned above. It was told by the geographer Yi-Fu Tuan in a book with a title that might seem odd at first glance: *The Hydrologic Cycle and the Wisdom of God*.[5] Tuan shows how the concept of the hydrologic cycle was articulated by natural philosophers writing in the tradition of natural theology from the seventeenth to the nineteenth centuries.[6] In Tuan's account, these writers sought to prove the existence and the providence of the Christian God by drawing attention to the wonderful design of Creation, and for this purpose the phenomenon of the circulation of water on earth served extremely well. The hydrologic cycle that they revealed bespoke the balance, harmony, and generosity of the Creator as manifest in his cosmological waterworks. Because of the religious orientation and motives of its proponents, I refer to this as the history of the "sacred" hydrologic cycle.

Although Tuan's history of the (sacred) hydrologic cycle involves occurrences and actors that also appear in the conventional history, the two stories very seldom overlap in the sense that they make very little reference to one another. They are, after all, dealing with different things. Tuan describes a hydrologic cycle that emerged from the desire to furnish proofs of benevolent design in nature; hydrologists describe the development of a concept that emerged from the desire to prove the scientific rationality of water. That the two stories hardly overlap is partly a consequence of their drawing different lessons from water and describing it in different (sacred versus scientific) terms. Another reason is that, unlike the scientific version, Tuan's does not presume that the sacred hydrologic cycle pre-existed its construction by the seventeenth- to nineteenth-century writers whose work he examines. Tuan regards the hydrologic cycle as less a natural system than a concept that was invented – a product of human "ingenuity," to use his term.[7] No wonder Tuan's study hasn't been cited in any of the conventional histories.

Finally, consideration of the term "hydrologic cycle" leads us to yet another history of yet another hydrologic cycle. (As the principle of potentiality, water is easily capable of sustaining such multiple identities.) This final history is rather my own. It tells the story of how and when the hydrologic cycle of contemporary fame (the one with which most of us are familiar today) came into being and how it managed to lay claim to the world's water, past, present, and future. But that is the subject of the next chapter and will make sense only after the tales of the first two hydrologic cycles have been told.

CONVENTIONAL HISTORY OF THE (SCIENTIFIC) HYDROLOGIC CYCLE

The hydrologic cycle is the most fundamental principle of hydrology. Water evaporates from the oceans and the land surface, is carried over the earth in atmospheric circulation as water vapor, precipitates again as rain or snow, is intercepted by trees and vegetation, provides runoff on the land surface, infiltrates into soils, recharges groundwater, discharges into streams, and ultimately, flows out into the oceans from which it will eventually evaporate once again. This immense water engine, fuelled by solar energy, driven by gravity, proceeds endlessly in the presence or absence of human activity.

– DAVID MAIDMENT, ED., *HANDBOOK OF HYDROLOGY*

Hydrology can boast of a modest volume of work written in English that is devoted to its origins and development. This work encompasses what is described here as the conventional history of the hydrologic cycle.[8] The dominant theme of this history is the story of the hydrologic cycle: its origins and its precursors in ancient thought; the intellectual detours and absurdities of those who espoused erroneous notions about the circulation of water; the identification of the earliest proponents of the "correct" view; and the steps leading to its completion, closing, and ongoing perfection through the exact quantification of all phases and scales of the cycle. The association of hydrology with the hydrologic cycle is thus cemented in the writing and the reading of this history. To put it in a nutshell, hydrology is described as having origins in ancient philosophy but is understood to have emerged as a true science with the quantitative, basin-scale studies of French and English proto-hydrologists in the seventeenth century. With these studies, as described in a publication of the US National Research Council, "the hydrologic cycle was firmly established, marking the beginning of scientific hydrology."[9]

It has been important for hydrologists, as natural scientists, to identify the hydrologic cycle as a *natural* process – something that takes place "in the presence or absence of human activity," as described in Maidment's *Handbook of Hydrology* cited above. The conventional history of the hydrologic cycle relates how, over the centuries, natural philosophers and scientists revealed the workings of this natural process. This kind of historical narrative is typical of traditional histories of science, which often assume the form of narratives of scientific progress. When written by practising scientists, as has most often been the case for hydrology, these histories perform the function of authorizing and contextualizing the work presently being done in the field. "Practicing scientists," writes historian of science Jan Golinsky, "are continually appropriating the work of their predecessors and orienting themselves in relation to it."[10] Most of the general histories described in this chapter are of this type, written with the leitmotiv of the *progress* of hydrological knowledge through the ages. Their purpose – as has been made explicit by the Committee on History and Heritage of Hydrology of the American Geophysical Union – is "to encourage the study of the classical history of hydrology in order to provide practicing scientists with a sense of continuity by knowing their professional roots."[11]

Although perfunctory reference is sometimes made to ideas about the measurement and the circulation of water in the ancient East, most writers in this genre locate the roots of the modern concept of the hydrologic cycle in ancient Western thought as it is reflected in classical literature.[12] Practically all the classical Greek and Roman philosophers and sages commented in one way or another on the nature and behaviour of water in general. A sense of water's cyclical flow is often detected in these early texts, a sense that is reflected in an often-quoted passage from *Ecclesiastes* 1:7, thought to date from the tenth century BC:

> All the rivers run into the sea; yet the sea is not full; unto the place from which the rivers come, thither they return again.

Read in context, the phrase is intended as a metaphor for the eternal flux of worldly existence, the only possible refuge from which is found in worshipping an eternal God. Nevertheless, as we learn from the conventional histories of hydrology, the *literal* question of "the place from which the rivers come" remained a subject of hot contention throughout ancient and medieval times, a question that is often put in terms of "the origin of springs."[13] According to conventional history, the long-standing dispute had to be resolved by the application of quantitative methods in order to pave the way for the "correct" understanding of the hydrologic cycle. Thus, the question of the origin of springs – and the methods developed by proto-hydrologists to prove that all spring waters are derived from rainfall – assumes critical importance.

In Chapter 4 I noted that Pliny the Elder, in addition to relating the socio-natural history of various contemporary waters, avowed that "water permeates the earth everywhere, inside, outside, above, along connecting veins running in all directions, and breaks through to the highest mountain summits."[14] This (to us) rather absurd notion reflects the commonly held belief that water *rose* from the sea or from the earth in underground channels to provide the fresh water that we see on the earth's surface. As late as the eighteenth century, it was widely held that rainfall could not possibly account for the volume of water issuing from springs, flowing in rivers, and pooling in lakes. First, there was the obvious – at least obvious to contemporary European observers – fact that most springs continued to flow, and rivers continued to run, during times of little or no precipitation. In fact, the greatest known river on earth, the Nile, apparently came from a place where there was practically no rain whatsoever. And even in humid climates, it would have been difficult to imagine that rainfall alone could have accounted for the flow of, say, the Danube River. Besides that, until the late seventeenth century, it was widely believed that precipitation could not penetrate the ground to a depth of more than one or two metres.[15]

The question, then, was whether the water issuing from springs and flowing in rivers was formed by condensation underground before emerging onto the earth's surface, or whether it came from the ocean via subterranean channels through which it was desalinated. Although some argued that rain and snow played an important part in furnishing the water flowing on the earth's surface, an astonishing number of ancient and more modern thinkers subscribed to one or another, or a combination, of the subterranean hypotheses. Indeed, for over two millennia, the dominant theory was that water was formed within, or found its way into, the interior of the earth, from where it was lifted by various means to immense reservoirs within mountains, from which it emerged to feed the streams and rivers that returned the water to the sea (see Figure 5.1).[16] The modern explanation, based on the argument that rainfall is adequate to provide for streamflow, slowly gained popularity from the end of the seventeenth century; however, it became universally accepted only in the twentieth century.[17]

The dispute over the origin of springs and rivers has been treated in depth by hydrological historians and is summarized here with an emphasis on *how* it is treated in these histories.[18] In *The Birth and Development of the Geological Sciences,* published in 1938, Frank Dawson Adams devoted a chapter to the "problem" of "the origin of Springs and Rivers – which was the subject of continuous speculation and controversy throughout the

FIGURE 5.1 A depiction of the widely held notion of the subterranean source
of springs and rivers, from Athanasius Kircher's *Mundus Subterraneus,* ca. 1664
The subterranean flow of water (originating at the sea bottom) is shown in darkened
channels. The cutaway view illustrates the common belief that water was channelled to
reservoirs beneath mountains from which it sprang forth, giving rise to springs and riv-
ers. | Frank Dawson Adams, *The Birth and Development of the Geological Sciences* (Baltimore:
The Williams and Wilkins Company, 1938), 437.

ages, but which may now be said to have been definitely and finally solved,
in all its chief features at least."[19] Because Adams' version of the "problem"
has been adopted by most subsequent historians of hydrology, it provides
the main source for much of this section.

From Adams we learn that ancient Greek philosophers such as Anaxagoras (ca. 500-428 BC) and Plato (428-47 BC) "imagined that there is
within the earth an immeasurable cavern filled with water in continual
motion, from which all rivers issued and to which their waters returned."[20]

Aristotle (384-22 BC) countered that such a subterranean reservoir, in order to accommodate all the waters flowing in the rivers, would have to be larger than the earth itself. He offered three complementary explanations: First, that a portion of this water comes from rain, which condenses in the atmosphere, falls to earth, and goes into the ground, from where it re-emerges in the form of springs. Much of the history of hydrology would require rewriting were this the only explanation offered by Aristotle. However, he argued that *most* of the water in springs and rivers was actually formed beneath the ground. The reasoning for this makes sense when we consider that water was understood to be one of the four classical elements. In Aristotle's words,

> It is unreasonable for anyone to refuse to admit that air becomes water in the earth for the same reason that it does above it. If the cold causes the vaporous air to condense into water above the earth, we must suppose the cold in the earth to produce this same effect, and recognize that there not only exists in it and flows out of it, actually formed water, but that water is continually forming in it too.[21]

Aristotle appears to have suggested that a third source of surface water comes from "the condensation of vapors which 'rose' from some source which is not stated."[22] In any case, instead of a subterranean reservoir, as Plato had imagined, Aristotle suggested that water in the earth rose to the interiors of mountains, from which it sprang forth to aliment springs and rivers. The flow characteristics of different springs and rivers were explained by variables such as the size and interior temperature of mountains.

Adams and other early historians of the earth sciences report that with one notable exception, the classical writers who turned their thoughts to the question subscribed to the Aristotelian view that the source of surface water was primarily a subterranean affair.[23] "The Roman philosophers," writes the mid-twentieth-century American hydrologist O.E. Meinzer, "in general followed the Greek ideas, and did not contribute much to the Greek hypotheses except erroneous details."[24] They often quote Seneca to provide typical illustrations of how the hydrologic cycle was understood in Roman antiquity:

> You may be quite sure that it is not mere rain water that is carried down in our greatest rivers, navigable by large vessels from their very source, as is proved by the fact that the flow from the fountain-head is uniform winter and summer ... So then, the earth contains moisture which it forces out ...

By nature, too, earth has itself the power of changing into water: this power it habitually exerts.[25]

Following the decline of ancient Rome, the standard histories report a long hiatus in the progress of hydrological thought: "Somehow," writes Asit K. Biswas in his *History of Hydrology*, "after Seneca, the origin of springs was not given much thought till the latter part of the fifteenth century."[26] Given the centrality of the development of the "true" or "correct" understanding of the origin of springs to hydrologic identity, this period of over one thousand years merits only minor attention.[27] During the Middle Ages, intellectual activity in Europe was dominated by the Christian Church, which explains why the earlier noted passage from *Ecclesiastes* 1:7 was quoted by practically every commentator throughout this period and beyond.[28] The prevalence of the subterranean thesis combined with the doctrine that "unto the place from whence the rivers come, thither they return again" to produce the widespread belief that the origin of springs and rivers was seawater that found its way underground and was conveyed by subterranean rivers to the tops of mountains. Medieval writers ascribed the removal of salt to processes of filtration and distillation, while the elevation of the water was explained by a wide variety of what today seem rather wild hypotheses.[29]

Among the proponents of the subterranean thesis, Adams and others have deemed the Jesuit Athanasius Kircher (1601-80) "especially worthy of mention."[30] Kircher was actually less a scientist than a science writer with a talent for making vivid diagrams, which is probably why he is deemed so worthy of mention. His 1664 book *Mundus Subterraneus* reported on the dominant view of things, offering an entertaining account of gaping holes in the bottom of the sea floor, subterranean passageways, and mountain reservoirs, and ingenious hypotheses for the elevation of sea waters to the tops of mountains. Adams reproduces a number of illustrations depicting a subterranean hydrologic cycle from Kircher's book (see Figure 5.1).

"Experiments to the Rescue"

It was not until the mid-seventeenth century and later that the subterranean theory was finally disproved and the foundations of modern, scientific hydrology were laid. As described by Biswas, it was "Experiments to the Rescue" of hydrological truth as men began to measure and compare phenomena such as precipitation and streamflow in a drainage basin.[31] The elements of modern scientific practice were then becoming established in the intellectual centres of western Europe: the formalization of experimental method, institutionalization of scientific societies such as the Royal Society in London and l'Académie Royale des Sciences in Paris, the regular publication of the results of new investigations in the journals of these societies, and – most notably for our purposes – the elevation of measurement and quantification to the status of a methodological prerequisite in all branches of scientific practice. Thus, the stage was set for the investigations that are deemed to have put the hydrologic cycle – and the science of hydrology – on a truly scientific footing. "The time had now come," writes Adams, "when close and critical observation was to replace conjecture and ingenious speculation, and when a true basis was thus to be laid for a final solution of many, if not all the problems concerning the origin of springs and rivers, with reference to which there had been such widespread differences of opinion down through the centuries."[32]

As noted, the most important contribution to this "final solution" is judged by most hydrological historians to have been found in field experiments conducted by Pierre Perrault. Perrault's hydrological reputation rests entirely on the single work he had published anonymously in 1674. Given the significance that has been attached to this work (and its importance to the discussion below), Adams' description is reproduced here in full:

> Pierre Perrault in his book entitled De l'Origine des Fontaines ... presents the results of the first serious attempt to actually measure the rainfall and determine its relation to the amount of water carried off by the rivers. Perrault selected and measured the drainage area of the Seine from its source to Aignay-le-Duc. He found this area to have a superficies of 121.50 square kilometres. He then determined by means of a rain gauge the average amount of water which fell as rain upon this area annually during the years 1668, 1669, and 1670 and found it to be 520 millimetres. Taking this as 500 millimetres, if all the rain which fell upon this area during the course of a year remained in place and if no portion of it was lost by evaporation or otherwise, on the last day of the year the area would be covered with water to a depth of 50 centimetres, which would represent a volume of 60,750,000 cubic metres. He then found the amount of water which was carried off the area by measuring the amount which passed through the Seine canal at Aignay-le-Duc each year and found it to amount to 10,000,000 cubic metres, that is to say about one-sixth of the total rainfall.[33]

By virtue of having quantified and correlated precipitation and river discharge in a basin, Perrault is judged by Adams, and almost all subsequent historians of the science, to be the founder of modern, scientific hydrology.[34] A contemporary of Perrault's and a figure of almost equal importance in the conventional hydrological literature is the physician and member of l'Académie Royale des Sciences, Edme Mariotte (1620-84). Several years after Perrault published his book, Mariotte made a more detailed investigation of the entire drainage basin of the Seine above Paris. Mariotte's book *Traité du mouvement des eaux et des autres corps fluids* was published in 1686 and translated into English in 1718. This work dealt for the most part with what is now known as hydraulics and hydrodynamics; however, it included sections on the origin of springs, the determination of velocities of running water, and the measurement of river or aqueduct discharge, which are questions that have been of particular interest to hydrologists. Mariotte's more detailed and exact experimental methods of quantifying precipitation and discharge have earned for him a lasting place in the annals of hydrological history. He is understood to have "verified" or "confirmed" the work of Perrault.[35] Thus, although Perrault is often identified as the first quantitative hydrologist, Mariotte is deemed "the most eminent hydrologist of the pre-eighteenth century era" and even "founder of ground-water hydrology, perhaps of the entire science of hydrology."[36]

Along with Perrault and Mariotte, Edmund Halley (better known for discovering the comet) is often credited with making the hydrologic cycle a scientific fact. O.E. Meinzer was the first of many hydrologist-historians to identify the beginnings of modern or scientific hydrology with the troika of Perrault, Mariotte, and Halley[37]: "A new epoch in the history of hydrology began in the latter part of the 17th century through the work of Pierre Perrault (1608-1680) and Edme Mariotte (1620-1684) and other French physicists, and of the English astronomer Edmund Halley (1656-1742). These men put hydrology for the first time on a quantitative basis."[38] This interpretation of the hydrological past is standard in today's hydrologic text- and reference books. The article on hydrology in the *Dictionary of Physical Geography* provides a more recent example:

> The study of hydrology is at least as old as the ancient civilizations of Egypt, because the provision of a reliable water supply is essential to the survival of man. However, the development of plausible theories concerning the circulation of water in the hydrological cycle did not appear until the seventeenth century. These were largely based on observations of rainfall and river flow in the Seine basin by Pierre Perrault and Edme Mariotte and

on the ideas of Edmond Halley who simulated evaporation from the Mediterranean and concluded that this could account for all surface drainage.[39]

The appending of Halley – whose interests in hydrology were peripheral to his work in astronomy – to this group results from his quantitative studies of evaporation. In 1687, he reported an experiment in a paper with the self-explanatory title "An Estimate of the Quantity of Vapour raised out of the Sea by the Warmth of the Sun; derived from an Experiment shown before the Royal Society; at one of their late meetings: by E. Halley." The experiment, carried out at a meeting of the Royal Society, took place in a field nearby Gresham College in London, possibly nearby where Christopher Wren had experimented with his rain gauges. Halley measured the amount of water evaporated from a pan during a fixed interval of time and extrapolated from this to estimate the volume of water that would be evaporated from the Mediterranean Sea. He then calculated the daily quantity of water flowing into the Mediterranean from its tributary rivers and found it to be "but little more than one-third of which is proved to be raised in vapour out of the Mediterranean in 12 hours time."[40]

On the basis of this experiment, Halley argued that the amount of water evaporated from the Mediterranean was more than adequate to supply the flow of the rivers feeding it. Despite the obvious difficulties of extrapolating from an experiment conducted in England to estimate evaporation rates for the Mediterranean Sea, and ignoring the effects of mass atmospheric transfer, from the perspective of the history of hydrology, the major significance of Halley's contribution was in having looked for proof of the atmospheric phase of the hydrologic cycle by quantifying evaporation and attempting to balance it with precipitation. Writing in 1974, Raymond Nace points out:

> In summary, Perrault and Mariotte started with water already in the atmosphere and from there described the disposition of precipitation. Halley sought and found the source of atmospheric water vapour and attempted to show a balance in the complete cycle of water movement. Owing to the crudity of measurements and estimates, the main contribution of the three men was the concept of the cycle rather than proof by accurate measurement. The proof was delayed a hundred years until John Dalton took up the problem.[41]

As mentioned in Chapter 4, Dalton rendered a visual representation of water's modern essence in the form of a diagram of the water molecule.

The fixity of this essence was strengthened by Dalton's proof that it – the chemical compound of hydrogen and oxygen – remained fundamentally unchanged during, and despite, its transformation from one physical state to another. However, Dalton's importance to hydrology, as suggested in the quotation from Nace cited above, is in proving the hydrologic cycle by *"accurate measurement."*

This proof came by means of working out what has since become known as the water balance, or hydrological balance budget equation, for a specific region. In 1802, Dalton published a paper titled "Experiments and Observations to determine whether the quantity of Rain and Dew is equal to the quantity of water carried off by the rivers and raised by evaporation, with an inquiry into the origin of springs." With this paper, Dalton aimed to settle once and for all the question of the origin of springs – which was evidently still in doubt in England at this time – by the following method: First, he estimated annual rainfall in England and Wales on the basis of readings from thirty rain gauges observed over several years and added to this an estimate of the annual contribution of dew made to the region. Next, he estimated total runoff of all the rivers of England and Wales by extrapolating from work done by Halley over a century earlier on the discharge of the Thames. Subtracting this total runoff from the sum of rainfall and dew, he was left with a figure of a very large volume of water that had to be accounted for by evaporation in order to produce a balance. His estimate of the quantity of water raised by evaporation in England and Wales was derived from reports of other researchers and from his earlier experimentation with an atmometer. His computation proved to be out of balance to the amount of seven inches, a discrepancy that Dalton accounted for by presuming an overestimation of evaporation.[42]

This discrepancy aside, Dalton's place in the history of hydrology rests on his having formulated the water balance for a specific region, which has been judged to have finally and irrefutably proven the concept of the hydrologic cycle. His contribution was given formal recognition in 2004 when the European Geosciences Union awarded the first John Dalton Medal for distinguished research in hydrology. The literature accompanying the medal provides a synopsis of Dalton's work and a fitting close to this section on the conventional history of the hydrologic cycle and its contribution to modern water:

> It took well over 2000 years for a reasonable understanding of the science of hydrology to evolve. Not until the pioneering work of John Dalton in about 1800 were all the mechanisms of the large scale hydrological cycle properly

determined ... The correct understanding of the hydrological cycle was established and from then to now the work in hydrology has gradually grown and improved with new techniques and the availability of more data. However in this field, vital to all our lives, we still owe a great debt to Dalton and the other pioneers who helped to form a basis for our current knowledge.[43]

THE (SACRED) HYDROLOGIC CYCLE AND THE WISDOM OF GOD

Water exists, then, in a closed system called the hydrosphere, and contemplating the hydrosphere and the hydrologic cycle is almost enough to make a sceptic believe in the omni-existent Gaia. The system is so intricate, so complex, so interdependent, so all-pervading, and so astonishingly stable that it seems purpose-built for regulating life.

– Marq de Villiers, *Water*

The statement above appeared in a prize-winning book on water that was published at the close of the twentieth century.[44] It combines the modern (scientific) concept of the hydrosphere and the hydrologic cycle with an older idea that the circulation of water on earth is "so intricate, so complex, so interdependent" that it could only have been designed purposefully by a divine being. These two kinds of hydrologic cycle are sometimes conflated, but they entail separate stories and describe rather different things. The conventional history of the hydrologic cycle makes practically no reference to the many works that presented the idea of the circulation of water as an argument for the existence and beneficence of God – a concept that we are calling "sacred hydrologic cycle." And even though the scientific formulation was eventually metamorphosed into the modern hydrologic cycle (the one with which we are familiar today), the sacred hydrologic cycle was almost entirely banished from hydrological discourse by the late nineteenth century.[45] Although portions of its history may be found in other works, the only dedicated effort to pull together a comprehensive study of this concept was undertaken by Yi-Fu Tuan.[46] His book *The Hydrologic Cycle and the Wisdom of God* describes how proponents of natural theology deployed evidence of the circulation of water on earth to argue their case for a divine architect.

Natural theology (also known as physico-theology) locates evidence of the existence and presence of God in the works of nature rather than in the supernatural revelation of biblical texts. Although it has roots in more ancient intellectual traditions, the Christian version was especially popular in England from the seventeenth to the nineteenth centuries, as represented by proponents including John Ray (1627-1705), John Keill (1671-1721), John Evelyn (1620-1706), William Derham (1657-1735), and William Palley (1743-1805).[47] In looking to the empirical world for concrete evidence of God's wisdom and providence, these men sought to rekindle the sacred in a world they felt had been secularized at the hands of Cartesian mechanical philosophy.[48] Against an apparently soulless, machine-like world devoid of final causes, English natural-theologians formulated a metaphysical compromise whereby matter, which they took to be "inert and stupid of itself,"[49] was imbued with a spiritual impulse that permitted a purposeful Christian cosmos. This "spirit of nature," as described by historian Carolyn Merchant, "was an incorporated principle that pervaded the universe, directing the parts of matter and their motions."[50] With this incorporated principle pervading the universe, knowledge from the most recent investigations of proto-scientists as well as ancient knowledge of natural phenomena could be adduced as proof of the divine plan of nature.

For the English natural theologians, the apparent balance and abundance that was suggested when they considered the circulation of water was deemed one of the most compelling arguments in their favour. Tuan's study illustrates "the popularity and the almost constant exploitation of the concept of the hydrologic cycle in physico-theology" from the late seventeenth to the mid-nineteenth century.[51] By 1900, the sacred hydrologic cycle had nearly stopped flowing. The growing influence of Darwinism meant that the wonders of nature could less easily be attributed to a divine architect but now required an explanation that accorded with the internal logic of nature itself.[52] The cyclic character of large-scale processes, established with James Hutton's late-eighteenth-century description of the geological cycle (uplift, subaerial denudation, deposition, lithithification, renewed uplift), had become something of a staple of the budding earth sciences by the late nineteenth century.[53] The circulation of water thereafter left its sacred course and was channelled through more secular, scientific language. But some sense of the magnanimity and abundance of water's circulation remained in even the most purified distillations of the concept: "The waters of the earth," reported T.H. Huxley in his 1877 edition of *Physiography,* "move in a continued cycle, without beginning and without end. From rain to river, from river to sea, from sea to air, and back again from air to earth – such is the circuit in which every drop of water is compelled to circulate."[54]

In contrast to the conventional history, Tuan eschews the urge to identify the origin of the concept. "Indeed," he writes, "it is not rewarding to decide on degrees of originality when we are concerned with a concept the roots of which are multiple, deep, and entangled."[55] Nevertheless, Tuan does use the term "hydrologic cycle" unselfconsciously, never considering that its coinage postdated by a considerable margin the written works he studied. Hence, our use of the term "sacred hydrologic cycle" to distinguish the concept of the circulation of water as it occurred and was represented in natural theology.

Much of Tuan's study is centred on what he takes to be one of the sacred hydrologic cycle's best and most popular expressions, found (among a wealth of other evidence of intelligent design) in John Ray's book *The Wisdom of God Manifested in the Works of the Creation,* first published in 1691. Ray was mainly interested in the biology of plants, but he assimilated knowledge from all aspects of natural history and philosophy to furnish arguments in support of his case. In the first two editions of the book, he subscribed to the subterranean thesis of the circulation of water, marvelling at how "Springs should break forth on the sides of Mountains most remote from the Sea."[56] However, by the time the third edition was published in 1701, the investigations of proto-hydrologists such as Edmond Halley had found their way into Ray's book.[57] These inclusions, Tuan argues, enabled Ray to assemble the main arguments for the hydrologic cycle. Let's consider a relevant (and entertaining) section from the third edition that Tuan regards as exemplary. In response to the question put by putative doubters, "Where is the Wisdom of the Creator in making so much useless sea, and so little dry land?" Ray replies with a lesson on the circulation of water:

> This, as most other of the Atheists Arguments, proceeds from a deep Ignorance of Natural Philosophy; for if there were but half the Sea that now is, there would be also but half the Quantity of Vapours, and consequently we could have but half so many Rivers as now there are to supply all the dry land we have at present, and half as much more; for the quantity of Vapours which are rais'd, bears a Proportion to the surface whence they are rais'd, as well as to the heat which rais'd them. The Wise Creator therefore did so prudently order it, that the Sea should be large enough to supply Vapours sufficient for all the Land, which it would not do if it were less than now it is.[58]

"The main elements of the hydrologic cycle," Tuan argues, "are all discernible in this paragraph: evaporation from the sea, the amount of vapour raised being proportional to the size of the evaporating surface; the transportation of the vapour to the land, its condensation over it to produce rivers ... The segment of the cycle that is not explicitly given is that the rivers eventually return the water to the sea, thus maintaining its level."[59] However, it might just as well be argued that although there are certain correspondences between the modern, scientific hydrologic cycle and Ray's argument, Ray has not exactly rendered the former here. Arguably, the hydrologic cycle has been constructed in Tuan's account, even to the point of inserting the missing "segment." Many of the other writers cited by Tuan provided arguments that correspond more or less to what we now understand as the hydrologic cycle. But theirs was a concept and a way of representing water that differed from the hydrologic cycle with which modern readers are familiar.

That Tuan failed to acknowledge his own layering of modern hydrological discourse on these works is surprising in light of the very different approach he took from that of the conventional hydrological historians. Not only the texts and quotations used but the orientation and positioning of Tuan's study is different from that of those who have written the conventional history of the hydrologic cycle to strengthen the identity of hydrology as a scientific discipline. Rather than seeing the hydrologic cycle as a fact of nature that awaited the application of correct method to be revealed to proto-hydrologists, Tuan is attentive to its constructedness: "For in its finer expressions the concept does honour to the ingenuity of the human mind if not indeed to the wisdom of God."[60] A couple of other particularities of Tuan's study might be noted. First, with the primary use of the sacred hydrologic cycle serving as evidence for the manifestation of God in nature, the distinction between "correct" and "incorrect" explanations – which plays such an important role in the conventional histories – is hardly relevant. So long as the principle of circulation is seen as giving rise to a "well-watered earth" by which man, animals, and plants are able to thrive, it doesn't really matter whether the explanation fits the model of the "true" scientific hydrologic cycle. Many of the works featured in Tuan's account actually espoused one or another variation of the subterranean circulation of water. Thus, the distinction that is so important to conventional histories of hydrology – the application of quantitative methods to prove the correct hydrologic cycle – is almost absent from Tuan's account; he relegates the conventional histories to a single footnote.[61]

Second, in attending to the constructedness of the sacred hydrologic cycle, the fact that it was conceived in a northern, temperate climate was seen as being significant. In proclaiming the wonders of God's sublunary water works, "scholars of northwestern Europe were helped in their delusion by the well-watered – and even drenched – landscapes they saw constantly about them."[62] As water was translated from Christian to Darwinian scientific catechism, some aspect of this delusion was carried over, as Tuan seems to be saying when he writes: "In 18th century England the idea of a well-watered earth *was* an unexamined article of faith to those who *have fallen* for the pervasiveness of the hydrologic cycle and to those who *have allowed* themselves to generalize from a very limited experience" (emphasis added).[63]

The Western imaginary of water had been (and to some extent still was, even as late as 1968) pervaded by a measure of ignorance of – or a contempt for – aridity. As Tuan observed, "We do not have to probe deeply into the history of geographical ideas before we become aware of the very late recognition of the extent of dry lands and of aridity as a climatic fact."[64] Although there are certainly other factors involved, Tuan's study shows how Western attitudes toward the dry lands are due in no small part to "the doctrine of the providence of God and the justification of that doctrine on the basis of the hydrologic cycle."[65]

CONCLUSION

In recognizing the northern temperate bias of the hydrologic cycle, Tuan hints at a broader critique of the Western imaginary of water that suggests a kind of hydrological Orientalism. The dominant (Western) apprehension of deserts and arid lands as barren, poor, uncivilized places that must be hydraulically re-engineered in order to be made civilized has been a motivating factor, or pretext, behind the colonial and neo-colonial materialization of modern water on several continents. This may be regarded, in part, as the effect of a discourse that normalizes copious volumes of liquid surface water, the history of which is traced in Tuan's study. The sacred hydrologic cycle is constituted in a way that makes deserts and dry regions seem anomalous: "The Wise Creator," to repeat John Ray's phrase, "did so prudently order it, that the Sea should be large enough to supply Vapours sufficient for *all the Land*" (emphasis added). The notion that "all the land" was well supplied with water was not unique to Ray. Whether out of ignorance or out of enthusiasm, the geographical fact of aridity was conveniently overlooked by proponents of the sacred hydrologic cycle.

From Ray on down to "the many reputable authors in the first half of the nineteenth century who trotted out the same concept and seldom bothered even to vary the rhetoric," the hydrologic cycle both reflected and effected a kind of faith in the universality of humidity.[66] To the extent that the scientific hydrologic cycle retains a measure of this abundance in its common form of representation (see Figure 1.4), it may be argued that it, too, naturalizes the presence of abundant surface water. The ubiquitousness of this form, moreover, suggests how the modern hydrologic cycle, like its sacred forebear, has contributed to the general misunderstanding that continues to characterize Western attitudes toward arid regions.[67]

There is a close relation between the sacred and scientific versions of the hydrologic cycle, as is suggested in the conflation of the two in the quotation from Marq de Villiers' *Water*, which opens the previous section. Furthermore, as Tuan points out:

> One purpose of this essay is to show that, for at least one hundred and fifty years (ca. 1700-1850), the concept of the hydrologic cycle in the "standard" form was a handmaiden of natural theology as much as it was a child of natural philosophy. It is not to be presumed, however, that theologians merely took over and simplified a concept of science and used it for their own didactic purposes. In the nineteenth century this may be been true, but towards the end of the seventeenth century and in the beginning of the eighteenth such a description of the development of the concept would have been misleading. Not only was there no sharp distinction then between natural theology and science but scholars who wrote on the theme of the water cycle within the context of a physico-theological treatise actually contributed to it.[68]

What is of greater interest to us here, however, is that the *histories* of these versions of the hydrologic cycle generally do not overlap. Tuan's study, which can justly be described as the most lengthy and thoughtful monograph on the concept of the hydrologic published in the English language, if not anywhere, has been completely ignored in the conventional history described above. None of the post-1968 writings that make up this conventional history even mention Tuan's study, nor do they attend to the physico-theological writings on the circulation of water. For his part, Tuan makes only passing reference to the conventional histories, attaching to them no particular significance: "A comprehensive history of the concept of the hydrologic cycle has yet to be attempted. Short summaries are available and the facts need not be repeated here. Our main interest, moreover, is in relating the cycle to a special conception of God's providence."[69]

One reason that the two histories do not overlap – and why no "comprehensive history of the concept of the hydrologic cycle" had (nor has) yet been written – is because there is no single concept of the hydrologic cycle. The sacred hydrologic cycle was unique. It certainly held "water" in the abstract but not exactly as a *quantity.* Although its later proponents made use of hydrological arguments derived from the early water balance studies by Halley and others, the sacred hydrologic cycle was not defined by, nor did it require, the measurement of water. Another important distinction between the sacred and scientific versions of the hydrologic cycle is that the circulation of water was not conceived of, or represented, as a distinct thing in itself by John Ray and the other natural theologians. The history of the hydrologic cycle as just such a thing is the subject of the next chapter.

6

The Hortonian Hydrologic Cycle

Water remains a chaos until a creative story interprets its seeming equivocation."

– Ivan Illich, H₂O and the Waters of Forgetfulness

Introduction

At the beginning of Chapter 5, we considered the history of the hydrologic cycle as it has been written by modern hydrologists. The approach taken in this chapter differs from this conventional history in that it does not consider the hydrologic cycle as something present in nature and requiring the application of correct scientific method in order to be revealed. Rather, it treats the hydrologic cycle as a way of representing water that was constructed in, rather than revealed through, scientific practice. The particular hydrologic cycle it describes will be referred to as the Hortonian hydrologic cycle, after Robert E. Horton, the American hydrologist who author(iz)ed it, and unless otherwise stated, references to "the hydrologic cycle" in this chapter should be read as the "Hortonian hydrologic cycle."

By means of *formalizing* water in the anglophone hydrological tradition, the Hortonian hydrologic cycle makes an important contribution to modern water. The particular form that Horton's hydrologic cycle imparts to water is explained by way of a historical reconstruction of its representation rather than by reciting past investigations into the nature of water itself. Accordingly, in this chapter we locate the origin of the hydrologic cycle in a set of historical circumstances entirely different from those with which it is usually associated. By this account, the hydrologic cycle – as distinct from earlier investigations into the circulation of water – is of relatively recent origin, coincident with the emergence of hydrology as a separate scientific discipline in the United States in the first half of the twentieth century. From the moment that it was first assembled – specifically, in a paper read by Horton in 1931 – this hydrologic cycle has been carried forward in the practice of modern hydrology, and projected backward in the writing of (conventional) hydrological history. Through this process of carrying the hydrologic cycle forward and projecting it backward, as well as invoking nature at the moment of its birth, the history of the Hortonian hydrologic cycle has been obscured, making it appear timeless and authorless, like modern water itself.

In what follows, I apply what historian of science Jan Golinsky describes as "constructivist history of science" to modern hydrology, with a focus on the development of the discipline in the United States: "By a 'constructivist' outlook, I mean that which regards scientific knowledge primarily as a human product, made with locally situated cultural and material resources, rather than as simply the revelation of a pre-given order of nature. This view of science has attained widespread currency in recent years."[1]

In the case of this study, a constructivist outlook directs our attention to the social ingredients of ideas and representations that take shape when quantitative scientific practice is brought to bear on the water process. Such an approach – because it deliberately attends to the social – runs the risk of sliding toward a radical, ontologically relativistic form of social constructionism whereby the only reality that can be acknowledged and spoken of is made by people.[2] That is certainly not the intention here. As David Demeritt emphasizes, attending to the way scientific knowledge gets produced and "demonstrating the social relations its construction involves does not imply disbelief in that knowledge or in the phenomena it represents."[3] We can agree therefore with Golinsky when he points out that "'constructivism' is more like a methodological orientation than a set of philosophical principles; it directs attention systematically to the role of human beings, as social actors, in the making of scientific knowledge."[4] Golinsky draws from the work of researchers in various fields, including sociology of scientific knowledge, philosophy of science, and history of science, to formulate this methodological orientation, of which a couple of aspects are particularly relevant to the discussion.

First, a constructivist history of science attends to the establishment and maintenance of scientific disciplines as influencing the production of knowledge. Scientific knowledge is understood and analyzed in the context of specific disciplines – the main social structures within which modern scientific practices take place. Moreover, scientific disciplines "cannot be observed and described as natural organisms existing in their own right. Rather, they are the result of deliberate activities oriented toward order and specific purposes."[5] As argued below, the Hortonian hydrologic cycle was deliberately brought forward to organize and assert discipline in the practice of hydrology.

Second, constructivist history attends to the stabilization of a scientific representation or statement in subsequent practice. As Bruno Latour puts it, "The status of a statement depends on later statements ... The fate of what we say and make is in later users' hands."[6] Latour has advocated a method of observing "science in action" that traces the stabilization of scientific facts through their incorporation in subsequent texts, emphasizing that: "To survive or be turned into fact, a statement needs the *next generation* of papers." Eventually, as a fact gets incorporated into the body of a scientific literature, its origins become obscure. To give a most relevant example, Latour asks, "Who refers to Lavoisier's paper when writing the formula H_2O for water?" The stabilized fact – in this case, H_2O – thus becomes "incorporated into tacit knowledge with no mark of its having been produced by anyone."[7] As we'll see, the same may be said of the hydrologic cycle.

HORTON 1931

All the components of the hydrologic cycle concept had been proven mathematically and were ready for assembly as a complete system as early as the first decade of the nineteenth century. By the latter part of the century, scientists such as Huxley had done much of the necessary work, yielding "a continued cycle ... [from] rain to river, from river to sea, from sea to air, and back again from air to earth."[8] It remained only that the concept be given a name and rendered in the form of a visual representation. This was accomplished by Robert E. Horton in a paper read at a meeting of the American Geophysical Union in 1931. "The Field, Scope, and Status of the Science of Hydrology" was signed from Horton's home in Voorheesville, New York, and published in *Transactions of the American Geophysical Union* the same year.[9] In addition to introducing English-speaking readers to the hydrologic cycle, the article included a diagram to illustrate it (see Figure 6.1).

The purpose of Horton's hydrologic cycle was to stake a disciplinary claim to water – all the water at, above, and beneath the earth's surface – on behalf of the emerging science of hydrology. Before Horton's intervention, the basic concept that he concretized as the "hydrologic cycle" was more or less understood by hydrologists and other scientists. However,

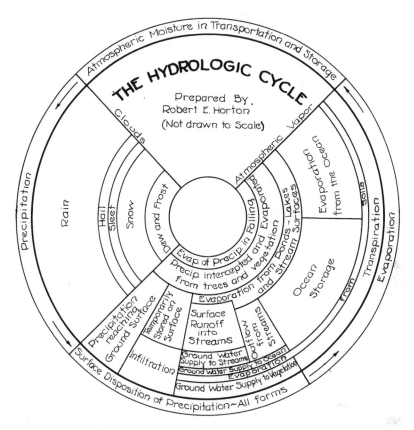

FIGURE 6.1 "The hydrologic cycle"
Robert E. Horton, "The Field, Scope, and Status of the Science of Hydrology," *Transactions, American Geophysical Union* 12: 193. © 1931 American Geophysical Union. Reproduced by permission of American Geophysical Union.

in the United States, as elsewhere, the term "hydrology" usually applied to only a portion of it, namely the study of underground-water phenomena.[10] Moreover, particularly in the United States, the study of hydrological processes in the atmosphere was usually considered the business of meteorologists. If hydrology was to make room for itself as an ambitious earth science, it required an ambitious unifying concept that would be capable of subsuming the widest possible range of hydrological phenomena. The idea of the cycle was popular in the earth sciences generally. A few decades earlier (in 1899), W.M. Davis had presented his famous "geographical cycle," which had been instrumental in organizing and systematizing geography in the United States. Davis' geographical cycle had had the effect of claiming the (study of) landforms for geography in anglophone academe.[11] Whether or not Horton sought to emulate Davis' success, his "hydrologic cycle" effected a similar acquisition for hydrology, bringing the entire hydrosphere into its remit while at the same time distinguishing hydrology from the other geosciences.

It is important to understand the institutional context in which Horton's paper was written to gain an appreciation of the significance of his presentation of the hydrologic cycle. By the late 1920s, hydrology had attained something of the status of a distinct and bona fide science internationally. In 1922, the Rome Assembly of the International Union of Geodesy and Geophysics had accepted a proposal to establish a separate International Section for Scientific Hydrology.[12] The following year, the International Association of Scientific Hydrology was established, with the adjective "scientific" being added to distinguish it from the commercial promotion of various mineral waters and thermal springs and "from the many charlatans and people misled by them who, with the help of all kinds of divining rods, undertook to find water, these people called themselves hydrologists; *the point was to be disassociated from them.*"[13] Hydrology's identity as a science nevertheless remained rather precarious. In some places, the term "hydrology" was still associated with the internal and external application of mineral waters to treat diseases.[14] And, as noted above, even among engineers and geologists, "hydrology" often applied only to subterranean hydrological phenomena. Those who considered themselves scientific hydrologists thus felt a need to, first, distinguish hydrology from its associations with mineral water enthusiasts, water diviners, and the like in order that it might properly be considered a science; and second, to define its field and scope so that hydrology might be distinguished from the other geosciences.

Meeting in Paris in 1924, the International Union of Geodesy and Geophysics defined the work of its new Hydrology Section in the following way:

> La nouvelle Section aurait le but de contribuer au développement des recherches scientifiques des phénomènes géophysiques lies aux fleuves, aux lacs, aux glaciers et aux eaux souterraines, qui forment le sujet de l'Hydrographie continentale ou de l'Hydrologie scientifique, nom peut-être préférable, afin d'éviter l'éventuelle confusion avec d'autres branches de la géophysique.[15]

For our purposes, the most significant aspect of this definition is that it limits hydrology to the surface and subsurface portions of what we now call the hydrologic cycle, leaving out all atmospheric hydrological processes.[16] The concept of the circulation of water had long been familiar to scientists and had given rise to sophisticated water balance studies in Europe and particularly in Russia beginning in the late nineteenth century.[17] However, the deliberate establishment of hydrology as the science that treats water through all phases of the hydrologic cycle seems to have been an American invention. In the United States, proposals to establish a separate Hydrology Section of the American Geophysical Union (AGU) had been rejected by the leadership on the grounds that there was insufficient "active scientific interest."[18] Finally, when the AGU underwent its own constitutional changes – and thus became a fully independent scientific society – in 1930, approval was given for a Section on Hydrology, with O.E. Meinzer as chairman and R.E. Horton as vice-chairman. At the union's 1931 meeting, Horton presented his paper on the "field, scope and status of the science of hydrology." With this paper, Horton introduced the hydrologic cycle, the main purpose of which was to provide a framework by which the new discipline of hydrology might become established and distinguished from the other geosciences:

> Defining science as correlated knowledge, it is true that a statement of the field, scope, and status of hydrology at the present time may be little more than a birth-certificate. Nevertheless, there is scattered through scientific and engineering literature a mass of quantitative results adequate to make a most respectable body of science, wanting only to be coordinated ...
>
> In one sense the field of hydrology is the Earth and so is co-terminous with other geo-sciences. More specifically, the field of hydrology, treated as a pure science, is to trace out and account for the phenomena of the hydrologic cycle ...
>
> Again, hydrology may be regarded as charged with the duty of tracing and explaining the processes and phenomena of the hydrologic cycle, or the course of natural circulation of water in, on, and over the Earth's surface. This definition has the advantage that it clearly outlines the field of hydrologic science.[19]

To varying degrees, something like the concept of the hydrologic cycle was implicit in the work of people who considered themselves hydrologists in the United States and elsewhere before Horton's intervention. Horton himself hints at this in citing texts and periodicals dealing with various aspects of hydrology that had been published in other languages, especially German, Russian, and French, beginning in the late nineteenth century.[20] Although Horton doesn't acknowledge it, it appears that Russian hydrologists in particular had been working with a very similar concept, "the rotation of water in nature," since the late nineteenth century.[21]

Within the United States, it could be said that the hydrologic cycle was implicit in the work of people who considered themselves hydrologists. Without actually using the term, for example, Meyer's 1917 American text *The Elements of Hydrology* was arranged in a way that corresponds to the sequence later given form in Horton's hydrologic cycle: "In developing the subject," noted Meyer, "the author has followed what from his viewpoint, acquired through about fifteen years of experience with the practical problems of hydraulic engineering, appeared a logical sequence."[22] This "logical sequence" took the shape of chapters on water in the atmosphere, followed by those on the condensation of water, precipitation, evaporation from water surfaces and land areas, transpiration, groundwater flow, runoff, streamflow, streamflow data, and finally, modification of streamflow by artificial storage. Another early American hydrology text associated hydrology with the "circulation of water above, on and within the earth's crust," noting that this "is as important and necessary in geological change and development as is the circulation of blood in the animal body or the circulation of sap to vegetable life."[23] Horton's signal contribution was to give substance to this "circulation of water," which to others merely "appeared a logical sequence."

As the main social structures within which modern scientific practices take place, scientific disciplines do not just materialize as reflections of nature's order but must be deliberately assembled. Before 1931, hydrology in the United States had been developed mainly as an aspect of civil engineering. Beginning around the turn of the twentieth century, the subject had been introduced as a separate technical study in several American engineering schools, and the first American texts on hydrology, such as those by Mead and Meyer, were written by engineers. Horton's paper – although written by someone who identified himself as a consulting engineer – represented a deliberate effort to raise hydrology from the status of a practical or applied field of study to a bona fide scientific discipline. Horton's 1931 paper has thus been deemed a key document in hydrology's "struggle for scientific recognition" in the United States.[24]

Horton's formulation of the hydrologic cycle accomplishes two necessary tasks associated with the establishment of scientific hydrology. First, it constitutes hydrology as a pure science in Horton's terms, because it represents the (abstract, objective) nature of water. As Horton defined it, the hydrologic cycle is "the course of natural circulation of water in, on, and over the Earth's surface."[25] By positioning it squarely in the realm of nature, Horton effectively applied what Bruno Latour has described as the "Modern Constitution" – the epistemological constitution dividing the world of nature from the world of society – to water.[26] For Latour, this constitutional separation is what allows science to represent and speak for nature. More will be said of this so-called Modern Constitution as it applies to water in Chapter 9. For now, I'll simply note that it was by asserting the hydrologic cycle as a natural phenomenon that Horton was able to make the claim for hydrology as a pure science: "As a *pure science,*" he proclaimed, "hydrology deals with the *natural* occurrence, distribution, and circulation of water on, in, and over the surface of the Earth" (emphasis added).[27]

In addition to establishing the hydrologic cycle as a thing that could be represented in diagrammatic form, Horton's paper gave it a basic mathematical expression:

There is a simple basic fact involved in the hydrologic cycle

Rainfall = Evaporation + Runoff

FIGURE 6.2 The hydrologic cycle as an expression of the
basic water balance equation for a basin

Robert E. Horton, "The Field, Scope, and Status of the Science of Hydrology," *Transactions, American Geophysical Union* 12: 190. © 1931 American Geophysical Union. Reproduced by permission of American Geophysical Union.

This "simple basic fact" is otherwise known as the basic water balance equation for river basins; it is frequently expressed as $P = E + R$, where P represents precipitation, E represents evaporation, and R represents runoff.[28] This equation is known to some as the Penck-Oppokov equation, after the Austrian and Russian scientists who are usually credited with developing it in the last decade of the nineteenth century.[29] Russian hydrologists in particular took a strong interest in the water balance concept and began to apply it around this time in their efforts to explain the phenomenon of endorheic (or closed-basin) drainage of the Caspian Sea.[30] In any case, by the 1930s, this equation had evidently become a well-established scientific fact.

Here, the term "fact" may be considered in the sense in which Ludwig Fleck elaborated it in his 1935 book *Genesis and Development of a Scientific Fact.* Fleck insisted that the production of scientific knowledge should be understood as a social enterprise. For Fleck, the social character of scientific facts is revealed by attending to the circumstances of their production within a specific scientific community: "A fact always occurs in the context of the history of thought and is always the result of a definite thought style."[31] Horton's hydrologic cycle was produced in the context of an interactive community (practitioners of hydrology) for whom the water balance equation was a "fact." By associating the hydrologic cycle so strongly with the water balance equation, he allowed the former to partake of the same facticity as the latter. The hydrologic cycle itself is therefore made a solid, scientific fact on which hydrology can be established as a pure science: "The central problem" of hydrology becomes that of "determining the physical process and principles and the quantitative relations involved in the hydrologic cycle."[32]

The second major contribution of Horton's hydrologic cycle was to define the (eventual) scope of hydrology. Horton charged the nascent science "with the duty of tracing and explaining the processes and phenomena of the hydrologic cycle," thus establishing definite boundaries around the science while orienting its work toward a specific purpose. The business of marking boundaries is always political, and Horton's intervention was no exception. By claiming the entire hydrologic cycle for hydrology, Horton was in effect intruding on the territory of another discipline. Before 1931, the atmospheric and other portions of what Horton called the hydrologic cycle fell squarely within the domain of meteorology in the United States.[33] Meteorologists were involved in "evaporation and run-off investigations," "river and flood work," and "mountain snowfall and water supply investigations," and these were even considered to be among the most important investigations undertaken under the aegis of meteorology.[34]

In the early years at least, hydrologists were sensitive to the intrusive character of their enterprise. On the occasion of the establishment of the AGU's Hydrology Section, its chairman, the groundwater hydrologist O.E. Meinzer, suggested that the work of the section be restricted to the nether portions of the hydrologic cycle: "The scope of the Section's activities," he proffered, "will not necessarily include everything that might be included within the science of hydrology. As the union includes a Section of Meteorology and a Section of Oceanography, it would seem that the work of the Section of Hydrology should relate chiefly to the study of the natural waters from their precipitation upon the land to their discharge into the ocean."[35] But the indivisibility of Horton's hydrologic cycle made it difficult to break up and parcel out among the different geosciences. By 1942, Meinzer had pushed hydrology's territory somewhat skyward: atmospheric water is still "primarily a part of the science of meteorology, but hydrology deals almost entirely with water of atmospheric origin."[36] By the end of the decade if not before, the entire hydrologic cycle was rather casually being claimed for hydrology, without any apology whatsoever.[37]

Since 1931, the hydrologic cycle has been reproduced by hydrologists and others in the United States and elsewhere who have carried it forward in scientific and popular literature. In the process, Horton's original contribution of naming and picturing it has disappeared from view. Today, the invisibility of Horton's handiwork allows the hydrologic cycle to be understood as part of nature instead of something with an origin and an author. At the same time, beginning around 1931, the historical roots of the concept of the hydrologic cycle began to be projected far into the past, further blurring its history and enhancing its apparent timelessness.

CARRYING THE HYDROLOGIC CYCLE FORWARD FROM 1931

Consolidating Hydrological Identity

Within the hydrological literature, the Hortonian hydrologic cycle has been carried forward since 1931 as a means of identifying the field and scope of scientific hydrology and providing "a conceptual framework for hydrological studies."[38] As already noted, the hydrologic cycle is represented diagrammatically and described as the "central concept" or "fundamental principle" in practically every textbook, handbook, and general article on hydrology published in the United States and beyond since 1931.[39]

Carrying the hydrologic cycle forward has been instrumental in upholding hydrological identity, at least in English-speaking countries. By most appearances, this identity has been relatively stable.[40] Nevertheless, hydrologists themselves have recognized the ambiguous status of their discipline, an ambiguity that arises from the perceived conflict between hydrology as a pure natural science and the practical, or applied purposes that often orient hydrological research.[41] In 1937, Horton called attention to "the fact that hydrology has largely grown up in the families of sister sciences, and the tremendous pressure for hydrologic research created by recent activities in soil and water conservation, have created a situation which seems to call for some pertinent discussion of the objectives and methodology of hydrologic research."[42] Horton's abiding concern for wresting hydrology from the status of an applied engineering science has persisted among many hydrologists until at least the recent past. "The present sorry state of hydrological science," declared the Canadian hydrologist V. Klemes in 1988, had arisen because hydrologists had allowed their discipline to be guided by practical and technical demands relating to water management rather than by the more pure, disinterested pursuit of knowledge about the behaviour of water; "the main reason is that, after all, many hydrologists do not have strong feelings about the science of hydrology and are only too easily seduced into performing the technological pirouettes required of them."[43] According to Klemes, in the late 1980s, hydrology still had not succeeded in establishing itself as a science. A year before Klemes made this statement, it had been observed that "the cultivation of hydrology as a pure science per se has not occurred."[44] And in 1991, the US National Research Council reported

> increasing concern among scientific hydrologists about the future and long-term vitality of their field ... owing, somewhat paradoxically, to the fact that throughout the history of this field applications have preceded science ... Because of the pervasive role of water in human affairs, the development of hydrologic science has followed rather than led the applications – primarily water supply and hazard reduction – under the leadership of civil and agricultural engineers.[45]

Nor has this increasing concern been limited to hydrologists working in North America. To quote British hydrologist J.E. Nash from 1992, "Hydrologists agonise over the status of their subject and seek, as they appear to have done for the last seventy years at least, the status implied by the term *scientific*."[46] For a scientific discipline that has, at least until recently,[47] exhibited something of an identity problem, the hydrologic cycle has lost none of its importance in upholding hydrology as "pure science," as Horton put it. This function is most obvious when it is insisted that the "mission" of hydrology needs to be "seen as improving the understanding of the water cycle" or, as put in more mathematical terms by James Dooge, "The business of hydrology is to solve the water balance equation."[48]

The hydrologic cycle, then, still serves Horton's original purpose as a means of framing and demarcating hydrologic science and establishing a distinctive hydrological identity. Perhaps the best illustration of this utility comes from the above-mentioned report by the Committee on Opportunities in the Hydrologic Sciences of the US National Research Council, the explicit purpose of which was "to establish an identity for hydrologic science as a separate geoscience."[49] Here the hydrologic cycle

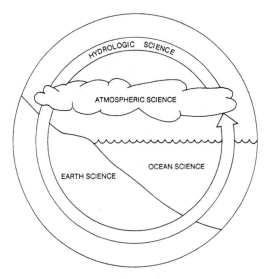

FIGURE 6.3 Hydrologic science = The hydrologic cycle
National Research Council, Committee on Opportunities in the Hydrologic Sciences, *Opportunities in the Hydrologic Sciences*, ed. Peter Eagleson (Washington, DC: National Academies Press, 1991) (from title page of the book). Reprinted with permission of the National Academies Press. © 1991, National Academy of Sciences.

is still the central concept of hydrology and "the framework of hydrologic science."[50] Indeed, the hydrologic cycle has graphically become the very identity of hydrology itself. As represented in the illustration on the frontispiece of the report, the hydrologic cycle and hydrologic science are *actually one and the same* (see Figure 6.3).

Picturing the Hydrologic Cycle

Perhaps the most *obvious* means by which the hydrologic cycle has been carried forward since 1931 has been through the reproduction of Horton's original idea of representing it in diagrammatic form. As a diagram, the hydrologic cycle has become a prolific image in scientific and popular discourse, such that it is often coeval with the concept of the hydrologic cycle itself. The American hydrologist Thorndike Saville put it in exactly these terms: "The generalized picture of the movements of water as it relates to the earth is called the *hydrologic cycle.*"[51]

   A brief discussion of the scientific diagram as a form of visual discourse is in order. The historian of science Bert Hall has written that, since the seventeenth century, visual images (naturalistic representations as well as diagrams) have gained authority in scientific practice as a consequence of having been accepted as "bearers of authentic information."[52] The authority of scientific images rests on structures of scientific discourse that are intimately associated with specific disciplines, or "thought communities":

> As a rule, we expect images that appear in the company of texts to explicate the material covered in the text in some fashion or other. Behind this lies the further assumption that both words and images are related to an external reality, a world "out there" whose description is one of the principal tasks of the scientist-writer whose texts we are reading. Yet it should be obvious by now that drawings cannot play so seemingly simple a role without an elaborate structure of institutional authority and personal credibility on the part of the author.[53]

The various parts of Horton's hydrologic cycle corresponded with different kinds of investigations that were underway in 1931 and were, in Horton's words, "wanting only to be coordinated."[54] Moreover, a scientific discourse in which the occurrence and movement of water – as water vapour, precipitation, groundwater, runoff, streamflow, and so on – could be accounted for quantitatively was clearly in place by 1931. There was, as Horton asserted, "scattered through scientific and engineering literature a mass of quantitative results adequate to make a most respectable body of science."[55] In other words, the readers of Horton's paper shared a language and set of rules that permitted a common hydrological discourse. This common discourse, together with Horton's immense personal credibility and the institutional circumstances in which his paper was presented made the hydrologic cycle and the diagram used to depict it intelligible and credible.

Nevertheless, the contemporary novelty of "the hydrologic cycle" and of the diagram by which it was represented meant that Horton had to, in effect, provide his readers with a kind of guide on how it should be interpreted:

> Beginning at the top and reading counter clock-wise the path of the water in the course through the hydrologic cycle may readily be traced. Figure [6.1] is, however, little more than a suggestive picture from which a wealth of detail is necessarily omitted. For example, on the diagram but little space is devoted to evaporation from the soil. This, however, is a very complex process. It involves the flow of vapor from the air into the soil, from the soil into the air, temperature differences between the soil and air, the diurnal and annual heat-waves flowing into the soil, evaporation from the

water-table, and also evaporation from capillary films in the non-saturated zone ... As a further illustration, the little segment labelled "Temporarily stored on the surface" includes water some of which goes into the soil as infiltration. In that case the detention or storage is often but a surface-film or thin layers in tiny pools. Temporary surface-storage ranges from such transitional phenomena to the extended and long-continued storage of water in lakes, swamps, marshes, as accumulated snow-layers in winter, as glaciers and polar ice-caps. Other details might be similarly amplified.[56]

Few of the subsequent descriptions and diagrams have been as detailed and as suggestive of the complexity that is avoided, or hidden, by representing water in the form of the hydrologic cycle. Nor have they been so explicit in acknowledging that the diagram is "little more than a suggestive picture." Nevertheless, so long as it was confined to texts and reference books circulating among hydrologists and students of hydrology, the hydrologic cycle remained relatively complex, and the difficulty of representing something as complicated as the behaviour of water by means of a single diagram or descriptive text was more or less acknowledged. For example, Meinzer's *Hydrology* – the first English-language hydrology textbook to be published in the United States after Horton's paper – featured in the frontispiece an illustration of the hydrologic cycle that provided considerable detail, particularly in the subsurface phases, and included only the barest hint of naturalistic features (see Figure 6.4).[57]

Meinzer's textual description begins:

The central concept in the science of hydrology is the so-called hydrologic cycle – a convenient term to denote the circulation of the water from the sea, through the atmosphere, to the land; and thence, with numerous delays, back to the sea by overland and subterranean routes, and in part, by way of the atmosphere; also the many short circuits of the water that is returned to the atmosphere without reaching the sea.[58]

This description represents something more convoluted than a simple cycle. Other descriptions of the hydrologic cycle appearing in the hydrology literature have done likewise. In Chow's 1964 *Handbook of Applied Hydrology*, it is stressed: "The hydrologic cycle is by no means a simple link but a group of numerous arcs which represent the different paths through which the water in nature circulates and is transformed."[59] The more recent descriptions of the hydrologic cycle within the hydrological literature have generally suggested less complexity. For example, compare

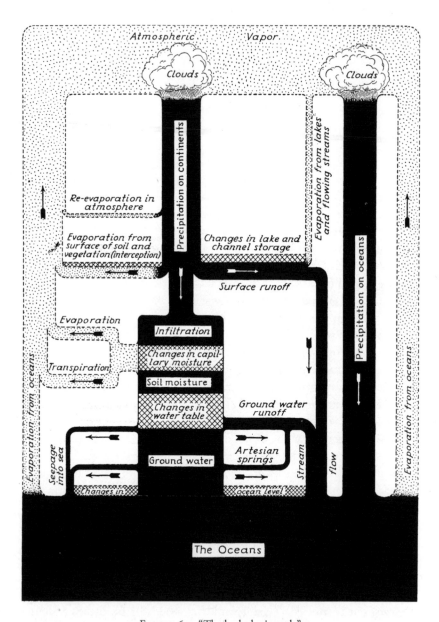

FIGURE 6.4 "The hydrologic cycle"

O.E. Meinzer, ed., *Hydrology* (New York: Dover, 1942) (from frontispiece of the book). Reproduced with permission of Dover Publications.

Chow's qualifications with the description of the hydrologic cycle found in Maidment's more recent *Handbook of Hydrology*, cited above.

Further, Meinzer's description of the "so-called hydrologic cycle" as "a convenient term" acknowledges its constructedness. A similar acknowledgement is made in some of the other earlier texts and articles by means of placing the term "the hydrologic cycle" in quotation marks or in italics.[60] However, in none of these is Horton's authorship – either of the term "hydrologic cycle" or the diagram – acknowledged, and even his role in having first identified hydrology as the science that treats of the hydrologic cycle is ignored.[61] In time, acknowledgement of the hydrologic cycle as a concept or term with a definite history disappears entirely. The term thus loses its quotation marks: the hydrologic cycle is no longer "so-called" but becomes instead "the natural pattern of circulation of water," "one of nature's grand plans," and "a great natural system."[62]

At about the same time that it became natural and ahistorical, the hydrologic cycle became far easier to *see*. The 1949 *Hydrology Handbook*, published by the American Society of Civil Engineers, partly acknowledged Horton's illustration by noting: "A comprehensive diagrammatic representation of [the hydrologic cycle] was published in 1931."[63] However, the illustration presented in the 1949 *Handbook* is far less complex than Horton's original and is described as a "somewhat less complete but perhaps more readily visualized diagram" (see Figure 6.5).

It is this type of simplified illustration that has served as the most common vehicle for disseminating the hydrologic cycle in popular discourse. In Figure 6.5 it is shown as the interchange of water between the ocean and a (continental) landmass, and employs numerous naturalistic features that make the concept more easily recognizable than in Horton's diagram. With only minor modifications, the 1949 illustration and commentary was reproduced from a paper titled "Basic Principles of Water Behaviour," presented at a 1936 conference organized by the US Soil Conservation Service.[64] Its author, in turn, had reproduced (again, with minor modifications) a slightly earlier diagram that had been prepared for a 1934 US federal government report on the need to rationalize the collection of hydrologic data on a national scale (see Figure 7.1).[65]

The relationship between the hydrologic cycle and the development of water resources by the US federal government is discussed in the next chapter. Here, we consider a couple of points about this descriptive style of diagram.[66] By virtue of its relative simplicity, its inclusion of stylized images of recognizable landscape features, and the ease with which it can

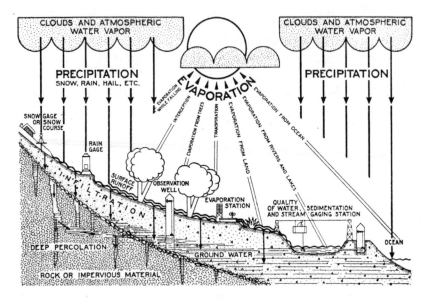

FIGURE 6.5 "The hydrologic cycle"
American Society of Civil Engineers, Hydrology Committee, *Hydrology Handbook* (New York: American
Society of Civil Engineers, 1949), 2. Reproduced with permission of the ASCE.

be read and understood, this type of diagram gained immediate popularity within and outside hydrological literature, such that it could be described as "the hydrologic cycle ... as conventionally drawn" as early as
1938 and "the usual form" by 1940.[67]

And so it remains today: perusing any standard earth sciences or geography textbook or publication (academic or otherwise) dealing with water
in general, one is likely to encounter a similar diagram.[68] There is, however,
an important difference between some earlier illustrations and the descriptive diagrams that circulate today; examples cited in this and the following
chapter are notable for their inclusion of the placement of hydrologic
instruments – snow gauge, rain gauge, observation well, evaporation station, and stream gauging station – at their corresponding places, or
phases, within the cycle. Some of these earlier diagrams even bore the title
"How the Water Cycle Is Measured" (see Figure 7.2). By featuring the
instruments by which the hydrologic cycle is measured, the mutually
constitutive relationship between the practice of hydrology and the hydrologic cycle is made rather apparent.

The immense popularity eventually attained by the hydrologic cycle
diagram is partly due to its scientific credibility, established in Horton's

time and confirmed by proofs of its quantitative integrity. In addition, however, there is an aesthetic quality in the idea and in the visual form of a cycle that acquires the power to convince by its very beauty and elegance. The persuasive power of such forms of representation is suggested by Hall, who, although describing technical illustrations of imaginative machinery in the Renaissance, offers an insight into the rhetorical effect of the visual depiction of the hydrologic cycle. Such images, he points out,

> persuade entirely through their effect on the viewer. The image ... seeks to reveal the hidden, inner workings of the machine to the viewer. The imagined machine will seem to work, and it will be seen to work by the viewer. In this way, the image draws the viewer into a process of verification that establishes the image's plausibility, and it creates its own authority thereby.[69]

There are obvious differences between the hydrologic cycle diagram and fifteenth-century drawings of imaginative machines. For one thing, depending on the climate in which we live, we may go outside and feel the rain, or dip a hand into a flowing stream, thus apparently corroborating parts of what we see in the picture with our personal experience. We can also measure these phenomena, thus confirming the picture with scientific evidence. But to the extent that the hydrologic cycle is presented as a kind of machine – an "immense water engine, fuelled by solar energy, driven by gravity," as Maidment puts it in his *Handbook of Hydrology* – to see it is, in effect, to believe it. As we know, the intellectual labour that preceded the crystallization of the Hortonian hydrologic cycle in 1931 was anything but simple. Moreover, each part, or phase, of the hydrologic cycle is astonishingly complex. In view of this complexity, the fact that a popular writer in the year 2001 could declare that "the hydrological system is simple to grasp" attests to the brilliance (and the success) of Horton's idea of representing the concept as a diagram seventy years earlier.[70]

As the science of hydrology became established as a bona fide discipline in the United States, Robert Horton was identified as the "father of American hydrology," the foremost leader of the science, and "the dean of American hydrologists."[71] But the reasons cited for Horton's fatherly status are not those we have given here. Rather, it is his elaboration of a quantitative relationship between the infiltration capacity of soils and the generation of floods by surface runoff that is most often cited as his major contribution to hydrology, especially his 1945 paper outlining a "hydrophysical approach to quantitative geomorphology."[72] But what was perhaps his greatest contribution – to materialize the hydrologic cycle as a distinct entity, to represent it in diagrammatic form, and to identify it with the discipline of hydrology – goes entirely unrecognized. Perhaps this should not be surprising. Hydrology, as a pure natural science fond of locating its roots in the august soil of classical natural philosophy, is hardly likely to proclaim its origins in a paper written in 1931 from Voorheesville, New York.

PROJECTING THE HYDROLOGIC CYCLE BACKWARD FROM 1931

Having been brought out in 1931, the hydrologic cycle was now ensconced by means of writing hydrological history. By this is meant that the hydrologic cycle was projected backwards into the works of ancient philosophers, seventeenth-century natural philosophers, and modern scientists alike. This project is described in some length in the previous chapter, under the rubric of the "conventional" history. Here, it remains only to highlight several key aspects of this history and to reconstruct the main sequence of events by which it was assembled.

There was a notable surge of historical interest in the controversy over the origin of springs around the time Horton wrote his 1931 paper. As already noted, despite the work of the seventeenth- and eighteenth-century proto-hydrologists, notions about the subterranean cycling of water remained popular, even among some scientists into the 1800s. These erroneous notions were finally discredited to everyone's satisfaction only around the turn of the nineteenth century.[73] Now that the matter had been "definitely and finally solved in all its chief features at least," wrote Frank Dawson Adams in 1928, it became "of interest to see where the observers of the past centuries strayed away from the path which would have led to the true explanation which they sought, and especially to determine what causes led them into error."[74] Although it was no doubt of interest, Adams might have added that it now became *possible* to see where the observers of the past had strayed from the path of truth. Adams' "The Origin of Springs and Rivers: An Historical Review," published in 1928, appears to have been the first paper published in English to deal with the controversy as a matter of historical interest. It was followed by Meinzer's "The History and Development of Ground-Water Hydrology," Baker and Horton's "Historical Development of Ideas Regarding the Origin of Springs and Groundwater," and Adams' chapter on the "problem" of "the origin of Springs and Rivers" in his book *The Birth and Development of the Geological Sciences.*[75]

However, as the hydrologic cycle was carried forward in hydrological discourse, the salient historical question was framed less in terms of the debate concerning *the origin of springs* and more in terms of the *development of the concept of the hydrologic cycle*. In Meinzer's 1934 account of the history of groundwater hydrology, he described its development in relation to the ancient dispute over the origin of springs (and never mentions the hydrologic cycle).[76] By 1942, however, Meinzer's chapter on the history of hydrology was centred entirely on the hydrologic cycle, which he introduced by noting: "The concept of the hydrologic cycle has become so generally accepted that it is difficult to appreciate the long history that lies back of its development and demonstration."[77] To cite another example, writing in 1938, Adams identified the famous French potter Bernard Palissy as "the first writer to recognize and insist that rain and the melting snows on the earth's surface were not one but the only source from which spring and rivers derived their waters."[78] Thirty years later, describing the same thinker, Biswas asserts: "Up to the time of Palissy very few persons had a correct understanding of the *hydrologic cycle*."[79] In general, after 1940, the main players – Aristotle, Vitruvius, Leonardo da Vinci, Palissy, Perrault, Mariotte, Halley, and others – whose work had been featured in the history of the dispute over the origin of springs in the late 1920s and 1930s, were recast in the new story of the "general evolution of the concept of the hydrological cycle."[80]

Interest in the history of the concept of the hydrologic cycle appears to have peaked in 1974, the year UNESCO sponsored a conference to coincide with the celebration of what it called the tercentenary of scientific hydrology.[81] The date was significant, as it marked exactly three hundred years since the publication of Perrault's monograph, *De l'origine des fontaines*. However, it was not the origin of springs per se whose history was written in the papers presented at this conference. As noted in the foreword to the published conference proceedings, "among the basic concepts of hydrology, the concept of the hydrological cycle may be considered as central. *Obvious as this cycle may seem nowadays,* it took a very long time to understand the mechanism correctly (emphasis added)."[82]

A very common practice in the process of writing the conventional history of the hydrologic cycle has been to locate its origin by identifying the first person to have the true, or correct, understanding of the modern concept. Reference has already been made to Bernard Palissy in this respect. Others have cited different writers: Meinzer observes that Book 21 of Homer's *Iliad* contains a statement suggesting the concept, but that Vitruvius was the first to state it clearly, and that Leonardo da Vinci was the first to understand it correctly.[83] According to Biswas, Theophrastus (ca. 372-287 BC) was the first to comprehend the hydrologic cycle, while Vitruvius merely restated his ideas a couple of centuries later.[84] Garbrecht claims that Xenophanes of Colophon (ca. 570-475 BC) was "the first advocate" of the hydrologic cycle, but that Anaximander of Miletus had described portions of the cycle a generation earlier.[85] Various other sages have been credited with being the first to describe the "correct" hydrologic cycle, ranging from the relatively renowned, such as the pre-Socratic Greek philosopher Anaxagoras of Clazomenae, to the rather more obscure nineteenth-century English hydrogeologist and expert on the thermal springs at Bath, one Edward Jorden.[86] With respect to the variety of these attributions, projecting the hydrologic cycle backwards into the writings of people for whom it did not exist is bound to be a somewhat haphazard affair.[87]

CONCLUSION: THE FATE OF PERRAULT

Perhaps the most interesting aspect of the projection of the hydrologic cycle backwards into history is twentieth-century hydrologists' identification of Pierre Perrault as the founder of the hydrological sciences. The legacy of Perrault offers another good illustration of Latour's observation that "the fate of what we say and make is in later users' hands."[88] Beginning in 1938 with Adams, the historical writings of modern earth scientists have almost unanimously identified Perrault as the foremost trailblazer of modern, scientific hydrology.[89] Larocque perhaps best summarizes the way Perrault's contribution is understood by stating, "Without Perrault's discoveries on springs and rivers, the entire concept of the hydrologic cycle would never have been developed."[90] And as if to stoke his legacy, at an international colloquium organized in 2001 on the origins and history of hydrology, a reproduction of Perrault's 1674 *De l'origine des fontaines,* was provided to every participant.[91]

Before the establishment of the hy drologic cycle in 1931, however, Perrault was more likely to have been ignored or even ridiculed than elevated to the status of founder of a discipline. There is, it may safely be said, no reference to his work by English or American hydrologists writing in the nineteenth century. And as late as 1926, A. Hallays, who wrote a history of the Perrault family, particularly of his more illustrious brothers, very contemptuously dismissed Pierre's scientific efforts.[92]

Twelve years after Hallays dismissed his work, Perrault was judged to be the key figure in laying the "true basis for a final solution of many, if not all the problems concerning the origin of springs and rivers" and, two years after that, he was the person to "put hydrology for the first time on a quantitative basis," a position he has held more or less ever since.[93] What accounts for this extraordinary improvement in Perrault's reputation? The fate of Perrault and that of the hydrologic cycle, it seems, are closely related. The birth of the hydrologic cycle in 1931 as the foundation of hydrologic science has had a marked influence on the way hydrologists have interpreted the past, including the significance of an otherwise obscure administrator of l'ancien régime who was summarily dismissed for dipping into tax receipts he had collected on behalf of the king.

7
Reading the Resource: Modern Water, the Hydrologic Cycle, and the State

Introduction

In this chapter, we consider the development of modern water in the United States in the first half of the twentieth century. Modern water, of course, was hardly unique to the United States at this time. As noted in the previous chapter, key moments in hydrological abstraction occurred elsewhere: in Germany, France, and, as discussed in Chapter 8, particularly in Russia and the former Soviet Union. As the internationalization of scientific hydrological discourse was effected through institutions such as the International Association of Scientific Hydrology after the First World War, the language of modern water was diffused around the world and translated into a plethora of large water projects undertaken by the modern state in Europe, North America, and the Soviet Union, and by the colonial state in many parts of the global south (see Chapter 3). "Hydronationalism," the drive to rectify rivers in the name of building national strength and economic development, became a common theme in many countries throughout the twentieth century, signalling an effect of modern water as well as a factor in its production.[1]

Nevertheless, the broad influence of American hydrological discourse, particularly outside the Soviet Union, justifies the focus that we give it here. The United States was to become a world leader in the propagation of modern water. Completion of the Hoover Dam (the world's first so-called super dam) in 1936 marked what many consider the beginning of a new era for water around the world. "The first of the world's great dams," notes Marc Reisner, "Hoover inaugurated an Age of Dams, which has spanned the past three-quarters of a century."[2] This American leadership was made evident in the dissemination of American hydrosocial models to other countries in the name of international development following the Second World War.[3] Immediately after the war, Michael Straus (President Harry Truman's commissioner of Reclamation) described the control of water "as a prerequisite of all development" while vaunting that "the American concept of comprehensive river basin development ... has seized the world imagination."[4] The influence of American hydrological discourse might also be inferred from an examination of hydrology textbooks published in English and other languages outside the United States in the 1960s and 1970s. Such an examination shows that one or another version of the Hortonian hydrologic cycle was commonly featured as a representation of the nature of water and a framework for the practice of hydrological science.[5]

The main purpose of this chapter is to show how modern water emerged in the United States in association with efforts by the state to gain control of the nation's waterways. In effect, by institutionalizing the quantification of stocks and flows of water on a national scale, the state took a major step in making water available for, and amenable to, management by state agencies. These agencies were eventually to succeed in controlling a vast portion of the water flowing in American rivers, thus materializing modern water and consolidating its identity as an abstraction of flow. A related aim of the chapter is to show how the hydrologic cycle concept fit into this process of materialization and abstraction. In 1931, the year of Horton's presentation of the hydrologic cycle as the conceptual framework for scientific hydrology, the country was on the verge of what was perhaps the most ambitious program of fluvial manipulation ever undertaken. The role of the hydrologic cycle in this undertaking was only minor; however, it drew from the wide ambition of the concept as a means of pulling together all water, at all scales, within a single integrated model. Particularly in its visual form, the hydrologic cycle was readily adapted to the needs of the state in the 1930s to make the nation's water visible – or "legible" – to central agencies. Ironically, although Horton's hydrologic cycle didn't play an important role in this chapter of US water history, the simple rendition of Horton's idea in the form of a diagram by these agencies was nevertheless to find its way into popular discourse.[6] .

The dialectical process by which modern water and the modern state were mutually strengthened in the United States began with the naming of water as a "resource" in the early part of the century.

In his 1909 paper titled "Water as a Resource," W.J. McGee, a leading
light of the early conservation movement and Theodore Roosevelt's main
confidante and spokesperson for matters pertaining to the exploitation of
the nation's waters, represents the cutting edge of progressive thought
about the correct way to dispose of the nation's water:

> No more significant advance has been made in our history than that of the
> last year or two in which our waters have come to be considered as a resource
> – one definitely limited in quantity, yet susceptible of conservation and of
> increased beneficence through wise utilization. The conquest of nature,
> which began with progressive control of the soil and its products and passed
> to the minerals, is now extending to the waters on, above and beneath the
> surface. The conquest will not be complete until these waters are brought
> under complete control.[7]

Why is McGee making this statement now? After all, water had been used
extensively for purposes of navigation, irrigation, industry, and kinetic
power for centuries. McGee was aware that prehistoric peoples of North
America had constructed "elaborate systems of irrigation" upon which
complex agricultural societies had been built, and he no doubt knew of
the great ancient hydraulic societies of Mesopotamia, China, India, and
Egypt.[8] In his own country, water had powered the American Industrial
Revolution in the east. The Erie Canal had been constructed some seven
decades earlier, providing a transportation resource between the Hudson
and Great Lakes-St. Lawrence drainage basins. Moreover, the Bureau of
Reclamation had been established in 1902 and had completed or initiated
dozens of projects (including the construction of numerous large dams
west of the Mississippi) by the time McGee wrote.[9] So why is water sud-
denly considered a resource in 1909?

Advances in hydroelectrical and industrial techniques around this time
are part of the answer.[10] Improvements in the design of turbines, the development of a dynamo (generator) capable of producing electricity on a
commercial scale, and the use of alternating current for the efficient distribution of electricity were all brought to bear in the latter part of the
nineteenth century, making it possible for business and the state to recognize a new dimension of resourcefulness in flowing water.[11] The historian H.V. Nelles describes "the mysticism of hydro development" that
gripped the United States and Canada, where the prospects of exploiting

hydroelectrical potential on a large scale fired the imagination of politicians, industrialists, and engineers alike.[12] These prospects were certainly in the minds of McGee and his colleagues in the conservation movement.[13] However, the deliberate naming of water as a resource needs to be seen in the broader context of scientific, economic, and political developments that occasioned this movement during the first two decades of the twentieth century. It was in this context that water became known to the state in a way that made it an object of calculation and subject to a particular kind of accounting and manipulation, which for McGee and his contemporaries was signified by declaring it a resource.

Many accounts of the early conservation movement stress the democratic ideal of serving the interests of the people over the interests of the corporations, highlighting the trust-busting rhetoric of its proponents. While the progressive social ideals of the movement are beyond dispute, historian Samuel P. Hays has argued that the populist rhetoric of the movement belied its scientific-technocratic dimension: "Conservation, above all," Hays argues, "was a scientific movement, and its role in history arises from the implications of science and technology in modern society ... The essence [of the movement] was rational planning to promote efficient development and use of all natural resources."[14] Rather than the question of resource ownership, Hays stresses, the conservation movement was mainly concerned with promoting a rational, efficient regime of resource use.[15] McGee's proclamation of water as a resource in 1909 can thus be regarded as a proclamation of the need for the scientific management of the nation's waterways, a need the scale of which could be satisfied only by the state.

The vision of water management that arose within the conservation movement envisioned river basins as complete units and called for the rational control of water within the basin so as to maximize utility. Hence, the concept of multi-purpose basin planning, by which "experts using technical and scientific methods... should adjust power, irrigation, navigation, and flood control interests to promote the highest multiple-purpose development of river basins."[16] This vision of multi-purpose planning required the centralization of authority (under the federal government) as the only means by which such rational planning could be put into effect. There were, of course, immense political hurdles to meeting this requirement (discussed below). In the meantime, however, the most important prerequisite for achieving scientific water management was to procure a quantitative account of the nation's water. Effecting the "complete control" of water required surveillance of the country's water resources, the capacity to account for its stocks and flows, and ways of representing these accounts that facilitated rational management. James C. Scott uses the term "legibility" to describe the means and apparatus by which the state has sought command of a vastly complex field, reducing it to a framework and to terms that are actionable by government agencies so as to bring things under control:

> Certain forms of knowledge and control require a narrowing of vision. The great advantage of such tunnel vision is that it brings into sharp focus certain limited aspects of an otherwise far more complex and unwieldy reality. This very simplification, in turn, makes the phenomenon at the centre of the field of vision more legible and hence more susceptible to careful measurement and calculation. Combined with similar observations, an overall, aggregate, synoptic view of a selective reality is achieved, making possible a high degree of schematic knowledge, control and manipulation.[17]

Scott describes how thoroughly the physical environment has been represented and refashioned by what he calls "state maps of legibility" – models of nature that "rather like abridged maps ... represented only that slice of [reality] that interested the official observer."[18] Borrowing Scott's term, in order for water to be brought under control, it had to be made legible. Knowing and representing "precisely how much water was running through the land" thus became a project to which the federal government devoted a great deal of attention.[19]

In 1888, Congress authorized the United States Geological Survey to undertake the first investigation of water in the west. As head of the agency, John Wesley Powell took it upon himself to coordinate a massive program of water measurement and accounting.[20] Powell appointed to the task Frederick H. Newell, another architect of the conservation movement who was known to American hydrologists as "the father of systematic stream gauging."[21]

Newell's assignment marked the beginning of the Hydrographic Branch of the Geological Survey, which rapidly evolved into a nationwide program dedicated to producing and disseminating data enumerating the country's water resources.[22] "The information thus obtained and widely diffused," Newell wrote subsequently, "laid the foundations for a presentation of the needs and opportunities of water conservation and furnished the facts for action by Congress."[23] In other words, on the basis of the hydrological data that was produced, government agencies – particularly the Bureau of Reclamation (of which Newell was named the first director) and later the

Army Corps of Engineers – were greatly assisted in the planning and execution of large-scale water development projects.

By 1911, McGee thought it reasonable and expedient to declare:

> Under the federal legislation and administrative operations, water is not only measured more accurately than in any other country but is steadily passing under control in the public interest ... The advance in this direction during the last decade has been especially rapid; and though apparently little noted, it is among the most significant in our entire history with respect to knowledge, use and administration of the natural waters.[24]

The relationship between the hydrographic needs of the state and the development of water science is suggested in noting the contemporary progress in hydrological discipline in the country's universities. In 1904, the first course on hydrology was taught at an American university and the first textbook of sorts was prepared to assist in the university instruction of hydrology.[25] In the first decade of the twentieth century, considerable advancements were made in the systematic production of hydrological data, pushed by the emerging practice of hydrology in the universities and pulled by the growing demands of the state for reliable hydrometric data. The specific nature of this data is especially worth noting. Like the resources of the land – mainly forests, soils, and crops – water was now becoming available to central planners as a quantum. McGee captured this moment in the instant of proclaiming water a resource:

> Our growth in knowledge of that definite character called science is notable – particularly in its ever-multiplying applications ... It is in harmony with the general development that the quantitative method is now applied not only to soil production in forests and crops, and to mine production and minerals in the ground, but finally to the rains and rivers which render the land habitable and the ground waters which render it fruitful.[26]

McGee stressed that thinking of water in quantitative terms marked a radical change: "The quantitative view of water, except in smaller measures," he pointed out, "is so new to thought that familiar units are lacking."[27] Before being regarded as a quantity, he noted, water "has been viewed vaguely as a prime necessity, yet merely as a natural incident or providential blessing. In its assumed plentitude the idea of quantity has seldom arisen."[28] We know that the quantitative approach to water had in fact been a feature of scientific (hydrologic) practice for some time. McGee's statement declaring water a resource might therefore be interpreted as defining the moment when the scientific mode of knowing water became official, in the sense of its adoption by the state. The thrust of the conservationists' argument was that the state must adopt a modern hydrological perspective so as to adequately read the resource and permit its rational exploitation for the general good. It thus became a preoccupation of the Geological Survey and other relevant federal agencies to standardize hydrologic methods, instruments, and measures while expanding the scope of investigation to encompass the full range of hydrological phenomena – from precipitation to infiltration, groundwater flow, runoff, streamflow, and evaporation.[29] In the meantime, despite lacking the "familiar units" of measurement, McGee came up with a means of quantifying water resources that he felt was suitable and used it to present a description of "our stock of water" on both national and global scales.[30]

Representing the so-called stock of water suggests another important reason that water was now considered a resource: Because water could now be seen to have a definite supply, it was recognized as having a definite limit. Water, in other words, was now susceptible to scarcity, which made it imperative that it be managed efficiently. So long as we could assume its plentitude and regard it as a "natural incident or providential blessing," the idea of scarcity could hardly apply to water, as was ever the case for proponents of the sacred hydrologic cycle. Now considering water in quantitative terms, McGee was compelled to warn, "Our growth in population and industries is seriously retarded by dearth and misuse of water."[31] The quantitative view of water renders it *naturally scarce,* an idea taken up in Chapter 10 when considering the crisis of modern water.

The "Complete Control" of Water Resources

The general principles of the scientific management of the nation's waterways were worked out during the time of the conservation movement in the early part of the twentieth century. But despite the best efforts of Roosevelt and his conservationist allies, political circumstances were not conducive to bringing the nation's waters under central control. Agency rivalry in Washington, political bickering, and outright hostility to conservationist policies in Congress prevented the movement from achieving much in the way of actual policy change in the development of water resources. With some notable exceptions, a backlash against the conservation movement during the years of Republican control of the White House from 1921 to 1933 held federally sponsored river-basin planning and development in check.[32]

It was not until the mid-1930s, under F.D. Roosevelt's New Deal programs, that new life could be breathed into these projects and McGee's vision of "complete control" could begin to be realized in the country's river systems. Improvements in technique in the first decades of the twentieth century provided the technical circumstances in which such a vision was possible. Changes in the design of turbines significantly increased the hydraulic head at which they could operate efficiently, greatly increasing the economic feasibility of using high dams for generating electricity. The efficiency of long-distance transmission was improved through the adoption of stepped-up voltage, which made it economical to transmit power to the cities from relatively remote sites. Meanwhile, increases in the cost of fuels (mainly coal) in the 1920s significantly raised the price of electricity generated from thermal power, making hydroelectricity all the more attractive. These factors contributed to substantial growth in installed hydroelectric capacity during the 1920s in the United States and elsewhere in the industrialized world. It was on the basis of this growth that, in his seminal treatise on the functional theory of resources published in 1933, the geographer and resource economist Erich Zimmermann devoted a chapter to "The New Era of Water Power."[33] In Zimmermann's functional terms, "One of the most valuable byproducts of the development of the electrical industry is the new significance which has been given to water power."[34]

However, it was the social and political climate in the United States in the 1930s that provided the circumstances in which the state could usher in this new era on a large scale. The Great Depression provided the social circumstances in which significant expansion of the role of the state to manage the national economy could take place. A defining feature of the New Deal era was the federal project to take command of the nation's waterways, monumentalized in agencies such as the Tennessee Valley Authority and in physical structures such as the Hoover Dam. Harkening back to the vision of the early conservation movement, a later edition of his book had Zimmermann describing the New Deal in rather crusading terms as a "revolutionary change," with huge implications for water:

> The most powerful blow struck in defense of water power has come from
> ... changes in human attitudes, attitudes which imperceptibly emerge from

the womb of social philosophy and take on tangible form in rewritten and reinterpreted laws and in revised public policies. In the United States these new attitudes with their tangible aftermath are largely identified with the New Deal. But that is not quite accurate. The roots of the new movement, the origin of the new way of viewing our natural and national endowment of basic assets, reach further back in time, to the conservationists of Theodore Roosevelt's day ...

The crux of the new philosophy is the realization that, in the long run, the magnificent achievements made under private capitalistic enterprise are endangered by inadequate regard for the durable basic assets, natural and cultural, on which all economic life depends, and for the conditions, forces, and processes of nature which underlie all human endeavor ...

In [this] scheme of things, water plays a vital part. If properly cared for, water is the bringer of life; if neglected, it can be the cause of disaster. And it has been neglected by the market-conscious leaders of our economy.[35]

Given this neglect, the argument for enhancing the role of the state in developing water resources now became compelling:

Here was a vast field in which governments – local, state and federal – could perform prodigious feats without stepping on the toes of private business. If government had not done this long before, it was due partly to incomplete realization on the part of both experts and laymen of this vital necessity, partly to the failure of penny-wise and pound-foolish Congressmen to appropriate the necessary funds. But now all this changed. Business was unable – in the Depression – to take care of the livelihood of large portions of the population. The government had to step in. Here was the golden opportunity to make up for decades, perhaps a century of neglect.[36]

The most comprehensive and ambitious thrust of federal resource strategy in the 1930s was to promote "a unified plan of water control" for the entire nation, based on the philosophy of multiple-purpose planning and integrated development of the nation's river basins.[37] Thus, the principles of multiple-purpose river development that had been advanced by the conservation movement over two decades earlier were partly realized in the New Deal era.[38] The president established a series of agencies to plan for coordinated national resource conservation and development: the National Resources Board, 1934-35; the National Resources Committee, 1935-39; and the National Resources Planning Board, 1940-43. "There must be," declared the National Resources Board in 1934,

national control of all the running waters of the United States, from the
desert trickle that may make an acre or two productive to the rushing flood
waters of the Mississippi ... There is no stream, no rivulet, not even one of
those tiny rills which cause "finger erosion" in the wheat fields of the Corn
Belt, that is not a matter of some concern to all the people of the United
States.[39]

Ultimately, a fully coordinated form of "national control of all the running
waters of the United States" would elude federal government planners.
Although some federal multipurpose river projects were realized (most
notably through the Tennessee Valley Authority), the mobilization of
particular (regional) political interests and inter-agency bickering had the
effect of fragmenting state authority among competing agencies and across
various special-purpose programs such as flood control, land reclamation,
and municipal water supply.[40] Thus, the "complete control" of the nation's
waterways became less a matter of coordinated federal programming than
the cumulative effect of diverse interests and state agencies (at federal,
state, and local levels) operating on the country's rivers.

It was, however, out of this moment of enthusiasm for national multiple-
purpose river basin planning in the 1930s that the hydrologic cycle was
translated from Horton's hydrological discourse to an administrative
register. Following Horton's 1931 paper, a diagram depicting the hydrologic
cycle appeared in the National Resources Board's 1934 report, the purpose
of which was to provide arguments for centralizing water planning: "The
key to the beneficial control and use of the waters of the country is to be
found in recognition of four unities," the report declares; "Unity of
Physical Factors ... Unity of Man's Interests ... Unity of Responsibility ...
Unity of Action."[41] Literally in the midst of these declarations of unity –
unifying all these principles, as it were – there appears a diagram of the
hydrologic cycle (see Figure 7.1).

This appears to be the first example of a descriptive hydrologic cycle
diagram, which, as discussed in Chapter 6, has become the most common
style. Given the immense popularity of this style, this particular diagram
might be considered a prototype of sorts. The simplified and naturalized
version of the hydrologic cycle allowed for ready visualization of what the
National Resources Board itself described as "the extremely complex and
diverse phenomena producing and affecting the water resources of the
country."[42]

As a means of making water legible for administrative purposes, the
hydrologic cycle was ideal. It had been presented by Horton as a means

Precipitation and the Hydrologic Cycle

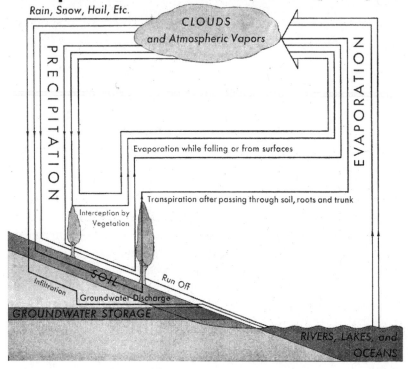

FIGURE 7.1 "Precipitation and the hydrologic cycle"
National Resources Board, *A Report on National Planning and Public Works in Relation to Natural Resources Including Land Use and Water Resources with Findings and Recommendations* (Washington, DC: US Government Printing Office, 1934), 262.

of expanding the scope of hydrology to include atmospheric and surface, as well as underground hydrological phenomena. Such a device served equally well the interests of the state to, as McGee had put it, bring "the waters on, above and beneath the surface ... under complete control."[43] The model thus provided a rhetorical device that by its very invocation proved the need for centralized collection of comprehensive, integrated, and standardized hydrometric data. Despite advances made to date, there was a perceived need to correct the vast "Deficiencies in Basic Hydrologic Data" that were *seen* by central planners to pertain throughout the country.[44] The National Resources Committee initiated a program for nation-wide coordination and standardization of hydrological data gathering,

HOW THE WATER CYCLE IS MEASURED

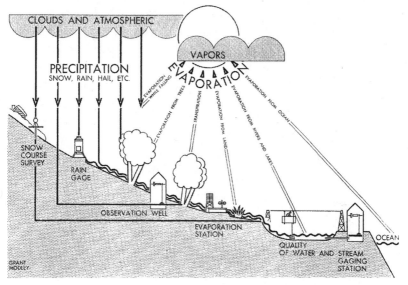

FIGURE 7.2 "How the water cycle is measured"
National Resources Committee, *Deficiencies in Basic Hydrologic Data: Report of the Special Advisory Committee on Standards and Specifications for Hydrologic Data of the Water Resources Committee* (Washington, DC: US Government Printing Office, 1937), viii.

whereupon the hydrologic cycle was (again) brought into use as a means of coordinating a vision of what was needed: "Water-conservation measures in the past," declared the committee, "have been hampered by lack of basic data concerning the hydrologic cycle."[45] Development of water resources depends on "exact knowledge of such factors as rainfall, snowfall, the flow of streams, the level at which water stands in the ground, the rate at which it evaporates or at which it transpires from trees and other vegetation, and the chemicals, suspended matter, or impurities which are found in it." More succinctly, "What we need to know" – a heading used in the document - is identified precisely, as "the hydrologic cycle."[46]

The diagram accompanying this particular report was titled "How the Water Cycle Is Measured" (see Figure 7.2). As noted in the previous chapter, a slightly altered version of this diagram was reproduced in the American Society of Civil Engineers' 1949 *Hydrology Handbook* (see Figure 6.5). The propagation of the hydrologic cycle in the United States can be regarded in part as a function of the way it has been traded back and forth between scientific hydrologists and state planners. As already noted about the teaching of hydrology in the universities, developing America's water resources has been characterized by a very tight relationship between the practice of hydrology and the knowledge requirements of the state. The adoption of the scientific hydrologic cycle by state planners, its altered appearance in government agency documents, and its reabsorption into scientific hydrology provides an illustration of the dialectical process relating bureaucratic interests with scientific practice. "Over time," writes V.K. Saberwal of hydrological discourse in the twentieth century, "one observes a two way process, whereby bureaucracies may use science to inform a particular rhetoric; at the same time, bureaucratic rhetoric comes to influence the scientific discourse itself, and, thereby, the very nature of science."[47]

CONCLUSION

"In the first few decades of the twentieth century," writes Donald Worster, "the command over rivers in the [American] West could hardly be said to have commenced."[48] It began in earnest in the 1930s and resumed with renewed vigour following the Second World War.[49] With only a tinge of hyperbole, Marc Reisner writes, "In that astonishingly brief twenty-eight year period between the first preparations for Hoover Dam [in 1928] and the passage of the Colorado River Storage Project Act [1956], the most fateful transformation that has ever been visited on any landscape, anywhere, was wrought."[50] Despite the limited success of federal multipurpose basin planning in the postwar years, agencies such as the US Army Corps of Engineers and the Bureau of Reclamation went a considerable distance toward realizing McGee's vision of complete control of the nation's rivers in pursuit of their mandates for flood control, irrigation, and hydroelectrical development. Water was thus internalized in the discursive machinery of the state, bent between levees and through turbines, spillways, and irrigation ditches to serve and to protect the people to the fullest extent that technology, public finances, and pork barrel politics would allow.[51]

Modern water can therefore be seen partly as an effect of what Worster terms the "immense ballooning of the state" with respect to its capacity for procuring hydrological data and managing water resources during this period.[52] At the same time, the materialization of modern water behind dams built by the US Army Corps of Engineers and other agencies helped produce a state that emerged as the most powerful in the world. The US

"hydrocracy" mushroomed, with the Bureau of Reclamation alone growing from two or three thousand to nearly twenty thousand employees
between 1930 and the late 1940s. After the Second World War, the further
enlargement of state water-control apparatus was rationalized by the contribution it made to the fight against Communism. The state, in other
words, materialized modern water, while modern water helped build the
state in a kind of reciprocal ratcheting process that contributed to making
America the foremost economic and military power in the latter part of
the century. The Columbia River, to cite an example, was set to providing
electricity for the production of aluminum in federal (Defense Plant
Corporation) factories used to build warplanes for the Second World War.
The river's contribution to the war effort was immense; "it is safe to say
that the war would have been seriously prolonged at the least without the
dams."[53] And by 1943, the Columbia – which was by far the largest source
of electricity in the world at the time – fed the reactors at the Hanford
complex as well as cooling them in order to produce the plutonium to
fuel the early American nuclear weapons program.

As noted in this chapter's introduction, the United States was hardly
the only country in which water was conscripted and put to work on a
massive scale in the mid-twentieth-century for the glory of the nation.
But the United States can fairly claim something of a leadership role in
this regard. By the 1940s, Americans were boasting of managing water
more effectively than any other country on earth.[54] By the late 1950s, it
was estimated that 10 percent of the entire national wealth was tied up
in "capital structures designed to alter the hydrologic cycle."[55] However,
in order to "alter the hydrologic cycle," the hydrologic cycle first had to
be made known, and the kind of water that it contained had to be spread
throughout the entire country by means of quantifying its flow. This accomplishment, as much or more than the physical manipulation of water,
constituted the realization of its "complete control" and the means by
which modern water came to dominate the rivers as well as the popular
imagination in the United States. In the next chapter, we consider how,
in this sense, modern water has been made to engulf the entire globe.

8
Culmination:
Global Water

*Any natural exposed surface may be considered as a unit
area on which the hydrologic cycle operates. This includes,
for example, an isolated tree, even a single leaf or twig
of a growing plant, the roof of a building, the drainage-
basin of a river-system or any of its tributaries, an
undrained glacial depression, a swamp, a glacier, a polar
ice-cap, a group of sand-dunes, a desert playa, a lake,
an ocean, or the Earth as a whole."*

> – ROBERT E. HORTON, "THE FIELD, SCOPE, AND STATUS
> OF THE SCIENCE OF HYDROLOGY"

*The water cycle is a global phenomenon. Therefore, water
resources are a global problem with local roots.*

> – RAYMOND NACE, "WATER RESOURCES:
> A GLOBAL PROBLEM WITH LOCAL ROOTS"

*The specific need for international collaboration in
hydrology stems from the very nature of the hydrologic
cycle.*

> – JAROMIR NEMEC, "INTERNATIONAL ASPECTS
> OF HYDROLOGY"

*Internationalism in science, insofar as it really does exist,
must be considered a social achievement, not the inevitable
consequence of some inherent scientific essence. It has to be
worked at.*

> – DAVID LIVINGSTONE, *PUTTING SCIENCE IN ITS PLACE:
> GEOGRAPHIES OF SCIENTIFIC KNOWLEDGE*

INTRODUCTION

M odern water is the presumption that any and all waters are to be considered apart from their social and ecological relations and reduced to an abstract quantity. This presumption culminates in the actual quantification of the world's water and its representation as a substantive fact, a culmination and fact that I call "global water." Global water has long been a possibility among those for whom it was natural to regard water as a quantity. When the early, mathematically inclined geographers, natural philosophers, and proto-hydrologists addressed themselves to questions of water in general (as opposed to the myriad waters of antiquity), global water was ever a sparkle in their beholding eye. In the early seventeenth century, for example, the German geographer Bernhardus Varenius (1622-50) suggested it might be possible "to compute what Quantity of Water the Earth containeth, and what Quantity of Land."[1] The mathematician John Keill (1671-1721) attempted to calculate the volume of water the ocean received from the world's rivers in what amounted to an early effort to compute a major component of the global water balance. Another example of an early attempt to quantify total river discharge into the oceans was made by the naturalist-writer Count Buffon (1707-88) in the mid-eighteenth century.[2]

For the most part, though, the quantification of global-scale hydrological phenomena seldom occurred to people before the early twentieth century, and global water remained a somewhat obscure hydrological preoccupation until after the Second World War. This was perhaps less a matter of considering water to be susceptible to such an operation than lacking adequate methods for such a purpose. But the most important requirement for establishing global water as a fact was the practice of a global hydrological discourse in which such a fact could thrive. As described below, global water balance studies had actually been developed in Russia in the first decade of the twentieth century, and estimates of the world water supply were published in the United States in the same decade by the likes of W.J. McGee. But these calculations represented a rather novel way of comprehending and representing water, one that required the right circumstances to take hold. As late as 1923, a leading American hydrologist had to admit (his ignorance) that "the position occupied by water in the economy of Nature is most remarkable ... yet it is nowhere told how much water there is in the world."[3]

The purpose of this chapter is to show the way in which this question of "how much water there is in the world" was eventually told to the satisfaction of hydrologists, and how this telling became an established fact.[4] Today, global water is commonly disseminated in tables, charts, and diagrams (for example, see Table 1.1 and Figures 1.3 and 10.1), and contained on and between the covers of books. Thus "Water of the World," "Water as a World Resource," "World Water Resources," "World Water," and simply "The World's Water" are common ways of conceptualizing and representing the waters of the world.[5] But there is more than a little artifice involved in these agglomerations; they came about as the issue of a well-planned and executed campaign to force all of the world's waters through the water balance equation and the hydrologic cycle.

GLOBAL WATER MATURES IN THE EAST

Global water materializes when the quantitative view of water described by W.J. McGee as "so new to thought that familiar units are lacking" gets projected onto the globe.[6] Propounding such a brand-new view himself, McGee was compelled to offer an assessment of the world water supply:

> The water of the world (or hydrosphere) is about 1/600 of the globe, or some 410,000,000 cubic miles. Nearly three quarters occupies depressions in the earth-crust as seas; about a quarter, or some 100,000,000 cubic miles, permeates earth and rocks as ground water; the remaining small fraction is gathered in fresh-water lakes and channels, accumulated in snow and ice, or distributed in the atmosphere as aqueous vapor.[7]

Nor was it a mere coincidence that McGee introduced his readers to global water in the same article in which he proclaimed water's resourcefulness. To become a resource is to become amenable to quantification and susceptible to scarcity. "Once a substance has been defined as a resource," notes resource geographer Judith Rees, "the question inevitably arises as to *how much is available* for use by man" (emphasis added).[8] McGee's interest in "the water of the world," moreover, came at a time of escalating geopolitical rivalry when America's relative share, especially for the purpose of generating electricity, became a matter of strategic importance.[9]

There are two basic methods of calculating global water. The one evidenced by McGee in the quotation above can be described as the static method.[10] Given the overwhelming preponderance of ocean water comprising the hydrosphere (about 97 percent), an approximation of the amount of water on earth can be derived by calculating the volume of the world ocean. As noted above, as long ago as the late seventeenth century, a method for estimating the total volume of the world ocean was intimated – although not implemented – by Varenius. Interest in calculating the relative volumes of ocean water and fresh water grew in the late nineteenth century, as evidenced for example in McGee's 1909 article. With the advent of echo-sounding surveys of ocean depths in the 1920s, calculations of ocean water became reasonably accurate. By the 1970s, a reliable calculation of the volume of the oceans had been worked out, largely abetted by a degree of coordinated international hydrological study that had taken place by this time.[11] Also by this time, many hydrologists were satisfied that estimates of the combined mass of glaciers and the volume of lake water on earth were reliable, though the volume of groundwater was, and remains, relatively unknown.[12]

This static method of calculating the world water supply, however, is usually considered less relevant and less accurate than the more dynamic water balance approach. Because there has generally been been less interest in the resourcefulness of ocean water than freshwater, and because freshwater is always in circulation, the static approach has long been thought to provide an inadequate account of the water resources available to society. What hydrologists consider the far more useful and accurate means of quantifying global water is to apply the water balance equation on a global scale. This approach takes into account the dynamic processes occurring in the hydrosphere that have the effect of constantly renewing water resources. The approach was developed in the laboratories and the literature of hydrologists working in Russia around the turn of the nineteenth century and matured in the Soviet Union in the following decades.

The basic water balance equation, as discussed in Chapter 6, was at least partly a Russian invention associated with the efforts of hydrologists to explain anomalous drainage phenomena in the Caspian Sea basin in the 1890s. It is generally acknowledged that the first to apply the idea of the water balance on the global scale (in 1905) was Russian hydrologist E. Ya. Brickner.[13] Following Brickner's work and applying the same basic method, some thirty-four global water balance studies by Russian, German, and other hydrologists in the years between 1906 and 1970 have been cited.[14] There is general agreement that by far the most sophisticated of these was made by Russian (later Soviet) hydrologists.

Several reasons may be given for Russian and Soviet leadership in this field. The influence of V.I. Vernadsky's pioneering work in global systems is likely to have sparked further development of global hydrology during the early years of the Soviet Union.[15] It was by building on the foundation of earlier Russian and Soviet hydrological investigations that hydrologists such as Gennady P. Kalinin, Mikhail I. Budyko, V.I. Korzun, and – perhaps the most renowned – Marc I. L'vovich (spelled L'vovitch in some publications) produced the world's most sophisticated and accurate methods of assessing the global water balance, beginning in the 1930s and 1940s. According to L'vovich, researchers working in the hydrology division of the Geography Institute of the Soviet Academy of Sciences developed methods that yielded hydrological information in circumstances where "hydrological data were very scanty."[16] The production of this information was made necessary by the demands of the Soviet system of central economic planning. As L'vovich and other Soviet hydrologists improved their techniques for producing information on the hydrology of the vast territory of the Soviet Union, they began to apply their methods on a global scale. Writing in 1974, L'vovich explained: "Because information available about sizable parts of the earth is less complete than the information available about other parts ... [the Russian] kind of approach ... has to be taken to hydrological studies of the earth's water balance and water resources."[17]

Thus, it was the work of Soviet hydrologists, including Kalinin and Korzun and his colleagues, but especially L'vovich, that set the standard for global water balance studies when these began to attain scientific credibility among Western hydrologists in the 1960s and onward.[18] Notwithstanding an early US study of the global water balance, American and other hydrologists who took an interest in global water in the 1960s relied for the most part on – ironically – Soviet sources for methods and data.[19]

GLOBAL WATER GOES WEST:
THE INTERNATIONAL HYDROLOGICAL DECADE, 1964-75.

If the International Hydrological Decade has done no more than force the countries of the world to look at their water on the world scale as a commodity they must inevitably share, and it has done much more than this, it will have justified the effort that has been made.

– REGINALD C. SUTCLIFFE, "INTRODUCTION,"
WORLD WATER BALANCE: PROCEEDINGS OF THE
READING SYMPOSIUM, JULY 1970

Before the 1970s, American and English hydrologists showed little interest in the global water balance.[20] This began to change with the work of an American hydrologist, Raymond L. Nace, whose reading of a 1945 paper by L'vovich sparked a life-long personal interest in global hydrology.[21] Nace, a staff hydrologist employed by the US Geological Survey who was able to read Russian, became a leading interpreter and proponent of global hydrology in the West. He authored numerous scientific and popular articles, as well as government reports that built on the work of L'vovich and other Soviet hydrologists, to present a picture of the availability of world water resources to English-speaking readers.[22] Nace's work came to fruition in the International Hydrological Decade (IHD), an international scientific program that Nace himself was instrumental in founding.[23] Through this international collaborative effort, global water gained a measure of credibility that it had been lacking earlier in Anglo-American scientific circles.

Systematic efforts to coordinate hydrologic practice on an international basis had already begun in the 1920s and helped provide an institutional foundation for the more refined collaborative exercises of the postwar era. From its inception, the International Association of Scientific Hydrology (IASH), founded in 1923, was dedicated to the international standardization of methods, measures, and instruments used to measure water in the hydrosphere.[24] Such coordination was hardly unique to hydrology; IASH was formed as a subsidiary association to the International Union of Geodesy and Geophysics, which had been established in 1919 to promote coordination of physical, chemical, and mathematical studies of earth and its environment in space.[25] Historians of science have identified such coordinating and standardizing activities as characteristic of the practice of "gathering the world together," a practice that has given scientists "the power to shape the way the world was put together, not least by their role in condensing the earth to the scale of a chart or an index or a catalog."[26] Although the production of charts, indices, and catalogues of world water would not become routine until decades later, this early work of the IASH helped lay the groundwork for the scientific calculation of global water in the postwar years.

After the Second World War, Nace authored a proposal that was adopted by the IASH and which eventually resulted in the International Hydrological Decade (IHD), a ten-year program (1965-74) of international scientific hydrological study that came under the aegis of UNESCO. Although a scientific exercise in the sense that it sought to improve the capacity to produce objective, scientific assessments of world water, the nature of the water that inspired these proposals was expressly that of a resource, and one that was definitely limited in quantity: "These proposals," wrote Jaromir Nemec, a Russian proponent of global hydrology, "undoubtedly corresponded to a specific need – in many countries the public and hence the governments were becoming aware of the scarcity of water and the difficulties of developing new sources of supply."[27] The main purpose of the IHD was to promote cooperation among hydrologists from different countries in the production of a full, quantitative assessment of the world's water resources.[28] As expressed by one hydrologist, the IHD was "man's first concerted attempt to take stock of his diminishing available resources of water in face [sic] of expanding population and rising standards of living and to coordinate world-wide research on ways of making better use of these resources."[29] Once again, the very act of thinking about water as a resource implies its scarcity while compelling its quantification.

The IHD facilitated dialogue among hydrologists and between international agencies with an interest in hydrology, advanced a measure of international coordination of hydrological education and standardization of methods, and identified important gaps in hydrological knowledge, particularly in developing countries. But most notable for our purposes was that it advanced international study of the global water balance, described by Nace as "an inventory of the total amount of water in the Earth system and its movement through the global hydrological cycle."[30] In particular, it brought the latest global water balance research of Soviet hydrologists to the attention of the international hydrological community and advanced discussion of how the latest computer and remote sensing techniques could be applied to produce improved data.

Thus, global water materialized as a moment of international scientific collaboration at a time when political concerns about water as a limited resource were coming to the surface. As it emerged during the course of the IHD, it appears that the scientific basis of global water was translated, at least superficially, into Hortonian terms. Russian/Soviet hydrologists, who had long been the world leaders in constructing the global water balance, adopted the terms "vodnyy tsikl" and "gidrologicheskiy tsikl," which were described by L'vovich as a "terminological inaccuracy that has become rather widespread and is related to the Anglo-American terms 'water cycle' or 'hydrological cycle.'"[31] But the change remarked upon by L'vovich was perhaps more than just a question of terminological inaccuracy. Soviet hydrologists also appear to have taken quite readily to the American mode of diagrammatical representation of the circulation of water. Kalinin's 1971 *Global Hydrology*, for example,

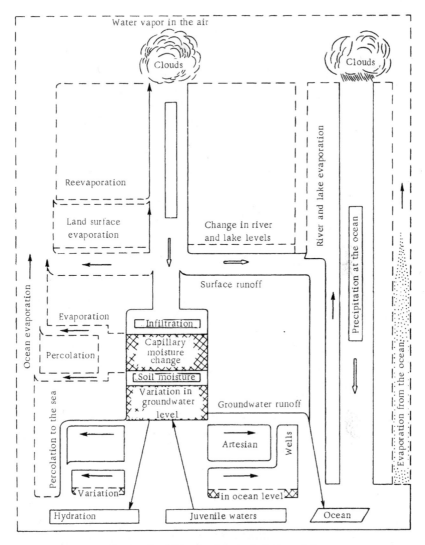

FIGURE 8.1 "Scheme of the hydrologic cycle"
Note the close similarity to the diagram shown in Figure 6.4. | G.P. Kalinin, *Global Hydrology*, trans. N. Kaner (1968; repr. Jerusalem: Israel Program for Scientific Translation, 1971), 8.

illustrates the hydrologic cycle with a diagram (see Figure 8.1) that is almost identical to the one appearing on the frontispiece of Meinzer's 1942 textbook and shown above in Figure 6.4. Horton's hydrologic cycle, it seems, held rhetorical powers that made it as irresistible to global hydrologists as it was to state planners.

The global water studies and data produced under the auspices of the IHD were published so as to reach the widest possible audience among professional hydrologists and water experts worldwide. The most significant accounting of global water to emerge during this time was *World Water Balance and Water Resources of the Earth*, prepared by the USSR Committee for the IHD in 1974 and published in English by UNESCO a few years later.[32] Drawing from the global water balance investigations of L'vovich and other Soviet hydrologists, this was arguably the definitive statement of world water and remained so into the 1990s.[33] Other global water resource assessments published in English around this time, which the IHD was critical in helping inspire and disseminate, include works by Nace, Kalinin A. Baumgartner and E. Reichel, and L'vovich.[34]

Despite these developments, it appears that the global dimension of hydrological thought was slow to pervade anglophone hydrological discourse. Writing in 1983, J.T. Houghton, observed:

> It is only recently that hydrologists have begun to think in global terms, largely because it is only recently that meteorological information with global coverage has become available and that theoretical models of global extent have been developed. This global thinking is at an early stage. Greatly improved data and much better formulations of physical processes are required before further progress can be made.[35]

It was not until the 1990s that the concept of global hydrology had become widely accepted among the community of Western hydrologists. In his preface to a 1997 textbook on global hydrology – the first of its kind – John C. Rodda, president of the International Association of Hydrological Sciences, wrote:

> Of course, a text entitled Global Hydrology would have been impossible to publish twenty or even ten years ago. Then the science was strongly dedicated to the river basin, with few hydrologists able to acknowledge that hydrology extends beyond the individual watershed towards a global dimension. But for a number of reasons this has all changed during the last decade and the science has "gone global."[36]

Although Rodda does not elaborate, part of the reason for hydrology's going global is no doubt due to improvements in the quality and quantity of data on parameters such as groundwater and surface streamflow, evapotranspiration, and precipitation in different parts of the world. Another factor in the globalization of hydrological discourse was the growth of an international network of water professionals and experts that promoted common approaches to accounting for and governing the world's water. Taking root in the 1970s (toward the end of the International Hydrological Decade), this network gained momentum in the 1980s, partly through the International Hydrological Programme, coordinated by UNESCO, which followed the IHD and had the general mandate of continuing the international collaboration among hydrologists that had begun under that earlier program. Other key developments were the establishment of the International Water Resources Association (IWRA) (a membership organization for water professionals founded in 1972); the holding of global water conferences, world congresses, and international meetings of experts by the 1980s; and the establishment of professional water publications such as *Water International* (published by IWRA) beginning in 1975, *World Water* in 1978, *International Journal of Water Resources Development* in 1983, and *Water Policy* in 1998.[37] As Ken Conca points out, this proliferation of membership organizations, publications, and international meetings in the 1980s and 1990s gave rise to a new discourse of "global water governance."[38]

By the early 1990s, global hydrology was no longer an oxymoron in the West. Global hydrology and the study of the hydrologic cycle on the global scale became a key component of the coordinated international scientific enterprise associated with the study of climate change. Furthermore, as seen in Chapter 10, it was on the verge of being taken up and repackaged for lay English-speaking audiences by writers who were increasingly concerned with the spectre of a looming worldwide water crisis.

Conclusion

When the landmark publication *Man's Role in Changing the Face of the Earth* appeared in 1956, the section dealing with water included nothing on the global water balance, on quantities of global water resources, or on the impending danger of water scarcity facing humanity.[39] Thirty-five years later, its sequel, *The Earth as Transformed by Human Action,* presented the latest global water data in the context of a general discussion of the scarcity of water as well as an article titled "Use and Transformation of Terrestrial Water Systems" (with L'vovich as lead author) that approached the topic using the framework of the global water balance.[40]

As we have seen, the proliferation of global water in the intervening period can be explained in large part by the deliberate coordination of hydrological practice on an international scale beginning after the First World War and given a substantial push in the late 1960s and early 1970s with the IHD. "Internationalism in science," to repeat the quotation from Livingstone that opens this chapter, "insofar as it really does exist, must be considered a social achievement, not the inevitable consequence of some inherent scientific essence. It has to be worked at."[41] But bringing global water to the attention of the public was not just a matter of the work of hydrologists; it also required, and was inspired by, growing concerns about the scarcity of global water resources that rose to prominence in the later part of the century. By the late 1990s, the work of hydrologists had become fused with these concerns to present the image of a mounting global problem:

> There is a growing concern over the future of the world's water resources. A series of world water assessments, starting with the classic work of L'vovich and the fundamental work by Falkenmark ... onwards to the latest freshwater assessments ... have convincingly shown that humankind is, through increased pressure on finite water resources, heading towards a water resources scarcity.[42]

We return in Chapter 10 to this question of whether we are headed toward a water resources scarcity, a concern that has been expressed widely in terms of a global water crisis. But first it is necessary to review the nature of modern water – this time from a less historical and more philosophical perspective. This will help us understand why global water, and modern water generally, is so potently capable of producing such a crisis.

PART 3
The Constitutional Crisis of Modern Water

9
The Constitution of Modern Water

We are so used to taking it for granted that we find it difficult to develop clear and specific ideas on the nature of water.

– Erich Zimmermann, World Resources and Industries: A Functional Appraisal of the Availability of Agricultural and Industrial Resources.

Indeed, we have become so accustomed to the presence of water in our daily life that it has been a long time since we have questioned its existence ... Invisibility is indeed the height of conquest.

– Jean-Pierre Goubert, The Conquest of Water: The Advent of Health in the Industrial Age

Introduction

In Part 2 of this book we considered the history of modern water. We now shift to a more philosophical register to account for – to apply Bruno Latour's terminology – its "constitution."[1] The articulation of modern water with people – the way it relates to people – now becomes a particularly important part of the argument. Although it was certainly produced in relation to social (particularly scientific) practice, modern water is apparently independent of humans. This apparent independence is at the core of modern water's constitution, and the constitution holds together only so long as the appearance can be sustained in hydrological and popular discourse. The increasing difficulty of maintaining water as a conceptual abstraction is now manifest in a host of problems and complications (co-implications) of water and people. These problems and

175

complications suggest that (modern) water is in crisis: the constitution is breaking down under the strain of having to acknowledge that water and society necessarily bleed into one another.

The crisis of modern water is described in Chapter 10. Here, we apply the concept of hybridity, again drawn mainly from Latour, to think about these water problems and complications. This concept was introduced in Chapter 2, which outlined the theoretical basis for this study. Essentially, hybridity captures the sense in which the things of our world are neither only social nor only natural but always both. As may be recalled, hybridity provides a way of accounting for things as internally related to social processes without succumbing to the fallacy that nature is merely a matter of social construction. Water might be considered a hybrid in the sense that in every instance of its involvement with people, it internalizes ideas, material practices, and discourses, as well as the unique properties that emerge from H_2O.[2] Through most of the modern period, the outcomes of this involvement were generally deemed salubrious, such that their hybridity went unnoticed. More recently, however, these outcomes are recognized as problematic, a situation that has drawn attention to their hybridity and to the difficulty of resolving them under the modern understanding by which nature and society are held as being separate entities.

The Modern Constitution

"Water has been critical to the making of human history," avers Donald Worster; "to write history without putting any water in it is to leave out a large part of the story. Human experience has not been so dry as that."[3] While fully agreeing with this statement, I focus on a slightly different idea here. Although water is undoubtedly critical to human history, Worster's terms may be somewhat altered: Human history has been critical to the nature of water; to talk about water without including its social and historical ingredients is to leave out a large part of the story. Water is never so pure as that.

One way of thinking about the impurity of water is through the work of Bruno Latour. Latour is well known outside the field of sociology of scientific knowledge, or "science studies" as it is often called, particularly for his collaboration with others in developing what is known as actor-network theory (ANT). Many researchers have taken up ANT as a way of thinking and acting that increasingly resists "the modern antinomy between nature and culture" and recognizes "the agency of 'non-human' actants."[4] In Chapter 2, we considered how a relational-dialectical approach might be considered sufficient for imploding this modern antinomy and how the concept of hybridity allows us to capture the enabling as well as the constraining properties of non-human nature (eg., H_2O) in the production of socio-nature. It is therefore not necessary that we engage directly with ANT here; instead, we are more interested in Latour's diagnosis of the modern predicament, as this may be used to diagnose the water crisis. The best expression of this diagnosis is found in Latour's book *We Have Never Been Modern.*

In the broadest of terms, Latour wants to get beyond what he sees as a fundamental intellectual impasse, a crisis arising from the inability of natural science, social science, or poststructural criticism to adequately recognize, analyze, comprehend, and respond to a host of contemporary problems and circumstances. Examples he gives of such problems include the hole in the ozone layer; the simultaneous global spread of the AIDS virus and the efforts by officials, chemists, biologists, patients, and industrialists to respond to it; the moral dilemma of keeping frozen embryos for later use; and the loss of forest biodiversity. These hybrid problems and events combine nature, society, and discourse in ways that elude the conceptual grasp of any one of the three traditional approaches. They are, as Latour argues, "simultaneously real, like nature, narrated, like discourse, and collective, like society."[5] To this list of problems we might very well add any of the water issues, or problems, that people face today. Water pollution, water scarcity, the inadequacy and uneven distribution of water and sanitation services – all these are combinations of the water process and human social processes. (If you thought water scarcity was a "natural" event, consider what such an event would mean without the presence of any people.)

To develop a stance from which to analyze these problems, Latour argues that it is necessary to put modern society under the microscope, much as social scientists in the past have studied premodern or amodern societies. Only once we have reflected on the fundamental intellectual commitments of our own society can we consider what it might mean to get beyond the dilemma we find ourselves in. Latour thus sets out to investigate what it means to be modern.[6] Being modern, he argues, means subscribing to a particular set of ideas – he calls this set of ideas a "constitution" – that define the relationship between nature and society. The most important of these ideas – the first clause of his Modern Constitution – is that human culture, or society, is separate from nature. "Nature and Society," as Latour puts it succinctly, "must remain absolutely distinct."[7]

To illustrate the peculiarity of this distinction (which we often consider quite natural), it is useful to recall that many non-European (i.e. indigenous) languages do not have a word "that translates even roughly into our 'nature.'"[8] Nature, as a realm defined by its separation from human culture, has become an important category only in modern Western thought: "Non-Western cultures," Latour points out, "have never been interested in nature; they have never adopted it as a category; they have never found a use for it."[9] Perhaps the most important use that we moderns have found for "nature" has been to provide the epistemological basis for the production of objective, scientific knowledge. And it is upon this knowledge that our technologies, economies, and many of our basic beliefs now rest. The irony of this conceptual separation of nature from human society is that it has proven so successful a foundation for the production of knowledge that we have come to believe that it is real; that is, that the world *itself* is actually divided into the natural and the cultural. As Erik Swyngedouw points out, "The separation [of nature and society under Latour's Modern Constitution] worked at the epistemological level, that is as a way of understanding the world, and as such has indeed managed to produce knowledge. The problem with this epistemological perspective, once it became hegemonic, is that it eventually turned from a dominant epistemology to a dominant ontology, that is a strong belief that the world was actually ontologically split into things natural and things social."[10]

Although this problem has always been present in modern society, the degree to which people now impinge on non-human nature makes it impossible to ignore. A second irony of the conceptual separation of nature from society is that it has allowed for an unprecedented degree of human involvement and intervention in the world, such that we now surround ourselves with proof of the fallacy of this very notion of separation. For a time, Latour notes, hybrids "posed no problem because they did not exist publicly and because their monstrous consequences remained untraceable."[11] By the late twentieth century, however, our subscription to the idea that nature and society are two categories was belied by the pervasion of hybrids in the environment and on the news – things, events, and problems that could not be slotted into either of the two categories but must be acknowledged as belonging simultaneously to both. It is upon recognition of the hybridity of the (by)products of modernity that a kind of crisis ensues. To again quote Swyngedouw, "The proliferation of 'hybrids' permits (and even necessitates) everyone (including scientists) to see the impossibility of an ontological basis for such a separation. Their very existence is proof of the flaw of such an argument."[12]

But Latour goes further by showing how the Modern Constitution effects a kind of resistance to – and is proof against – the reality of hybridity and socio-nature. Drawing from studies in the history of science, he shows that it is in the event of producing a scientific fact that an apparent gap is opened up between nature and the representations made by the people who reveal its secrets.[13] He describes the opening of this gap in terms of "mediation": mediation is "an original event [that] creates what it translates as well as the entities [i.e., nature and society] between which it plays the mediating role."[14] At the same time, Latour avers that it is in the mediating events that non-human things are mobilized to produce durable societies (through, for example, invention, commerce, the arts), that the same gap appears to manifest between these things and the society that mobilizes them: "Despite its human construction[, society] infinitely surpasses the humans who created it, for in its pores, its vessels, its tissues, it mobilizes the countless goods and objects that give it consistency and durability ... as demonstrated by the work of mediation."[15] Thus, every effort that is made to fix our problems by shoring up the purity, integrity, and sustainability of nature only drives us more deeply into the modern intellectual predicament. Instead, Latour wants us to recognize that everything in which we are involved takes place in what he calls the "Middle Kingdom," a realm that is neither natural nor cultural but out of which nature and culture are spun. "At last," he touts, with this (his) writing of the Modern Constitution, "the Middle Kingdom is represented. Natures and societies are its satellites."[16]

To take a relevant example: Under the Modern Constitution, I might describe drinking a cup of water in terms of my intervention in the hydrologic cycle in order to sustain my precious health. Projecting this explanation on a larger scale, I might say that a city alters the natural flow of the hydrologic cycle by its water intakes, treatment, distribution, and sanitation systems in order to sustain a sanitary urban environment. Under the influence of Latour's critique, however, I might say that this description itself has the effect of opening a gap between me and the hydrologic cycle – or between an urban water system and the natural flow of water – which then produces the idea of separation. From Latour's perspective, it may be conjectured that in drinking a cup of water, I am not intervening in nature in order to sustain myself; I am, rather, sustaining my nature in the event of drinking water, while the nature of the water process is sustained in exactly the same event. Through this event, it can thus be acknowledged that the water process and I are actually one; I *am* the process by which water flows through my body. The event of my drinking water is therefore neither natural nor cultural but an occurrence in the Middle Kingdom. To arrive at the natural and the cultural, I would have to spin these out – or abstract them – as satellites of the event.

It has to be admitted that Latour's constitutional metaphor is useless when it comes to questions involving the disarticulation of society into groups or regions that might benefit more than others from its arrangements.[17] Its usefulness is instead in revealing the contradictions that characterize modern thought and the real difficulties that this thinking has produced in the general relations between humans and the rest of the world: The constitution, Latour points out, allows us to regard nature as external (he uses the term "transcendent") to human society, while simultaneously permitting the constant mixing and mediation of things natural and things social, a move by which nature is immanent to social life. As for the other half of the formula, the constitution upholds that "society is our free construction; it is immanent to our action," while simultaneously allowing for the constant enrolment of non-human nature into the fibre of social life.[18] The result is what can only be described as a kind of deceit giving rise to a host of troubling hybrids that cannot be recognized as such. Those who subscribe to the constitution, he argues,

> are going to be able to make Nature intervene at every point in the fabrication of their societies while they go right on attributing to Nature its radical transcendence; they are going to be able to become the only actors in their own political destiny, while they go right on making their society hold together by mobilizing Nature. On the one hand, the transcendence of Nature will not prevent its social immanence; on the other, the immanence of the social will not prevent the [body politic] from remaining transcendent. We must admit that this is a rather neat construction that makes it possible to do everything without being limited by anything.[19]

The result of this transgression is the proliferation of hybrids which, because of the constitutional imperative that "Nature and Society must remain absolutely distinct," we fail to – indeed, we cannot – recognize for what they are. As Latour stresses, "The essential point of this modern Constitution is that it renders the work of mediation that assembles hybrids invisible, unthinkable, unrepresentable ... *the modern Constitution allows the expanded proliferation of the hybrids whose existence, whose very possibility it denies.*"[20]

To bring this brief description of Latour's Modern Constitution to a conclusion, it is the very proliferation of hybrids that has now reached the point at which the constitution itself is threatened, precipitating an intellectual crisis. The constitution, Latour argues, "has collapsed under its own weight, submerged by the mixtures ... The diagnosis of the crisis with which [he began his] essay is now quite clear: *the proliferation of hybrids has saturated the Constitutional framework of the moderns.*"[21]

ADDING WATER TO THE MODERN CONSTITUTION

How, then, might Latour's Modern Constitution be applied to water?[22] The Modern Constitution suggests an important distinction between what it is to be modern and what it is to be premodern or amodern, a distinction Latour characterizes in terms of the "Great Divide."[23] By this distinction, premodernity (as well as thriving or emerging non-Western or non-modern cultures) is regarded rather contemptuously by moderns as failing to subscribe to the fundamental constitutional divide between nature and culture. Thus, in our anthropological and historical investigations of other cultures, we recognize a seamless coextension of society and the lifeworld, between nature and culture, that can hardly be applied to the investigation our own (modern) society: "It is impossible to do with our own culture – or should I say nature-culture? – what can be done elsewhere with others," says Latour. "Why? Because we are modern. Our fabric is no longer seamless ... For traditional anthropologists there is not – there cannot be, there should not be – an anthropology of the modern world."[24] Of course, writing "an anthropology of the modern world" is more or less what Latour is up to. By revealing the contradictions (one might say the sleight of hand) of the Modern Constitution, this very distinction by which we moderns hold ourselves to be different from all others, is shown to be false, suggesting that we might indeed look in the mirror and apply the same kind of critique. (Hence, we have never been modern, for at the hidden core of the Modern Constitution lies the seamlessness of socio-nature.) For our purposes, the significance of this conclusion is to suggest that although it may appear to have been rendered asocial, modern water is as deeply embedded in the social fabric of modern Western culture as the tanks of southern India or the water temples of Bali are embedded in those cultures – and that modern water may be analyzed as such.

We can begin by considering in what sense modern water is (dis)embedded from and in the social fabric and the implications of this paradox. In representations of modern water considered so far – including H_2O, the hydrologic cycle, and global water – there is little or no evidence of people, including the people who made it possible to represent water in such an abstract manner. We know that this absence is only apparent – although appearing perfectly natural, modern water is nevertheless thoroughly imbued with human action – a fact the production of which has opened up a gap between water and society. Latour's elaboration of the Modern Constitution disentangles the paradox by highlighting the socio-material scientific practices in which the "facts" of nature are fabricated. As Latour puts it, "Even though we construct Nature, Nature is as if we did not construct it."[25]

Such a constitution allows water facts to perform a kind of political work by appearing to speak for themselves. Thus, "Water Facts for the Nation's Future" – to use the title of a landmark assessment of the American hydrological sciences in the late 1950s – reinforce the hegemony of modern water while serving as an argument for devoting more resources to the technical practice by which such facts are produced.[26] "Progress toward ... the gathering of water facts," its authors point out, "is well under way and is continental in scope. Today, it is routine to work out the potential supply of a river for a city that may be hundreds of miles distant. Water engineers regularly calculate the capacity of reservoirs to store flood waters. Water geologists determine the yield of groundwater reservoirs."[27] The thrust of the report is to show the growing need for such data and recommend greater public spending on hydrological-data-gathering programs.[28] Of course, such data are indispensable to modern society. The cycle by which water facts provide the epistemological means for controlling water, which in turn produces a growing need for water facts, flows naturally from modern water. Once the control of (modern) water became the dominant paradigm for the development of water resources, the need for water facts became indispensable, as without them, the entire edifice of control was unthinkable.

Water facts also do political work of another sort by establishing a discourse and a set of hydrosocial relations that cannot possibly be gainsaid on their own terms. The fact that a certain quantity of water is known to be present in a river or a lake or an aquifer makes this water *available* to the agencies and the people who know water in this way. Making water available, whether for industry, for domestic use, or for ecosystem services, is a political act, the first step of which is accomplished by merely knowing and representing water as a fact. Quantities of water appear to speak for themselves by presenting themselves to us as something that is available. Such, it may be recalled, was the effect of declaring water a resource in the early twentieth century, a declaration that, as we saw in Chapter 7, flowed directly from the *idea* of water as a quantifiable substance.

These ideas find complementary themes in Heidegger's work on what he called the modern technological mode by which people reveal and represent things. "Modern science's way of representing," Heidegger argued, "pursues and entraps nature as a calculable coherence."[29] The quantification of water in the perfect coherence of the hydrologic cycle is a suitable example of this mode of revealing things, to which Heidegger gave the name *Gestell* (Enframing). *Gestell* "concerns nature, above all, as the chief storehouse of the standing energy reserve ... The rule of Enframing ... demands that nature be orderable as standing-reserve."[30] Heidegger's critique of modernity has two implications that are relevant here: First, he points out that *Gestell* makes it impossible to entertain other modes of revealing: "It drives out every other possibility of revealing."[31] Second, once revealed in this way, as Michael E. Zimmerman points out, "the technological disclosure of entities as raw material compel[s] humanity to erect a world consistent with that disclosure: the world of total mobilization."[32] *Gestell* thus reveals the world as given over to the possibility of control or mastery.

If we return to the Modern Constitution with this lesson of Heidegger's in hand, we can see perhaps more easily how the constitution allows us to have our way with water and to imagine that we can get away with it intact. By making water naturally available, the constitution gives us licence to abstract, adulterate, drain, dam, divert, and contain water – intellectually and materially – without concerning ourselves unduly with the products of these operations. Perhaps it would be more accurate to say that it *gave* us licence to think and do these things, as the proliferation of some of the more obviously troublesome hydrosocial hybrids has lately occasioned a great deal of consternation. Some of this concern is discussed in Chapter 3, for example, water pollution and the social and ecological impacts of large dams since the 1970s.

These concerns appeared as something of a challenge to the practice of scientific hydrology, committed as it was to the intellectual purification of water. "To meet the criticism from within and without," stated the eminent hydrologist J.E. Nash in 1992, "we must consider and decide for ourselves, what hydrology is, where it lies in the spectrum of human knowledge end endeavour."[33] As the proliferation of hydrosocial hybrids posed serious problems for hydrology as a science, it also threatened the integrity of modern water. Instead of appearing to speak for themselves, the facts of water now began to speak *to* the severed relations, the organisms, ecosystems, economies, and human cultures that had been stranded by modern water. It is slightly disorienting to have to reconcile modern water, which still seems so natural to us, with the actual effects of letting it loose in the world. Thus, for example, the residual notion of hydroelectricity as a perfectly clean or free sort of energy is incongruent with the (sometimes devastating) socio-ecological effects of its production.[34]

Some epistemological commitments necessary to upholding modern water's constitution can be distilled from the various strands of the discussion above. Borrowing from Latour's metaphors, I will put these commitments in terms of four clauses. The constitution, I might note, begins to unravel as soon as we become conscious of having made these commitments – especially the first:

Clause 1 – Water and society are separate and must remain distinctly so.

Clause 2a – We carry on as if water were a natural fact, while producing this facticity in social practice. Social practice, in other words, is dissolved in water, allowing things such as H_2O, global water, and water scarcity to form a perfectly transparent solution.

Clause 2b – We carry on as if society and water fell into separate categories, while depending on the water process for virtually every aspect of social production. Water, in other words, is dissolved in society – human history is wet (to paraphrase Worster); human society, "in its pores, its vessels, its tissues, ... mobilizes" water, to quote Latour.[35]

Clause 3 – This double contradiction "makes it possible to do everything [with water] without being limited by anything."[36] In other words, society can do what it will with water, while maintaining the (mis)apprehension that society itself remains unaffected.

Clause 4 – The result of this transgression is the proliferation of water hybrids, but because of the first clause, these are not recognized as such. Things such as water resources, water scarcity, and water purity combine social and hydrological reality but are seen and responded to as though they were perfectly natural.

THE CONSTITUTION AND THE HYDROLOGIC CYCLE

"Man and the hydrological cycle" is a story in itself.

– RAYMOND NACE, "WATER OF THE WORLD: DISTRIBUTION OF MAN'S LIQUID ASSETS IS A CLUE TO FUTURE CONTROL"

Having specified modern water's constitution will prove useful in the next chapter, when it comes to discussing the crisis of modern water. Here we consider how this constitution relates to the hydrologic cycle.

The hydrologic cycle is itself constituted by two moves that correspond to the constitution. First, the hydrologic cycle effects the banishment of people so as to produce an objective representation of nature. The hydrologic cycle is, as Horton defined it, "the course of natural circulation of water in, on and over the Earth's surface."[37] Such a move, it will be recalled, was necessary to establish hydrology's credentials as a natural science, a "pure science," in Horton's terms. The hydrologic cycle is thus consistent with the first clause of the constitution, by which the world of nature is divided from the world of society, a constitution that enables science to speak for, and therefore to represent, nature.

However, although the hydrologic cycle was constituted as "a great natural system," it was surreptitiously adulterated by the presence of humans from the very start, a move that corresponds to clause 2a of the constitution.[38] Humans can be found lurking in even the most ostensibly uncontaminated sections of Horton's paper, defiling even the purity of the water balance equation:

There is a simple basic fact involved in the hydrologic cycle:

Rainfall = Evaporation + Runoff

(Inasmuch as most persons think of evaporation in a more restricted sense, it is better to define runoff as equal to rainfall minus water losses. Water-losses are of three kinds, all evaporative in their nature: (a) Interception; (b) transpiration; (c) direct evaporation from soils and water-surfaces.)[39]

Horton's use of the term "losses" to describe various kinds of evaporation is revealing. Evaporation is not a loss to the hydrosphere, to non-human nature, or to the hydrologic cycle. It may be considered a loss only by those for whom the *available* water flowing in rivers, stored in lakes, or held in aquifers is what really counts – in other words, those for whom water is a resource.[40] The use of the term "losses" to describe those phases of the hydrologic cycle perceived as being of no immediate use to humans has been very common in hydrological discourse.[41] "Losses" reveals a contradiction of the hydrologic cycle; namely, that its naturalness is nevertheless suffused with human intention. This contradiction is perhaps at the root of hydrology's occasional bouts of identity crisis, as discussed in Chapter 6. (Hydrology's identity problems are perhaps unavoidable for a "pure science" that excels at producing facts that are indispensable to the state.) The hybridity of the hydrologic cycle cannot but give rise to a certain amount of confusion among those for whom the natural and the social must remain categorically distinct.

The hydrologic cycle is perfectly legal under the constitution of modern water in the sense that it internalizes both the water process and hydrologic practice. Thus, "hydrology" is defined in a typical American hydrology textbook as "the science that treats of the various phases of the hydrologic cycle."[42] Here the hydrologic cycle exists independently in nature, as something that has been discovered in scientific practice and that water scientists can "treat of" objectively. But, on the very next page of the book, we find that the "hydrologic cycle ... provides the groundwork upon which the science of hydrology is constructed."[43] The circularity of this formulation – the hydrologic cycle providing both the groundwork on which the science of hydrology is constructed as well as the object of its investigations – illustrates its simultaneous transcendence and immanence, the constitutional paradox of modern water.

As intimated by research reviewed in Chapter 3, the ghostly presence of people in the hydrologic cycle is now manifest in a way that makes it more and more difficult to sustain the constitution. In Horton's time, however, and until quite recently, this separation authorized all manner of human interventions in the hydrologic cycle without presenting a constitutional challenge. Thus, the hydrologic cycle served as a framework in which humans could safely situate their interventions without fear of producing radical socio-ecological dislocation. To paraphrase Latour, it "made it possible to do everything with water without being limited by anything." As the conceptual framework of scientific hydrology, the hydrologic cycle constituted a group of experts formally schooled in the abstraction and accounting of water. This expertise has been instrumental in the "improvement" of rivers, the impoundment of runoff, and all other hydrologic manipulations contributing to the monumental twentieth-century project of altering the hydrologic cycle. So long as this intervention could be seen as a matter of improving something that remained external to society, the project could proceed without risk to water, people, or the constitution that held them apart.

The sense of this risk-free manipulation of water is reflected in the contribution of Harold E. Thomas, a groundwater hydrologist working with the US Geological Survey, to a landmark 1955 symposium on "Man's Role in Changing the Face of the Earth":

> For the hydrologist, there is a need to know as accurately as possible the modifications that man makes in the hydrologic cycle – past, present, and future – in the hope that man can progressively increase his ability to modify the hydrologic cycle to his advantage. By working with nature, adapting his needs to the natural cycle or adapting that cycle to his needs, man can obtain the greatest beneficial use of the water resources.[44]

In effect, what Thomas does here is situate "man" both outside and inside the "natural cycle," the only (im)possible position from which man is able to "modify the hydrologic cycle" without having to acknowledge the process – and the politics – of hybridization.[45] The "complex pattern of circulation" that ensues is thus naturalized so as to make invisible the expert discourses, the interests, the winners, and the losers that might be entailed in such apparently innocuous features as "withdrawal," "diversion," "reservoir storage," "artificial recharge," and "pollution" (see Figure 9.1). Cleaving to the Modern Constitution, the hydrologic cycle allows us to maintain the fiction of our constitutional separation from nature and water, while perfecting the politics of mixing. With the simultaneous presence and absence of people, the hydrologic cycle can be depicted in a way that features major structural works such as reservoirs and dams without raising political questions about the social production of water (see Figure 9.2). Even the instruments that hydrologists have used to measure precipitation, streamflow, evaporation, and so on can be shown (as in Figure 6.5) without having to acknowledge that it is by virtue of such instruments and the disciplined coordination of the measurements they record that the hydrologic cycle has been constructed in the first place – the recording of such measurements is understood merely as a matter of observing and representing the nature of water.

By the 1970s, awareness of the unintended consequences of human impact on the physical environment began to produce a change in the way people and the hydrologic cycle came together. For the most part (and as suggested in the quotation by Thomas above), this relationship had been seen as benevolent; it was for humanity to "obtain the greatest beneficial use of the water resources." [46] Now, however, the agency of humans was increasingly regarded as having the potential to do harm. Thus, by the mid-1970s and onward, the problem was often put in terms of human "impact" or "influence" on the hydrologic cycle.[47] During this (ongoing) episode of the history of the hydrologic cycle, the constitution has held full sway. Instead of tending toward a more relational understanding of

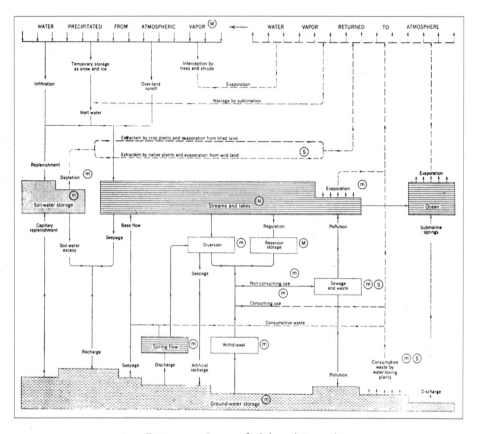

FIGURE 9.1 Can you find the politics in this picture?

H.E. Thomas, "Changes in Quantities and Qualities of Ground and Surface Water," in *Man's Role in Changing the Face of the Earth,* ed. W.L. Thomas, C.O. Sauer, M. Bates, and L. Mumford (Chicago: University of Chicago Press, 1956), 545. © University of Chicago, 1956. Reproduced with permission. The source of this diagram offers no explanation for the circled "M"s, "m"s, and "s"s.

water and society, our thinking has remained for the most part in the groove of human impact on the hydrologic cycle. Although the appearance, in other words, of socio-environmental problems such as hybrids might have brought a challenge to modern water and its basic constitution, this hasn't been the case, as evidenced by the hydrologic cycle's remarkable resilience in the face of our incursions.

Even today, the hydrologic cycle retains a kind of waterproofness – an impermeability – to social depredations that makes it resistant to the social nature of water. Its identity as something that may be modified and improved with impunity is definitely a *thing* of the past. However, rather

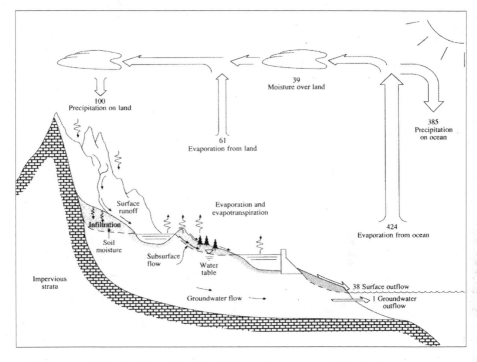

FIGURE 9.2 The hydrologic cycle, with global annual average water balance given
in units relative to a value of one hundred for the rate of precipitation on land
V.T. Chow, David R. Maidment, and Larry W. Mays, *Applied Hydrology* (New York: McGraw-Hill,
1988), 3. © 1988 McGraw-Hill. Reproduced with permission of the McGraw-Hill Companies.

than challenging the hydrologic cycle itself, the strength of modern water's
constitution is such that it has, if anything, become naturalized to an even
greater extent. Today, the objective is no longer to modify and improve
the hydrologic cycle as it was in 1955 but to maintain its integrity:

> A working definition of sustainable water use ... is the use of water that
> supports the ability of human society to endure and flourish into the in-
> definite future without undermining the integrity of the hydrological cycle
> or the ecological systems that depend on it.[48]

The thesis conveyed by these pages is simple. Nature is the source of water;
therefore our ability to support additional human lives on planet Earth
depends upon the protection of nature and the continued operation of the
water cycle.[49]

CONCLUSION

The project of modifying the hydrologic cycle has everywhere given rise to social and ecological problems that were unforeseen by those for whom water was taken to be separate and distinct from society, in accordance with Clause 1 of modern water's constitution. Because we have made water what it is while maintaining that it is natural (Clause 2), we have imagined it possible to do anything with water without being limited by anything (Clause 3). The proliferation of hydrosocial hybrids such as water pollution, canalization of rivers, salination of soils, aquatic ecosystem degradation, groundwater depletion, alteration of the hydrologic regime, and river diversions – all made possible under the pretense of the risk-free exploitation of modern water (and the hydrologic cycle) – has come back to haunt us in various unexpected ways (Clause 4). But because of the first clause, "Water and society are separate..." these hybrids are not recognized for what they are. The urge to purify the realm of water as well as the realm of society is reflected in the most recent, but inevitably futile, calls to respect, protect, and sustain the integrity of the hydrologic cycle.

Like nature itself, the hydrologic cycle has often served to bolster someone's social and political agenda. Whether as an argument for God's wisdom and providence, a means of constituting an exclusive body of knowledge workers, a framework for legitimizing and coordinating state intervention in the nation's rivers, or an argument for applying greater efficiency in a post-industrial economy, the hydrologic cycle has served as a powerful ally. But the effectiveness of such an ally is ensured only so long as its constitution holds. As a corollary of modern water, the hydrologic cycle functions only if it is seen as categorically separate from and independent of the ambitions of those who variously seek to define, measure, modify, or protect it. Upon the first clause of modern water's constitution rests its legitimacy and its utility.

In the next chapter, we consider how this constitution has begun to unravel. The scandalous (to use Latour's term) problems to which the constitution has given rise – the effects of translation and mediation that it has so long authorized – are proliferating and becoming obvious in a way that makes modern water less and less invisible all the time.

10

Modern Water in Crisis

The water crisis is real. If action isn't taken, millions of people will be condemned to a premature death. According to the World Water Development Report, a UN survey[,] ... population growth, pollution and climate change are conspiring to exacerbate the situation. Over the next two decades, the average supply of water per person will drop by a third. Heightened hunger and disease will follow. Humanity's demands for water also threaten natural ecosystems, and may bring nations into conflicts that – although they may not lead to war – will test diplomats' skills to the limit.

– PETER ALDHOUS, "THE WORLD'S FORGOTTEN CRISIS"

Make no mistake: the world water supply is in crisis, and things are getting worse, not better. In spite of the many grandiose plans made by the United Nations and other international bodies since the 1970s, the basic issues have yet to be tackled in practical terms. The situation will continue to worsen until effective action is taken on a worldwide basis.

– ROBIN CLARKE AND JANNET KING, THE WATER ATLAS

INTRODUCTION

For the purpose of establishing a date, the publication of Peter Gleick's *Water in Crisis: A Guide to the World's Fresh Water Resources* in 1993 marks the moment when a certain wave swept ashore, flooding the popular imagination as well as the academy with a conviction that something was fundamentally wrong with the world's water. Today, it is widely felt that the world faces an ever deepening water crisis.[1] In this chapter, we consider this crisis as it became known in the early 1990s. We also consider historical precedents from earlier in the twentieth century, particularly the apprehension held by many Americans in the 1960s that the calamity of water scarcity loomed on the horizon. This precedent is important because the *idea* of the global water crisis of the 1990s emerged mainly from the pens of American writers and was founded on presuppositions and concerns similar to those that gave rise to the crisis of the 1960s. The main difference – apart from expansion of the crisis to the global scale – is in the kinds of solutions proposed: In the 1960s, it was presumed that a water crisis could be averted only by exploiting new supplies. By the 1990s, the supply option was becoming less and less feasible, and it was recognized that the solution was to use available water much more efficiently. These differences, although undoubtedly important, nevertheless mask an underlying similarity. Both these historical incidents of concern about water crisis rest on what we have described as modern water. As it was presented in the 1960s and again in the 1990s, the water crisis presumed the naturalness of modern water. In this chapter, we consider how modern water itself has created the conditions for its own crisis and how, instead of a water crisis, what we are facing is *the crisis of modern water*. Framing the problem in this way suggests that we need to change the way we think about water, which is the subject of the concluding chapter. Here, the main idea is to identify the unmarked presence of modern water in the conventional crisis literature.

To describe the water crisis as an "idea" might seem odd or even perverse to many readers. After all, things such as droughts, groundwater depletion, and inadequate or non-existent rural and urban water services are distressingly real for billions of people. The point here is not to question these realities but, rather, to investigate in what exactly the "water crisis" consists. This investigation involves a critical examination of the idea of water crisis as it has been presented in the relevant literature. To be sure, this literature has served the vital purpose of drawing the world's attention to urgent and emergent water problems as well as to the need for radical change in water management. As discussed in Chapter 3, the identification of a new water paradigm in the 1990s signalled a moment of enlightenment whereby the twentieth-century model of procuring supplies of water to meet growing demands was recognized as no longer viable.[2] As researchers and managers recognized the bankruptcy of the supply model, the imperative of utilizing water resources more efficiently was put forward with such strength of argument and conviction that governments and international agencies had to take notice.[3] The concern here is not with the need for improved water-use efficiency but with the way the water crisis was presented as a means of proving this need. It will be argued that the strength of modern water is such that the water crisis of the 1990s was framed in its terms, using the idiom of abstract water to construct the crisis. The irony is that this framing of the crisis actually works against realizing the new water paradigm, an argument taken up in Chapter 12.

With a few exceptions, there has been little critical examination of the so-called water crisis.[4] Although the social dimensions of other environmental issues projected on the global scale – including climate change, desertification, deforestation, and biodiversity loss – have been examined from critical perspectives, the water crisis has been mostly taken for granted as an inevitable and obvious consequence of growing human populations, increasing affluence, and limited physical water supplies.[5]

FOR AN INSTANT CRISIS, JUST MIX POPULATION AND WATER

When it comes to water, nature has dealt a difficult hand.

> – SANDRA POSTEL, *PILLAR OF SAND: CAN THE IRRIGATION MIRACLE LAST?*

The root of the argument that the world is facing a crisis of modern water is in recognizing that modern water does not mix well with people, especially in their abstract, statistical guise as "population." Modern water, as per its constitution, is a combination of social practice and the water process that nevertheless purports to have nothing whatsoever to do with people. Modern water would not present much of a problem if only it could be contained on the remote side of the levee separating water from society. But as is so typical of our contemporary predicament, the levee is full of holes, and modern water has got mixed with people in ways that make it more and more difficult to sustain. After water's having spent around three hundred years incubating as a social leper, when it is reintroduced to society, especially in its culmination as global water, a kind of constitutional crisis necessarily ensues.[6] And as a constitutional crisis, the problem can finally be resolved only by reconstituting water.

To put this slightly differently, all that is required for a crisis is to re-combine (modern) water with people. An article very aptly titled "Troubled

Waters," appearing in the British *Observer Magazine* in 1993, provides an illustration of this explosive combination:

> All land-bound life has to share one ten-thousandth of the planet's water. Less than three percent of the world's water is fresh, and more than three-quarters of that is frozen, mainly at the poles. Ninety eight percent of the rest lies deep underground.
>
> The tiny fraction that remains should still, in theory, be more than enough. Every year, about 27,000 cubic miles of rain fall on the continents, enough to submerge them under two and a half feet of water. But nearly two-thirds of it evaporates again, and two-thirds of what is left runs off in floods. Even the remaining 3,400 cubic miles of rainfall could still sustain more than double the world's present population – if only it would fall evenly where people live. But while Iceland gets enough rain every year to fill a small reservoir for each of its quarter of a million inhabitants, Kuwait, with seven times as many people, scarcely gets a single drop to share between all of them.
>
> In all, 26 of the world's countries – including many of those in Africa and the Middle East – get less water than they need. Over the next 30 years another 40 nations are expected to join them, as their populations outstrip rainfall. The number of people affected is expected to grow tenfold from the present 300 million to three billion – one-third of the projected population of the planet.[7]

Here we have modern, global water presented (in the first paragraph), which actually appears to constitute a very serious problem all by itself. Such is the inevitable outcome of what has been described as "the gloomy arithmetic of water" when presented as a tiny fraction of a fraction of a fixed quantity.[8] We have already considered in Chapter 7 how merely by virtue of its quantification, modern water is distinguished by a general susceptibility to scarcity. The very quantification of the world's freshwater *in relation to* the total amount of water on the planet yields a ratio that ineluctably produces a rather startling picture. In the fifteenth edition of *Encyclopaedia Britannica,* we read:

> Quantitative studies of the distribution of water have revealed that an astonishingly small part of the Earth's water is contained in lakes and rivers. Ninety-seven percent of all the water is in the oceans; and, of the fresh water constituting the remainder, three-fourths is locked up in glacial ice and most of the rest is in the ground.[9]

Obviously, the mere quantification of the world's water establishes the condition by which scarcity becomes manifest. Let's describe this as the latent scarcity of modern water. Modern water cannot help but be latently scarce, a condition that is very commonly conveyed in representations of the world's water (see Figure 10.1; see also Figure 1.3). It is this latent scarcity that lends greater force to statements such as that by leading water authority Sandra Postel: "When it comes to water, nature has dealt a difficult hand."[10]

A statement such as this is the antithesis of the idea of the well-watered earth propounded by adherents of the sacred hydrologic cycle, as discussed in Chapter 5. But neither the natural, latent scarcity of water nor the well-watered earth reflects water's reality so much as the way different natures (and natures of water) can be invoked to support different arguments or points of view. The point in making water out to be naturally scarce is to drive home the fact that the hydrosocial predicament that many people face is deadly serious, just as the point in stressing its natural abundance is to argue for the providence of the Creator. These arguments may be considered equally valid, but they produce different outcomes. To naturalize abundance suggests that incidents of poverty and dearth must be attributed to social causes, be they religious impiety or a failure to redress economic disparities. To naturalize scarcity suggests that the very same problems must be attributed to more material or technical causes, be they climate change, poor water management, or a failure to implement economic incentives to promote the efficient allocation of water. To say that nature has dealt a difficult hand when it comes to water is not untrue. However, we need to be aware of the sense in which this truth has the effect of determining a particular and apparently commonsensical way of responding to such difficulties.

This is what Latour means when he says that nature can have the effect of paralyzing politics. Alluding to the Modern Constitution of "a two-house politics in which one house is called politics and the other, under the name of nature, renders the first one powerless," he puts this point as follows: "From now on, whenever people talk to us about nature, whether to defend it, control it, attack it, protect it, or ignore it, we will know that they are thereby designating *the second house of a public life they wish to paralyze.*"[11] Latour identifies globalized environmental discourse as the site on which this operation is most manifest: "Where 'global thinking' is concerned[, environmentalists] have come up with nothing better than nature already composed, already totalized, already instituted to neutralize politics."[12] But nature alone, or the latent scarcity of water, is not sufficient grounds

FIGURE 10.1 "The world's water supply"
This diagram appeared in a special 1993 issue of National Geographic on water. The caption reads: "If all earth's water fit in a gallon jug, available fresh water would equal just over a tablespoon – less than half of one percent of the total. About 97 percent of the planet's water is seawater; another 2 percent is locked in icecaps and glaciers. Vast reserves of fresh water underlie earth's surface, but much of it is too deep to economically tap." |
M. Parfit, "Sharing the Wealth of Water," in "Water," ed. M. Parfit, R. Conniff, J. G. Mitchell and W. S. Ellis, special issue, *National Geographic* 184, 5A (1993): 24. Image credit: Chuck Carter/National Geographic Image Collection. © National Geographic Society.

for declaring a crisis. Such a declaration requires that we add a certain kind of people to the equation: when it comes to water, nature has dealt a difficult hand – *to people* – and to some people more than others. Returning to the "Troubled Waters" article referred to above, it is when we add these people to the equation – particularly those of the twenty-six countries lacking sufficient water and of the forty nations expected to join them –

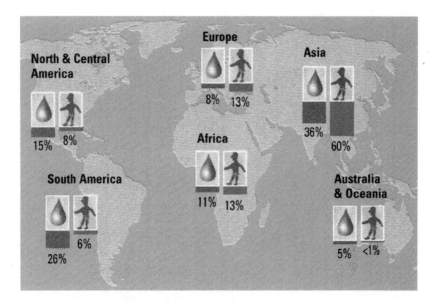

FIGURE 10.2 A juxtaposition of two abstractions: "Water availability vs. population"
(Percentages indicate proportions of the world total.)
UNESCO (United Nations Educational, Scientific and Cultural Organization), *Water for People, Water for Life: The United Nations World Water Development Report* (New York: United Nations Educational, Scientific and Cultural Organization and Berghahn Books, 2003), 69. Reproduced with permission of Berghahn Books.

that a full-blown crisis takes shape. In all, as we are told, the number of people affected is expected to grow to some 3 billion. Modern water does not constitute a crisis all by itself; it requires modern people. Just as the water presented here remains a fixed, abstract quantity, so, too, do the people involved; they are as just as inert as the water against which they are juxtaposed (see Figures 10.1 and 10.2). Neither the water nor the people compared in this juxtaposition are alive; they relate to each other only in an external sense, that is, as a mathematical ratio. From a relational-dialectical perspective, we can argue that people and water are also *internally* related in the sense that water can change the very nature of human society, while human society can change the nature and disposition of water. But so long as modern water (and modern people) prevail, these internal relations remain perfectly invisible.

The map shown in Figure 10.2 provides a crude illustration of the juxtaposition of water supply and population commonly drawn in the water literature. The figures shown here indicate proportions of the world

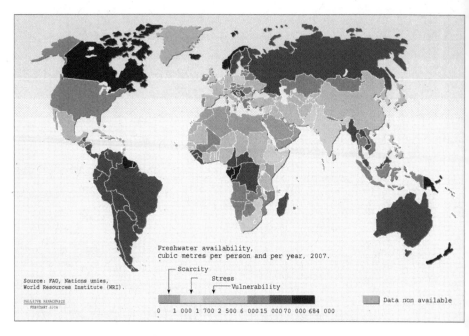

FIGURE 10.3 "Freshwater availability, cubic metres per person and per year, 2007"
Reproduced from *Vital Water Graphics* website, http://www.unep.org/dewa/vitalwater/index.html,
published by the United Nations Environmental Programme.

total. More commonly, this *comparison* is shown as the availability of water
per capita, per country, as illustrated in Figure 10.3.

Let us describe the people who give rise to this water crisis as Malthusian man – one dimensional, consuming, procreating, biological units
whose relation with water is as fixed and determinate as the statistical
methods by which it is made known. Modern water needs to be juxtaposed
alongside Malthusian man in order for its latent scarcity to be revealed
as part of its nature. This juxtaposition cannot help but produce an immediate water crisis. As described succinctly by researchers at the Worldwatch Institute, "Wherever population is growing, the supply of fresh
water per person is declining."[13]

Before we consider its genealogy more fully, it should be stressed that
the water crisis is not only a matter of dwindling supplies. As populations
have grown, as industry has expanded, and as humans have concentrated
themselves in urban regions, water pollution has added a qualitative dimension to the dismal arithmetic of the population-water equation. Water
pollution is a real problem. But the particular way that we understand

the problem – like the way we understand the water crisis generally – is conditioned by modern water. In a famous study of the social construction of pollution, anthropologist Mary Douglas elaborated on the manner in which "dirt" should be considered "matter out of place."[14] Being properly "in place" or "out of place" is, of course, socially relative. Recognizing this is not meant to downplay the seriousness of water pollution but to show that it can be understood in different ways. Modern water – as noted by Hamlin – makes the idea of pure water natural and renders everything else an adulteration by adventitious substances.[15] Water pollution thus becomes, literally, a matter involving water, rather than a hybrid of water and society. Hence, water pollution has often been considered as a water problem rather than a social problem, and the crisis to which it contributes is "the water crisis" rather than the crisis of modern water.

The Perpetual Twentieth-Century American Water Crisis

Today, man is losing his race with the growing need for water.

> – President Lyndon B. Johnson, welcoming
> address at the Water for Peace Conference, *1967*

Modern water contains the seeds of an apparently insoluble dilemma. Virtually everywhere modern water – especially in its culmination as global water – has been mixed with people, a paradox ensues. Professor Raymond Furon of the Faculté des Sciences in Paris described this paradox rather forcefully in a 1963 study, perfectly titled *The Problem of Water:*

> There are only 20,000 cu. km of fresh water available each year on the Earth's surface, which would imply that a world population in excess of 20,000 million could not be supported; this is the future foreseen for the year 2100. It is not my intention to predict how coming generations will die, but to show that we are on the threshold of a world water shortage, and that in the year 2000 we shall have only the sea to drink. The problem affects the entire world.[16]

Furon's startling assessment suggests a prescient warning of the global water crisis. His pioneering application of global water as a means of establishing the epistemological basis of a global-scale water catastrophe was to be taken up most convincingly by North American water experts a generation later. Meanwhile, "the problem of water" was recognized by some of Furon's contemporaries, especially in North America, who looked ahead and saw the collision of modern water and Malthusian man on their home turf.

In North America, the dilemma can be traced back to the early twentieth century. W.J. McGee, who in 1909 offered American readers an introduction to the idea of water as a resource and to the "quantitative" view of water, was compelled by the same token to warn that the nation's "growth in population and industries is seriously retarded" by a dearth of water, a situation that called for extraordinary measures to bring water under "complete control."[17] Resourcefulness and scarcity, as we have noted, are two sides of the same coin. Little more than two decades after McGee's declaration, as it was about to embark on a program approximating McGee's vision of total control, the US Department of the Interior stressed, "The western third of the United States grapples always with one stern fact ... Out west there is a shortage of water."[18]

The idea of a generalized dearth or shortage of water makes no sense unless we presume the presence of people who are constitutionally incapable of being satisfied with what is available to them. Like "losses" in the hydrologic cycle, the idea of such a general dearth or shortage reveals the presence of such people in what is otherwise presented as a fact of nature. Thus, the presence of Malthusian man is naturalized in every statement asserting or predicting a general dearth of water, water shortage, water scarcity, or water famine. Modern water and Malthusian man grew to fantastic heights in the United States in the twentieth century, the combination of which produced a recurring national water crisis. Concerns about general water shortage rose with the experience of drought on the Great Plains in the 1930s, water shortages in New York City in 1949, and a decade-long drought in the Southwest, from 1942 to 1952. A growing "feeling of impending disaster" was mitigated somewhat by redoubled efforts to manipulate the hydrologic cycle – by artificially inducing precipitation as well as by the usual means of building bigger reservoirs, extending control of water over larger basins, constructing inter-basin water diversions, promoting artificial groundwater recharge, and so on.[19] But these measures, if anything, only exacerbated the underlying problem in the long run by entrenching and materializing modern water to a greater degree than ever. So long as the underlying conditions – modern water and Malthusian man – continued to gain strength, concerns about general water shortages were sure to reappear.

These concerns were indeed very apparent shortly after the Second World War, when there arose a nation-wide anxiety over water.[20] By the late 1950s, the country's most senior hydrologists were warning, "Every estimate indicates that this country's need for water in the next several decades is bound to grow enormously ... It is no wonder then that strains on water resources are already showing up, and that these stresses are certain to grow."[21] A national report on the state of the hydrological sciences in 1962 found, "There is a continual appeal for water studies, bolstered by recurrent grim warnings about water problems by 1980, by 2000, and so on."[22]

By the mid-1960s, as the data from these very studies became more widely available, the postwar economy boomed, the major river basins in the continental United States approached full regulation, water pollution became a matter of national concern, and an exceptionally protracted drought struck the northeastern states, water-scarcity worry reached new heights of rhetoric: "The United States of America, the richest and most powerful nation in the world, is running out of its most indispensable commodity," wrote US congressman Jim Wright in *The Coming Water Famine*, published in 1966. "That commodity is clear, usable water."[23] In another popular book, *The Water Crisis*, published the following year, Senator Frank Moss reflected, "There is general agreement that the nation is in the midst of a perilous water crisis."[24] Indeed, by the late 1960s, many Americans had worked themselves into a fit of anxiety over water scarcity.[25] As Canadian author Richard Bocking observed (with understandable alarm) in the early 1970s: "A stream of books, magazine and newspaper articles, and political pronouncements have proclaimed 'The Water Crisis' one of the most serious problems to be faced by Americans in the very near future."[26]

In projecting this "crisis," the gloomy arithmetic of the water resources available to the nation was set against an equally gloomy prognosis of future demand for water, predicated on the assumption of a rock-solid correlation between population growth, rising living standards, and growing water demands. The prevailing assumption was that the water resources of the United States could support a maximum population of some 230 million, beyond which "our standard of living starts to suffer."[27] After the Second World War, in addition to providing the usual data on the availability of water resources, hydrologists in the United States became more involved in compiling and publishing data on water use.[28] Because the available trend data showed a positive correlation between growth in gross domestic product and increases in water withdrawals and consumption, a

presumption was made that rising living standards led inevitably to pro-
portionate per capita demands for water. That hydrologists would be called
upon to give predictions of future water demands might seem mistaken
in light of the need to consider water's social nature. But recalling the
hegemony of modern water and its constitutional divorce from society, it
is perhaps not surprising that those deemed expert in its quantification
would naturally be presumed most qualified to give projections of its
future use.[29]

These projections were, of course, invariably wrong – something to
which we return below. For now, let's just note that the spectre of general water scarcity was, and remains, underwritten by the presumption
that with respect to their hydrological needs, modern people are as inflex-
ible as modern water. Senator Moss relied largely on the work of the renowned American hydrologist Raymond Nace for the gross hydrological data underpinning *The Water Crisis*. He also relied on Nace to substantiate the correlation (vital to Moss' argument) between population, living
standards, and the need for greater supplies of water: "Water need per
individual in industrial societies is greater than has been supposed, by
nearly ten times," he quotes Nace as saying, "and the population that can
be supported with acceptable living standards undoubtedly is much
smaller than some writers have calculated."[30] The conclusion was obvious,
at least for Moss and most of his contemporaries: "A rising standard of
living is not possible without increasing water use."[31] Nor, it might be
added, despite solid evidence to the contrary, has this view of a rigid correlation between water demands and living standards changed much, at
least in some quarters.[32] "Increasing affluence," declared a popular report
in 2005, "inevitably means more water consumption: from needing clean
fresh water 24 hours a day, 7 days a week and basic sanitation service, to
demanding water for gardens and car washing, to wanting Jacuzzis or
private swimming pools."[33] Such is the strength of modern water.

As with earlier expressions of concern about water dearth and shortage,
the American water "crisis" of the 1960s was fuelled by the ambitions of
the state, allied with politicians and corporate leaders, all of whom stood
to aggrandize themselves by manipulating water supplies to address the
perceived imbalance between supply and demand. Politicians such as
Wright and Moss provided moral support for the hydraulic engineering
agencies of the federal government in calling for redoubled efforts (and
expenditures) to make wise (and full) use of the nation's rivers. The wave
of hydrological hysteria that seized Americans in the 1960s attracted attention to even more distant supplies, and it was thus that the notion of

diverting rivers from Canada to slake the growing American thirst became a matter of earnest speculation in US political circles. Thus, the amalgam of modern water, Malthusian man, and US water politics gave rise to the remarkably enduring, and not unreasonable, concern shared by many Canadians that it is only a matter of time before the United States comes after Canada's water resources. But that is another story.[34]

THE GLOBAL WATER CRISIS OF THE 1990s

Despite ebbing and flowing since the 1960s, the American water crisis has maintained a grip on the popular imagination into the present century.[35] Meanwhile, the gloomy arithmetic of water continued to be summoned by authors, experts, and planning agencies with an eye to future projections of scarcity on a global scale. For example, there were Furon's dire warnings in *The Problem of Water* cited above.[36] Several years later, the Club of Rome's famous *Limits to Growth* study warned that among constraints to food production, "second in importance only to land, is the availability of fresh water." The study warned that "the approach of the increasing demand curve for water to the constant average supply" would be reached in some places "long before the land limit becomes apparent."[37] A decade after that, the US Department of Agriculture yearbook proclaimed: "The energy crisis of the 1970s will take a back seat to the water crisis of the 1980s and 1990s."[38] For as long as modern water had gone global, the spectre of a global water crisis was only a matter of combining it with the world's (growing) population.

That this particular concoction failed to ignite much concern before the 1990s is perhaps because of the diversion of the world's attention to other matters. The late 1980s marked a general shift in public concern toward environmental problems operating on a global scale.[39] The 1987 *Report of the World Commission on Environment and Development* (the Brundtland Commission) popularized a decade or more of scientific research and environmental thought that had treated the planet as a single system and the most relevant scale for investigating a growing number of environmental problems. With these problems now regarded as global in nature, the Brundtland Commission and other leaders of world opinion seized on the need to coordinate environmental management on the same scale. A special issue of *Scientific American* appearing in 1989, titled "Managing Planet Earth," indicated the emerging global orientation of much scientific and environmental-managerial discourse:

It is as a global species that we are transforming the planet. It is only as a global species – pooling our knowledge, coordinating our actions and sharing what the planet has to offer – that we may have any prospect for managing the planet's transformation along the pathways of sustainable development. Self-conscious, intelligent management of the earth is one of the great challenges facing humanity as it approaches the 21st century.[40]

The globalization of environmental concern galvanized around several issues of systemic and cumulative global change, especially global warming, depletion of the stratospheric ozone layer, deforestation, and losses of biological diversity. Water remained a relatively minor concern in the 1980s. In fact, around the time of the Brundtland Commission's report, Swedish hydrologist Malin Falkenmark complained bitterly that the international environmental and policy community was failing to acknowledge a looming water catastrophe: "It is highly alarming to note," she stressed in an article titled *The Massive Water Scarcity Now Threatening Africa*, "that endemic water scarcity seems to be a hidden international issue."[41] Falkenmark pointed out that the Brundtland Commission had paid little attention to water while elevating climate change, deforestation, and biodiversity loss to the top of the international environmental agenda. This oversight was repeated, albeit less egregiously, at the monumental United Nations Conference on Environment and Development in 1992, where a failure to give adequate recognition to the water problems of poor countries was described by Falkenmark in terms of the "water blindness" of the world's decision makers.[42] Falkenmark suggested three related causes for this blindness, a discussion of which is useful in clarifying some of the arguments below: First was the world's – or, rather, the international environmental elites' – preoccupation with what were perceived as global issues. For Falkenmark, water scarcity was (and remains) a problem specifically endemic to certain parts of the world, and was and is not a global issue. The places particularly affected, she argued, were regions of the developing world (particularly sub-Saharan Africa) where poverty prevented the adoption of techniques that would otherwise enable people to overcome water scarcity and problems associated with interannual fluctuations.[43] Significantly, she took aim at "northern scientists and professionals" and, in presenting her views on the water crisis, completely ignored the global water balance data as well as the global water crisis literature of the 1990s. Second, Falkenmark argued that the world community of environmental experts failed to see endemic water scarcity as a problem because their thinking was conditioned by experience in wealthy temperate regions, where "water scarcity is generally no serious problem, and supply of water tends to be seen as a technological issue."[44]

Third, the water blindness of the northern environmental and policy community was abetted by general indifference to problems that either appeared not to affect them directly or called for measures that were unlikely to provide them with much political or economic benefit. The appropriate means of addressing the water needs of people in poor countries, Falkenmark had long argued, was to apply "a modified approach" that consisted of exploiting the pre-runoff portion of the hydrologic cycle instead of concentrating on conserving the runoff portion – that is, building dams – which was the favoured – that is, most capital-intensive – Western approach.[45] In addition to being of no benefit to the construction and engineering consortiums of the industrialized world, Falkenmark's modified approach eschewed the economically inspired solutions that were increasingly in vogue among water resource planners, economists, and policy experts influenced by contemporary neoliberal perspectives: "The view held by influential water engineers and economists – biased by the present conditions in the southwest USA and claiming that the water-related future of arid lands is just a problem of setting the right price on water" – was singled out for criticism.[46] In the mid-1990s, by which time this view had become a global creed, Falkenmark pointed out that this preoccupation with water economics had diverted attention from the central question of access to clean water and sanitation for people in low-income countries.[47]

The Western world was to be cured of its "water blindness," though not exactly in a manner of which Falkenmark would approve. By the mid-1990s, no one in the international environmental community could possibly have been unaware that the world was embroiled in a global water crisis. What Sandra Postel had described in 1992 as a "sleeper of a problem" had been brought to a boiling issue a few years later, as reflected in the title of the contemporary academic paper *Global Water Crisis: The Major Issue of the 21st Century; A Growing and Explosive Problem.*[48] By the end of the decade, the United Nations Environment Programme reported that a survey of leading environmental scientists had identified water shortage as one of the two most worrying problems facing the world in the new millennium. The other was global warming. In 2000, the cover story of an issue of *Harper's Magazine* was titled "Running Dry: What Happens When the World No Longer Has Enough Freshwater?"[49]

What accounted for this rather rapid shift from "water blindness" to the major issue of the twenty-first century? One explanation was that the problem was now made out to be global in nature and thereby touched the sensitivity – and the interests – of people in the wealthy industrialized world in a way that Falkenmark's presentation of water problems as endemic to poor countries could not. Through its presentation as a global issue in the 1990s, water was poured into the mould of global environmental concern that had been forged in the previous decade.

A few "key documents" brought the matter of global water scarcity to the attention of decision makers and the general, particularly Anglo-American, public.[50] One of the first and most influential of these was Postel's *Last Oasis: Facing Water Scarcity*, published by the Washington, DC-based Worldwatch Institute in 1992.[51] Another was Peter Gleick's edited collection *Water in Crisis: A Guide to the World's Fresh Water Resources*, published by the California-based Pacific Institute for Studies in Development, Environment, and Security the following year.[52] These studies published the essential data and established the discursive space in which a host of others could elaborate on the water crisis in academic journals, through the mass media, and in forums of government agencies, water experts, and business leaders such as the World Water Council. As indicated above, both these studies could be read as indictments of modern water, especially in their critique of the engineering and supply-oriented paradigm of water management. But in presenting the water crisis by appealing to modern water, these studies also inadvertently helped to sustain it.

"Last Oasis"

The thesis of Sandra Postel's *Last Oasis: Facing Water Scarcity* is that the world has entered a new water era, one marked by water scarcity.[53] At the core of this new era is an acknowledgement that we can no longer presume to resolve water problems by means of engineering more and improved water supplies: there are fewer and fewer unexploited supplies available, and the water remaining is needed to sustain aquatic ecosystem health. Securing enhanced water supplies has had, Postel argues, the effect of hiding, or masking the scarcity that characterizes (is latent in) water: "In a sense," she suggests, "masking scarcity is a principal aim of water development, the collection of engineering projects and technologies that give people access to and control over nature's supply."[54] With the option of developing new supplies rapidly disappearing, our last oasis is found in improving the efficiency with which available water supplies are used. Much of the book (parts 2 and 3) outlines strategies for improving water-use efficiency, stressing the need for "major changes in the way water is valued, allocated, and managed."[55] These strategies are considered in the next section. For now, let's consider how the general condition of water scarcity is put forward.

The cover of *Last Oasis* features a composite photographic image of an anonymous, desiccated lakebed with a windswept sand dune in the middle range and barren mountainscape in the background. The textual introduction to global water scarcity starts by conjuring the image of our "strikingly blue planet" from space. It is "hard to believe scarcities could arise in the midst of such amazing water wealth."

> The total volume of water, some 1,360,000,000 cubic kilometres, would cover the globe to a height of 2.7 kilometers if spread evenly over its surface. But more than 97 percent is seawater, 2 percent is locked in icecaps and glaciers, and a large proportion of the remaining 1 percent lies too far underground to exploit.[56]

Of the 500,000 cubic kilometres of precipitation that falls on the earth every year, it is pointed out, most falls into the oceans. After these reductions, we are left with only 40,000 cubic kilometres as the earth's renewable freshwater supply. Furthermore, fully two-thirds of this amount "runs off in floods, leaving about 14,000 cubic kilometres as a relatively stable source of supply."[57] Presented in this way, it has to be admitted that 14,000 cubic kilometres of water seems like a very small amount indeed. The data constituting this world water predicament are derived principally from the work of the Soviet hydrological studies discussed in Chapter 8.[58] *Last Oasis* effectively disseminated global water in a very well-informed yet popular format. In addition, data on regional water balances derived from L'vovich and other Soviet scientists, together with demographic data, provide the empirical basis for constructing the important concepts of water stress and water scarcity, which are described in this way:[59]

> One of the clearest signs of water scarcity is the increasing number of countries in which population has surpassed the level that can be sustained comfortably by the water available. As a rule of thumb, hydrologists designate water-stressed countries as those with annual supplies of 1,000-2,000 cubic meters per person. When the figure drops below 1,000 cubic meters (2,740 liters per person a day), nations are considered water-scarce – that

is, lack of water becomes a severe constraint on food production, economic development, and protection of natural systems.[60]

"Today," writes Postel, "26 countries, home to 232 million people, fall into the water-scarce category."[61] Several methods of calculating water stress and water scarcity have been presented in the water crisis literature.[62] Most quantify human populations and available streamflow and present these as a ratio over a specific territorial unit – most commonly, the territorial state (see Figure 10.3). As fixed indices (ratios of water supplies per capita), water stress and water scarcity presume a rigid set of hydrosocial relations, rooted in what is described as the "population-water equation."[63] Postel is cautious about using the term "crisis" to describe the situation. Nevertheless, when fed through such an equation, water and people yield what appears to be an irrefutable scientific argument for declaring a state of crisis, a declaration that was made with less hesitancy by others.

"Water in Crisis"

A year after the publication of *Last Oasis*, the water crisis was announced unequivocally in a landmark publication edited by Peter Gleick and titled simply *Water in Crisis: A Guide to the World's Freshwater Resources*.[64] This publication achieved widespread distribution and was considered by many to be an essential, if not the definitive, statement of the condition of freshwater resources on the planet.[65] The success of the book has since given rise to a series of reports authored mainly by Gleick and published every two years – *The World's Water: The Biennial Report on Freshwater Resources*.[66] As with the *Last Oasis*, we are confronted with water scarcity before opening the book; the front cover features a photograph of the dry bed of Russia's Aral Sea, with the dusty hulks of scattered ships in the foreground. The first page of Gleick's introductory chapter presents the latent scarcity of modern water by alluding to the famous poem "The Rime of the Ancient Mariner," by Samuel Taylor Coleridge, who

> effectively described the principal characteristic of the earth's water resources when he wrote, "Water, water everywhere, nor any drop to drink." Ninety-seven percent of all the water on earth is salt water – unsuitable for drinking or growing crops. The remaining 3% is fresh water, comprising a total volume of about 35 million km³. If this water were spread out evenly over the surface of the earth it would make a layer 70 m thick. Yet almost

all of this fresh water is effectively locked away in the ice caps of Antarctica and Greenland and in deep underground aquifers, which remain technologically or economically beyond our reach. Less than 100,000 km³ – just 0.3% of total fresh water reserves on earth – is found in the rivers and lakes that constitute the bulk of our usable supply.[67]

The first substantive article of *Water in Crisis* is "World Fresh Water Resources" by Igor A. Shiklomanov. Then based at the State Hydrological Institute in St. Petersburg, Shiklomanov appears to have inherited Marc L'vovich's informal title as "the dean of Soviet [now Russian] hydrology."[68] In a world in which popular concern about water issues has given some hydrologists an unprecedented measure of prominence, it may be said that Shiklomanov is world famous. Chosen by the United Nations to lead the massive collaborative inventory of global water known as the *Comprehensive Assessment of the Freshwater Resources of the World,* his work is citied in the vast majority of studies dealing with global water issues over the past fifteen years.[69] In the award-winning book *Water,* author Marq de Villiers describes Shiklomanov as "a formidable figure in the water world" and even "the central figure in the water universe."[70]

The global water data presented in Shiklomanov's chapter (the same as those presented in Table 1.1 above) provides the essential numbers that allow Gleick to warn that water is in crisis.[71] As in *The Last Oasis,* these numbers present an obvious problem: The available water resources just don't add up to the projected demands, especially in the developing world.[72] These projected demands were calculated mainly on the basis of estimated economic development and population growth, reflecting a method only slightly more sophisticated than a simple juxtaposition of modern water and Malthusian man.[73] Stressing that the number of countries with inadequate water resources per capita will grow to unacceptable levels unless extraordinary measures are taken, Gleick is brought to conclude: "The world's population cannot continue to grow indefinitely. It must be stabilized as quickly as possible ... The problem of population must be tackled directly."[74] "The world's population" is, of course, as powerful an abstraction as "the world's water." The moral, political, and economic difficulties associated with such a prescription are legion, and they may be regarded as symptoms of the constitutional problems associated with realizing such an abstraction. This prescription, moreover, holds Malthusian man and modern water in a perpetual embrace. Combined with other solutions put forward in the concluding chapter of *Water in*

Crisis (mainly improving water-use efficiency and applying economic instruments to the allocation of water and water services), it has the effect of maintaining a fairly rigid correlation between population, living standards, and demands for water.

CONCLUSION

A water crisis is inevitable whenever the quantification of abstract water is brought into relation with the quantification of abstract people. In this chapter, we have seen how this modern juxtaposition produced a perennial water crisis in the United States that waxed and waned (but mainly waxed) throughout the twentieth century and how the same basic crisis was projected on the global scale in the 1990s.

It is also inevitable that a neo-Malthusian resource crisis takes on a different appearance when either people or water are regarded less in the abstract and more in terms of the concrete, historical circumstances in which they become what they are. Recently, the investment of population with the clothing of real people seems to have had just such an effect. Since publishing *Water in Crisis* in 1993, Gleick has admitted that every one of the "projections and estimates of future freshwater demands ... made over the past half century ... [including those of Shiklomanov that Gleick himself used in 1993] have invariably turned out to be wrong."[75] The reason for this consistent error is the presumption of a fixed mathematical relation between population and water use. That in changing historical circumstances, people would find new ways relating to water, discover new forms of resourcefulness in water, and apply new techniques to mediate their relations with water seems not to have been within the range of possibility foreseen by mid-twentieth-century hydrologists. Their projections, Gleick admits, "routinely, and significantly, overestimated future water demands because of their dependence on relatively straightforward extrapolation of existing trends."[76] Around the same time that he made this admission, Gleick appears to have begun refraining from using the term "crisis" to describe the basic condition of the world's water.[77]

The tenacity of modern water is such that, almost invariably, these changing hydrosocial relations are understood as, and are reduced to, a matter of efficiency. Thus, for Gleick, the fault of hydrological forecasters of the twentieth century was not their inability to see through modern water but their failure to factor in improvements in water-use efficiency in all sectors.[78] Of course, there is nothing wrong with water-use efficiency or with adopting measures to increase water-use efficiency; this is, after all, what "the changing water paradigm" of the twenty-first century is all about. But in a sense, the new paradigm is not all that new. Merely changing the ratios of people to water – or the rates at which people use water – leaves the new paradigm constructed still of modern water. It could thus be argued that instead of addressing the water crisis, the new paradigm of water efficiency only postpones it. The possibility of going beyond efficiency to consider the more basic question of *how people relate to water* requires a different sort of paradigm change, one based on an understanding that what we face is less a water crisis than the crisis of modern water. This kind of paradigm change will be the subject of the concluding chapter. In the meantime, we will consider how the possibility of this change is somewhat hindered by the dominant response to the water crisis.

II

Sustaining Modern Water:
The New "Global Water Regime"

The beginning of the twenty-first century has been marked by efforts to coordinate and sustain a worldwide response to the water crisis. These efforts began in the 1990s and they reflect how the water crisis was framed in that decade. As discussed in Chapter 10, the declaration of a global water crisis drew from and reinforced the presumption of external relations between water and people; that is, that the relation between people and water should be understood in quantitative terms. Although some degree of appreciation for the social dimensions of water has characterized these efforts, they have for the most part been limited by a view of water as an abstract resource. Instead of developing the potential of water's polyvalence, the predominant response to the water crisis has been to find new ways to sustain modern water.

In this brief chapter, we consider the response to the water crisis led by a coalition of powerful actors making up what the critic Riccardo Petrella has described as "a kind of global high command for water."[1] The main player in this "global high command" is a quasi-nongovernmental organization known as the World Water Council (WWC). The WWC was convened in 1996 by the World Bank, the United Nations Development Programme, representatives of the water services industry, professional water associations, and water policy experts. In 1997, the WWC gave itself the mandate to develop a report that it called the *World Water Vision,* the purpose of which was to "offer relevant policy and ... recommendations for action to be taken by the world's leaders to meet the needs of future generations."[2] The ambition of the *Vision* was broadened through an ongoing series of high-level meetings known as the World Water Forum.[3] *The Vision* and a companion document, *World Water Security: A Framework for Action,* were presented at the Second World Water Forum held in 2000:

> The first report, the *World Water Vision,* framed the global water challenge as a case of inadequate supply in the face of greatly increasing demand. Without dramatic technological innovations, institutional change, and substantial new investment, the world in 2025 was projected to face an even more sizable "water gap" than that of today, when an estimated 1.3 billion people lack access to safe drinking water and 2.6 billion lack access to adequate sanitation.
>
> The second report, *World Water Security: A Framework for Action,* presented a blueprint for achieving that vision. The Framework called for dramatically expanded investment in water supply infrastructure, primarily by mobilizing the private sector through incentives such as privatization and full-cost pricing of water. The Framework also called for more effective water governance based on a paradigm of Integrated Water Resources Management.[4]

In these reports, water is framed as a scarce resource and an economic good that must be managed in an economical and integrated fashion. Together with the machinery by which they are diffused, these doctrines constitute what political scientist Ken Conca has called a new "global water regime ... a set of norms – prescriptive rules and standards of appropriate behaviour meant to govern water-related actions on a global scale."[5] We can read Conca's critique of this "global water regime" as a statement of the need to overcome the limitations imposed by modern water. So long as we remain beholden to modern water, entertaining a simultaneous variety (and a politics) of waters is constitutionally impossible:

> Attempts to create a broadly cooperative international approach to managing water – to govern water globally, so to speak – seem doomed to founder on more fundamentally contested questions. Should it be the privatized, supply-oriented vision of the [World Water] Forum? Or the grassroots, watershed-scale vision of the Forum's most ardent critics? Or an updated version of the state-led model of infrastructure expansion and water as a public good that so many governments have historically favored?[6]

Putting this problem in terms of our argument, it might be said that *modern water* cannot withstand these "more fundamentally contested questions." Ultimately, what is being questioned – and what makes the new global water regime "doomed to founder" – is the presumption of modern water that it is built on.

Our conventional understanding of the water crisis reinforces the presumption of modern water and thus gives greater credibility to the "global water regime." "In recent years," we learn from the preface of *World Water Vision,* "it has become evident that there is a chronic, pernicious crisis in the world's water resources."[7] "The world," as is proclaimed on the back cover, "is experiencing a water crisis." The term "water crisis" is used fourteen times in the foreword, preface, and executive summary of *Vision.* And we are informed by the WWC's president:

> From its inception the World Water Council has understood the dimensions of the world water crisis. As the world population increased and urbanisation and industrialisation took hold, the demand for water kept rising while the quality continued to deteriorate. Water scarcity afflicted many more nations, and access to clean drinking water and sanitation remained poor. A decline in public financing and a rise in transboundary water conflicts made these problems worse. But awareness of the problems was limited to the few on the "inside," in the water sector. We start the new century with a water crisis on all accounts. A concerted effort and extraordinary measures are needed to face the challenges head on.[8]

The only possible solution to such a crisis "is the development of a shared vision on world water for the long term."[9] Let's now consider the two basic ideas upheld in this shared vision to show how they reinforce modern water.

Water as a Scarce Resource and an Economic Good

Central to the new global water regime is the idea of maximizing water-use efficiency and reducing water to an economic good. Given the fixed constitutional relationship between modern water and people, and with the difficulties of actively stabilizing the abstraction known as "the world's population," only two possible solutions to the water crisis remain available without jettisoning modern water altogether. The first is to augment supplies of water in the face of greatly increasing demands, particularly in rapidly growing urban regions of the developing world. This, in turn, calls for vast increases in infrastructural investment, which at a time of putative state failure in the water sector translates into the need for substantial new private investment. The other solution is to increase the efficiency with which available water resources are used. As noted, the latter notion is the gist of the "changing water paradigm" and "the last oasis" to which humanity must turn in this era of growing water scarcity.[10] Increasing private investment in water supply and improving the economic productivity of water have become the twin hydrological hopes of state planners, water experts, and corporate leaders alike. Both solutions presume a slightly different, more flexible, set of relations between people and water than those that pertained in the old state-hydraulic paradigm, but they are nevertheless thoroughly modern in the sense that the fundamental relation is unchanged – water is still a resource (and increasingly an economic good) to be supplied and consumed or used in the most rational, economical fashion possible. Moreover, both solutions – increasing private investment to augment water supplies and economizing water use – complement the framing of water as an economic good. And in accordance with the end-of-history ideology of the day,[11] the only possible means of economizing water is presumed to be via the hidden hand of the market. The commitment to the market and to private sector solutions that is entailed in the *World Water Vision* is perhaps best indicated by its subtitle: *Making Water Everybody's Business*.

The economization and privatization of water and water services have, of course, occasioned a great deal of debate, as discussed in Chapter 3. The point to be stressed here is that the single most powerful response to the crisis of modern water – that is, the response proffered by the most powerful actors in the water policy world – has been to attempt to salvage it by means of deliberately rendering water an economic abstraction built on the presupposition of naturally scarce water resources, or the latent scarcity of water. Despite the appearance of making it more versatile, scripting water as an "economic good" effectively sustains modern water in the face of crisis. Prescribing the universalization of the commodity form of water is intended to induce the efficiency gains that would be presumed to follow from stripping water of all but its economic identity. But such a scripting only shifts the aspect of the predominant hydrosocial relation from public, or state-owned, resource to commodity. In other words, merely substituting one of the "things" that modern water makes possible for another has only the effect of strengthening its unidimensionality. Modern water is prerequisite to pricing water and making it into an economic good. The creation of markets for buying and selling water requires that it be reduced to the commodity form – that it be reduced to that which can be presented in terms of price. By the same token, the commodification of water strengthens modern water by fixing hydrosocial relations in terms of water-consuming subjects and water-commodity objects.

Concerted efforts by the World Bank, the regional development banks, national governments, and the water industry to promote a new, global hydro-economics predate the publication of *World Water Vision* by more than a decade.[12] Perhaps the signal – and most frequently cited – moment of the rebranding of water as an economic good occurred at the International Conference on Water and the Environment held in Dublin in January 1992. This conference has been described as "a seminal frame-shifting event" and deemed of critical importance in setting the agenda for subsequent international water discourse and policy.[13] The conference adopted the following principle: "Water has an economic value in all its competing uses and should be recognized as an economic good."[14]

This has been by far the most frequently cited and controversial of the four so-called Dublin principles (the other three are discussed below). It could be argued that not since McGee's declaration of water as a resource in 1909 had water been so deliberately forced through the frame of an abstract concept. The timing of this statement – virtually at the moment of the emergence of the global water crisis – suggests how the crisis helped produce the circumstances in which the commodification of water became such a compelling means of fixing water.[15]

INTEGRATED WATER RESOURCES MANAGEMENT

Treating water as an economic good was not the only solution to the water crisis proposed by water experts in the 1990s. The idea of getting the price of water right has been a central theme, but only one theme of integrated water resources management (IWRM), an overarching prescription for global water governance that gained international currency in the 1990s and has become the dominant water policy approach. As Conca observes, IWRM "has become *the* discursive framework of international water policy – the reference point to which all other arguments end up appealing."[16] IWRM is nevertheless an extremely difficult concept to pin down. Typically, it is understood to have several characteristics, including recognition of the cultural and ecological as well as the economic values of water;

integration of different sectors or water uses (such as agricultural, industrial, and municipal uses) in an overall management strategy; integration of water management at multiple scales (local, national, transnational); and inclusion or participation of all relevant stakeholders in water management.

However, as with the idea of water as an economic good, IWRM floats in modern water – modern water is the unmarked or presumed essence of *what water is* that makes IWRM possible and that makes it possible to imagine that such a concept may be applied to the world's water and the world's people. But the point I want to stress here is that by its own logic, modern water also dooms IWRM to a kind of futility: The most often-cited brief definition of integrated water resources management is found in the aforementioned report *World Water Security: A Framework for Action.* "IWRM," it declares, "is a process which promotes the coordinated development and management of water, land and related resources in order to maximize the resultant economic and social welfare in an equitable manner without compromising the sustainability of vital ecosystems."[17]

The presence of modern water in this formulation is clear where IWRM is identified as "a process that promotes the coordinated development and management of water." However, the urge to formulate a single, coordinated process for developing and managing the world's water has produced what must be recognized as a completely unworkable concept. Asit K. Biswas has recently critiqued IWRM as a "vague, indefinable and unimplementable concept" that has nevertheless achieved widespread popularity in the past fifteen years because water professionals required a "paradigm for management, *which will solve the existing and the foreseeable water problems all over the world*" (emphasis added).[18] The problem, as Biswas points out, is that a paradigm with such universal ambition suffers a critical breakdown when brought to bear on specific water issues. "Not surprisingly," Biswas writes, "even though the rhetoric of integrated water resources development has been very strong in the various international fora during the past decade, its actual use (irrespective of what it means) has been minimal, even indiscernible in the field."[19]

The problem that Biswas identifies is perhaps best illustrated by quoting from one of the foundational representations of IWRM: "Integrated water resources management," declares *Agenda 21,* a UN plan of action for sustainable development in the 21st century, "is based on the perception of water as an integral part of the ecosystem, a natural resource, and a social and economic good."[20] Consistent with the constitution of modern water, each of these ways of perceiving discloses water as *one thing.* But by the same token, (modern) water can hardly be all of these things at once. This difficulty is apparent when considering all four of the principles adopted at the 1992 International Conference on Water and the Environment (the Dublin principles) together:

- Fresh water is a finite and vulnerable resource, essential to sustaining life, development, and the environment.
- Water development and management should be based on a participatory approach, involving users, planners, and policy makers at all levels.
- Women play a central part in the provision, management, and safeguarding of water.
- Water has an economic value in all its competing uses and should be recognized as an economic good.[21]

Within these principles are numerous contradictions. How can recognizing water as an economic good be reconciled with its essential function of sustaining life? Does adopting "a participatory approach, involving users, planners, and policy makers at all levels" not create a formula for protracted conflict? Does the third principle mean that the needs and views of women should trump those of men in matters regarding the provision, management, and safeguarding of water? How, for that matter, are such needs to be defined, especially in circumstances where power relations prevent the interests of women and children from being heard? What is obviously missing here is recognition that water is, literally, a political matter and that any regime of water management needs to be founded on this recognition of water's fundamentally social nature rather than a reformulation of its abstraction:

> Although the concept of IWRM holds the promise of reconciling goals of economic efficiency, social equity and environmental sustainability it is becoming clear that there is no consensus on how to weigh these priorities, or on how best to ensure their realization. Dominant visions of IWRM promote a view of a technical optimality to be achieved by good science, rational and neutral problem-solving, and negotiations between well-intentioned and well informed stakeholders. They obscure the reality of the (hard) choices and tradeoffs that have to be made. State and donor-driven water reforms take precedence over the necessity of empowerment as a means of redressing past injustice or unbalanced power relationships.[22]

Because the new paradigm retains modern water, the unfolding of an unresolvable global debate over the question of what water *is* under the new global regime has characterized water politics in recent years. If water "has an economic value in all its competing uses and should be recognized as an economic good," as declared in the fourth Dublin principle, how can this be reconciled with the perception of water as a social good or as an integral part of the ecosystem, as declared in *Agenda 21*? Is water a *natural substance* to which all people are entitled as a function of their fundamental human rights, or a *product* that may be priced and traded? Is it a *national* resource, to be conserved and safeguarded by the territorial state, a *global commons* to which all of humanity has some claim, or a *transnational commodity* to be disposed of in accordance with market principles? Is it the *lifeblood of ecosystems* or a *resource* in the functional sense? As Conca argues, the global water regime, grounded as it is on such ambivalence, is "doomed to founder" on "chronic conflict and controversy":

> For some, IWRM represents the death of the idea of rivers as resources and substitutes an imperative for comprehensive planning to balance economic, ecological, and social considerations. For others, it constitutes an effort to perfect rather than abandon the rivers-as-resources idea – to shift river development projects and water itself from a state-supplied public good to a private economic good, subject to the disciplining rule and valuation techniques of the market. As a result, the IWRM arena has been marked by struggles over public versus private authority, conflict over market versus nonmarket bases for resource valuation and allocation, and tensions between the territorially fixed character of the state and the transnationally fluid character of contemporary global capitalism.[23]

CONCLUSION

It is telling that practically all sides engaged in the politics of water have used "the water crisis" as a rationale for advancing their respective points of view. The World Water Council's parlaying of crisis into its particular *Vision* has been described above. Meanwhile, opponents of the commodification of water and privatization of water services are just as adamant about the water crisis: "There is simply no way to overstate the water crisis of the planet today," avers Maude Barlow in one of her many interventions against the corporatization of water.[24] The same could be said for, and has been said by, others who are primarily concerned with protecting the health of freshwater ecosystems in the face of a growing suite of anthropogenic assaults: "This water crisis," reports the International Rivers Network, "has significant ramifications for the world's river systems and all who depend upon them."[25] That such different – and often opposing – sides appeal to the water crisis might be taken as a hint that there is much more than just water at stake here, and that the crisis involves the *idea* of water as well as the matter of scarcity. For the most part, however, the political debate itself attests to the strength of modern water's constitution. Instead of *a crisis of modern water,* the water crisis is presented as a set of circumstances that demands that modern water be strengthened. The irony is that it can hardly be sustained in this climate of competing claims. Water – or, rather, modern water – cannot be an economic good, a state-owned resource, a public good, and the lifeblood of rivers all at once. When all these various waters are channelled through "the water crisis," what emerges is a paradox. No wonder, as Conca observes, "We seem, therefore, to be at an impasse."[26]

PART 4
Conclusion: What Becomes of Water

12

Hydrolectics

Water is not about water. Water is about building people's institutions and power to take control over decisions.

> – Sunita Narain, head of the Centre for Science and Environment, on the occasion of accepting the 2005 Stockholm Water Prize

Let me put it bluntly: Political ecology has nothing to do with nature.

> – Bruno Latour, *Politics of Nature: How to Bring Sciences into Democracy*

Introduction

This chapter brings together several strands of the discussion above to outline a practice of social hydrology that I call "hydrolectics." Consistent with a relational-dialectical outlook, hydrolectics conceives of a water process out of which particular instances of water get fixed, or instantiated, in social relations. Hydrolectics thus complicates the science of abstract water with the idea that we cannot have knowledge of water except in relation to our own circumstances and modes of knowing. In every case, it is the relation that defines the essence of what water is. This primacy of relation is obvious in the literal and material sense in which every cell of our body contains water and functions by virtue of water. The very process that we call "life" is so conditioned by water that it becomes impossible to separate these two processes except by fixing them in abstractions such as "people" and "H_2O." Thus, knowing and identifying water is necessarily a product of engagement, with engagement itself being the real – that is, relational – substance of both the knower and the known.

That we live and think by virtue of engagement with and participation in the water process means that we cannot identify water as something apart from ourselves except through the violence of abstraction. With respect to water and ourselves, there is no inside and no outside; water is always in our life process and so long as we live, we are always part of the water process. This statement may be taken as the hydrological imperative of a general ontology of process. As A.N. Whitehead puts it, "We are in the world and the world is in us."[1]

An ontology of process might seem a rather impractical stance on which to build any sort of practice. But it simply means that for all practical purposes, every instance of water is secondary to the process of engagement that makes it part of our world. From the drilling of a well to the discovery of H_2O on Mars by researchers at NASA, water is a collaboration between ourselves and the environment rather than a self-identical thing in itself. And to the extent that we make practical use of water – by pumping water from a well or perchance by living on Mars – the water process occurs *through* us. Water problems, therefore, are never just water problems; to imagine them in such a way is to deprive ourselves of the potential that exists in the water process.

Every relevant instance of water is realized by someone, whether a state planner, a well driller, a consumer, a researcher at NASA, or a child playing in the rain. Water is therefore conceived not as a self-identical object but as a process whose identity is formed in social relations. The possibilities for water are thus open to social change. In this sense, hydrolectics draws attention to the *politics of waters* instead of *water politics*. The politics of waters begins with the assertion that water is never simply "neutral stuff"; it is never merely given.[2] The fact that water always becomes what it is in accordance with a particular kind of engagement means that it always becomes *for someone or something*. Wherever water is declared a resource, the nature of its resourcefulness is defined by the particular uses for, and interests in, water held by those who declare it thus. Much of this study has examined how, throughout the era of modern water, a particular idea of the resourcefulness of water has been naturalized by government agencies, private corporations, and water managers and experts who have held the authority to generalize particular kinds of hydrosocial engagement – and particular waters. In this concluding chapter, I want to suggest a practice founded on the idea that *water itself is political* – that what is always at play is not just the question of who gets it, but the question of *what water is*.

ALIEN WATER

Several years ago, as an earlier draft of this book was being prepared, Canadians were learning (or being reminded) of the appalling water conditions that pertain in many of the country's remote Native communities. In late October 2005, it came to light that the water treatment and distribution system of the Kashechewan reserve on the shore of James Bay was contaminated with potentially deadly *Escherichia coli* bacteria. Most of the community's 1,200 residents were evacuated to cities in central and southern Ontario pending a solution to the problem. In the ensuing media coverage of the event, it was reported that for more than two years before this outbreak, federal government officials had been advising people in Kashechewan to boil their tap water before using it. Now, in the wake of an infamous *E. coli* outbreak that killed seven people in Walkerton, Ontario in 1999, the discovery of *E. coli* in Kashechewan was rightly deemed a national disgrace.

The plight of the residents of Kashechewan makes an argument for water's social nature and for an approach to water problems that recognizes social conditions as the key ingredient. Although most acute, the problem faced by these people was hardly unique. At the time the incident was reported, thirty-eight other Native communities in Ontario were under official "boiling water advisories," and for First Nations communities in Canada as a whole, the number was more than one hundred.[3] The residents of Native communities all across the country were thus compelled to use bottled water or water imported by other means (e.g., by tanker truck) from hundreds if not thousands of kilometres away for their basic needs. Nor was this situation only a recent occurrence: In 1988, the Science Council of Canada had reported that fully one-half of all on-reserve homes in Canada had no piped water. And in 1994, an outbreak of water-borne hepatitis B. in the community of Pukatawagan in northern Manitoba reached the attention of the public when a group of its residents began to walk three hundred kilometres south to the provincial capital to protest the deplorable water conditions in their community.[4] As with the revelation of *E. coli* in Kashechewan, the Pukatawagan story drew national attention to the water crisis facing Canada's First Nations communities. Then, as now, Canadians expressed their shock and disgust, and politicians promised to fix the problem.

The irony of the water crisis in Canada's northern Native communities is striking. Although a couple of the reserves identified as having inadequate drinking water were located in the southern, more densely populated part of the country, almost all were in more remote northern regions. There is perhaps no place on earth where "natural" water resources are, in proportion to the population, more plentiful and of better quality than in northern Canada. Even when including the more heavily populated parts of the country, Canada boasts a per capita supply of river runoff greater than almost every other country, and our share of the world's lake water is second to none.[5] With its vast hydrological resources and relatively miniscule human population, there could hardly be a less likely place for severe water problems than northern Canada. And yet, for the people involved, these problems are of a magnitude similar or equal to those facing communities in parts of the world that have been designated as "highly water stressed" when measured in terms of water resources per capita.[6]

The stories of Kashechewan and Pukatawagan illustrate how the water crisis has less to do with water than modern water leads us to believe. As the Indian socio-environmental activist Sunita Narain has declared, "Water is not about water." This assertion strikes us as paradoxical – like a water crisis in northern Canada – because we regard "water" in the first instance as something that we act upon. In the way of thinking conditioned by modern water, water can always only be "about water." That water is, or may be, about something entirely different from that to which we have become accustomed is suggested in the second part of Narain's statement: "Water is about building people's institutions and power to take control over decisions."

Taking "control over decisions" is relevant to the water crisis in Canada's north. Of all Canadian communities, the residents of northern reserves have been among the least likely to have a say in the water systems on which their collective and individual health depends. Although First Nations band councils have primary responsibility for the construction, operation, and maintenance of these water systems, they are designed, built, and operated in accordance with federal (or in some cases provincial) government standards. Furthermore, they consist of technologies that have been developed in industrialized, urban environments. The choice of technologies, as well as the construction, operation, and maintenance of these systems, is largely a matter of remote control from Ottawa, governed by what can only be described as foreign knowledge and expertise.[7] The result is a water so alien to people in these communities that many of the residents do not trust it for drinking, or even for bathing. Instead, many use expensive bottled water or take their water directly from lakes and rivers.[8]

The irony that people living in northern Canada feel they must use bottled water for their daily needs suggests a fundamental, and tragic, disconnect in hydrosocial relations. It suggests a mode of engagement with water that is hardly on the terms of the people who actually live in these places. The water of such places is not merely given – rather, it has become in accordance with programs, plans, and technologies that emanate from another place and another, entirely different, set of hydrosocial relations. Although the government agencies involved are undoubtedly sincere in their efforts to furnish water for the First Nations peoples of these communities, no expenditure of goodwill – or money alone – can substitute for allowing the people the capacity to engage, and produce, water for themselves. Until they gain a greater measure of power and control over the means by which water comes into their lives, such people are ever susceptible to water crises.[9]

A central argument of this book is that defining the water crisis in terms of diminishing quantities of available water obfuscates the bigger picture – namely that every issue involving water is realized in a specific social and cultural context. The sense in which water scarcity is the product of social and economic circumstances by which some people are allowed access to water while others are not has already been discussed (especially in Chapter 3). Defining the water crisis in terms of a world running out of water plays most strongly into the hands of those who would benefit from the provision of water supplies, especially in wealthy markets. It is no accident that tales of nightmarish water are often purveyed most emphatically by the companies that sell the stuff in bottles. Among people in places where such extravagant provision is unaffordable or inappropriate, the situation calls for an attitude that recognizes the need to engage more equitably and carefully with the water resources available. To do otherwise is to revert to the supply-oriented paradigm that has been discredited by virtually all water managers and experts.

So long as we accept uncritically the notion that water is about water, we acquiesce to fixing its nature in a particular web of language, knowledge, law, management, and material infrastructure that does not serve social purposes very well in either water-rich or water-poor regions. In a summary of recent scholarship on water issues, hydrologist Malin Falkenmark notes that water "is recognized increasingly as an essential component in the dynamics of poverty; poor water management can indeed create and perpetuate poverty. Not only is secured access to water essential for poverty alleviation, but water development is closely linked to food production and hunger alleviation, and to energy development."[10] Resource economist

David Brooks has observed how development professionals "have come to recognize that, if efforts to improve the quantity and quality of water supply are to be successful, not only must they be technically sound and economically feasible, they must also deal directly with poverty alleviation, local empowerment and ecological protection."[11] Some water-management thinkers have begun to criticize what they call "hydrocentricity," the mistaken notion that problems in the provision of water services to people can be solved entirely within the water sector itself.[12] In sum, it is evident that a growing cadre of researchers is beginning to operate with a far more complex idea of water; they are – as geographer Leah Gibbs, who has researched meanings and values associated with water in the Lake Eyre basin of central Australia, concluded – quickly coming "to learn that water isn't just water."[13]

As we have seen, modern water demands that water problems be framed in such a way that water itself is made the issue. Accordingly, addressing such issues and problems must be a matter of "water management," that is, performing some operation on water, traditionally by engineering new supplies and more recently by devising ways to allocate available supplies more efficiently. Instead of conceptualizing *water* as something to be manipulated and governed in the abstract, hydrolectics begins with the premise that *waters* are diverse, are made and are made known through different modes of social engagement. This approach suggests an alternative kind of hydrological science, one that fits philosopher Catherine Hayles' description of "a science understood to flow from historically specific interactions." Hayles points out that such a science "implies that we know the world because we are involved with it and because it impacts upon us. While such an understanding of the scientific enterprise does not guarantee respect for the environment, it provides a conceptual framework that fosters perceptions of interactivity rather than alienation."[14]

The Hydrosocial Cycle

How might a hydrosocial science "understood to flow from historically specific interactions" be grounded? If, as discussed in Chapters 5 and 6, a science of abstract water was organized around the concept of the hydrologic cycle, is there a similar framework around which hydrolectics might be considered?

The concept of the hydrosocial cycle might very well serve such a purpose. Already introduced in studies of the political ecology of water and discussed in Chapter 3,[15] the hydrosocial cycle offers an approach to studying and managing flows of water as phenomena that necessarily involve a mix of hydrological and social processes. "Just as the investigation of the circulation of money and capital illustrates the functioning of capitalism as an economic system," writes Erik Swyngedouw, "the circulation of water – as a physical and social process – brings to light wider political economic, social, and ecological processes."[16] For Swyngedouw, specific instances of the hydrosocial cycle are illuminated by tracing the circulation of water, particularly in urban environments: "These flows of water ... carry in their currents the embodiment of myriad social struggles and conflicts. The exploration of these flows narrates stories about the city's structure and development."[17]

The concept of the hydrosocial cycle can be elaborated here to provide a general framework for hydrolectics. We can thus consider the hydrosocial cycle as having ontological and epistemological relevance. The case for the ontological relevance of the hydrosocial cycle is suggested in the fact that practically every body of water on the planet bears traces of human involvement in the form of minute quantities of anthropogenic substances such as chlorinated organic compounds.[18] And almost everywhere it falls, snow is laden with particulates and other residues of human activity.[19] The flows of water on the earth's surface, moreover, are radically affected by people: In the Northern Hemisphere, some 80 percent of river discharge is regulated, or controlled, by dams.[20] Combined with this vast scale of human diversion and regulation of streamflow on the earth's surface, the systematic and global effects of anthropogenic climate change on hydrological processes mean that these processes are thoroughly and unavoidably involved with human ambition. The very nature of the circulation of water on earth, in other words, has to be described in *social* as well as *hydrological* terms. Under these circumstances, the "the course of natural circulation of water in, on, and over the Earth's surface," which Robert Horton had termed "the hydrologic cycle" in 1931, ought to now be expressed in a different idiom: it is now the case that the *hydrosocial* cycle characterizes flows of water in the hydrosphere more accurately than does the *hydrologic* cycle.[21]

The claim for the reality of the hydrosocial cycle is, moreover, buttressed by recent work in water science: "Evidence now shows that humans are rapidly intervening in the basic character of the water cycle," reports the framing statement of the Global Water System Project, an international research effort that facilitates integrated study of the "biogeophysical and social dimensions of the water system."[22] This statement may be understood

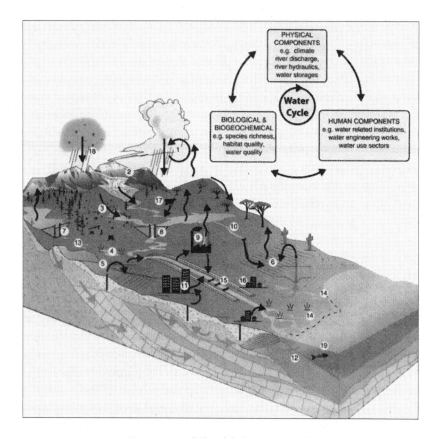

FIGURE 12.1 "The global water system"
C. Vörösmarty, D. Lettenmaier, C. Leveque, M. Meybeck, C. Pahl-Wostl, J. Alcamo, W. Cosgrove, H. Grassi, H. Hoff, P. Kabat, F. Lansigan, R. Lawlord, and R. Naimann, "Humans Transforming the Global Water System," *EOS* 85, 48 (2004): 509. © 2004 American Geophysical Union. Reproduced by permission of the American Geophysical Union.

in two ways: Clearly, as the authors point out, anthropogenic interventions, such as climate change, basin-scale water balance changes, river flow regulation, sediment fluxes, chemical pollution, microbial pollution, and changes in biodiversity, are "transforming the contemporary global water system."[23] But at the same time, the authors of the paper have themselves transformed the water cycle (see Figure 12.1). The water cycle is now understood and represented as the integration of physical, biological, biogeochemical, and human components of a more comprehensive system. Now it is the "water system" that represents the nature of water, a nature that is highly complex and highly social.

The hydrosocial cycle makes it unreasonable (and irrelevant) to isolate flows of water from social processes and vice versa. Whereas the hydrologic cycle "proceeds endlessly in the presence or absence of human activity," [24] the hydrosocial cycle frames a science whose field is defined by relations between the hydrological and the social. The hydrosocial cycle can be considered a way of conceptualizing, envisioning, and accounting for the necessary correspondences between hydrological and social processes. To elaborate on the comment made by Swyngedouw and cited above, analyzing physical flows of water in particular places "narrates stories" about social structure and cultural norms. [25] Therefore, in contrast with "the global water system" depicted above, the hydrosocial cycle can hardly exist as an abstract concept – it doesn't exist apart from the process of tracing and narrating specific flows. In this way, drawing the hydrosocial cycle might be considered a performance that yields practical knowledge of the social-historical nature of water.

Drawing the hydrosocial cycle can thus be regarded as a means of producing critical knowledge of the social nature of waters. The examples and images presented here are drawn to reflect my position on water politics in Canada. Readers are invited to consider similar examples drawn from their own places, experiences, and positions. Figures 12.2 and 12.4 illustrate two different *kinds* of water, as defined by the social structures and physical infrastructures by which they are made available to people. Figure 12.2 shows how the use of a public drinking fountain sustains water as a public good, while simultaneously producing a kind of public/citizenship – or "body public" – in which all members of society have equal access to water services. The fountain, the provision of high-quality water, and the public itself are sustained by the vested interests of fountain-users in maintaining these services. The interruption of this cycle by the strategic placement of a commercial bottled-water vending machine (Figure 12.3) is illustrated in Figure 12.4 to show how the diversion of water through private channels has the effect of producing a different kind of access, with the corollary of producing individual consumers rather than a body public. One socio-political effect of sustaining the flow of water through commercial vending machines (and similar means of securing private supplies of water) is suggested by considering how people who procure such private supplies might be less willing to fund public water infrastructure and facilities through their taxes. The general trend of the dereliction of public drinking fountains in many parts of North America might thus be analyzed as a function of this widespread change in the hydrosocial cycle. [26]

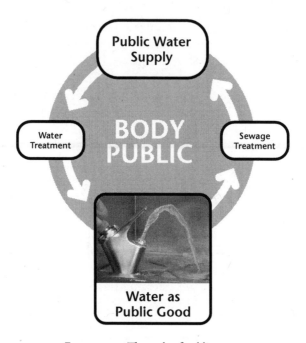

FIGURE 12.2 The cycle of public water
and the production of a body public: a drinking fountain
Photograph by the author.

As argued in Chapter 6, the hydrologic cycle has long provided a means of defining the field of scientific work known as hydrology. The suggestion here is that the concept of the hydrosocial cycle might provide a common basis for work in the political ecology of water. To this end, researchers have already begun to explore ways of formalizing a connection between this concept and critical investigations of water. A series of special sessions were organized at the annual meetings of the American Association of Geographers around the themes of "Water, science, humans: Adventures of the hydrosocial cycle" and "Water, science, humans: Advancing the hydrosocial cycle" in 2008 and 2009 respectively. The call for papers for the 2009 conference sets out the basic framework:

> Considering water as a socio-physical process makes it impossible to abstract water from the social circumstances that give it meaning and from the social and ecological relations that get consolidated in material flows of water. At the same time, attending to the social dimensions of these flows provides a means of analysing water's political nature so as to promote/facilitate change

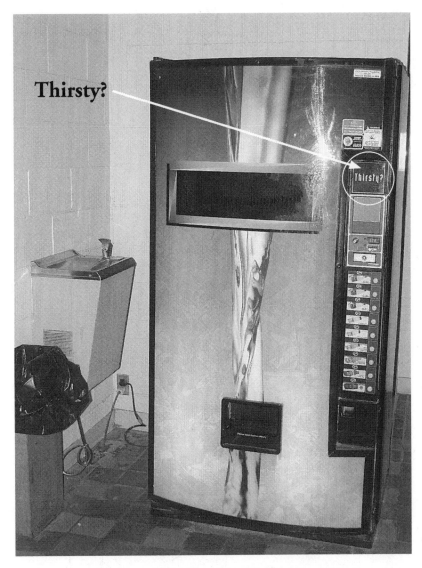

FIGURE 12.3 Interruption of the cycle of public water:
a drinking fountain blocked by a bottled water vending machine
Photograph by the author.

in what is often taken to be a fixed set of circumstances. The hydrosocial cycle thus offers a framework around which researchers interested in the social and political nature of water might rally as well as an approach to investigating the social production of hydrological knowledge.[27]

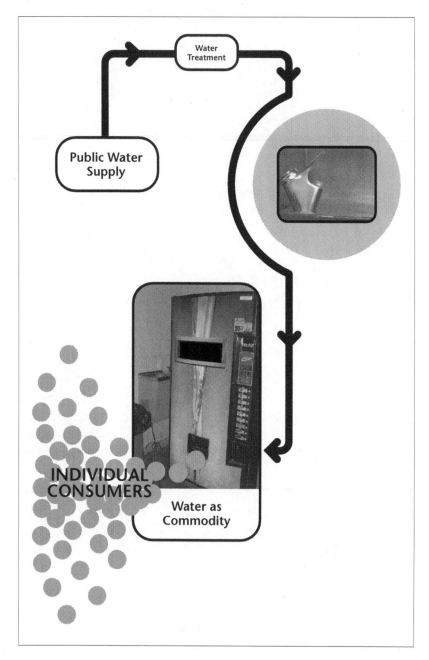

FIGURE 12.4 Diversion from the public water cycle
to a private vending machine and the production of individual consumers
Photographs by the author.

Hydrolectics may be applied practically and analytically. This distinction is somewhat arbitrary, as the way we analyze a problem or thing is internally related to the way we respond to it and vice versa. Although analytical and practical hydrolectics overlap, they may be considered here separately. To offer a rough approximation, practical hydrolectics is the business of changing waters and the social relations that sustain them, whereas analytical hydrolectics is the study of such relations and changes.

Practical hydrolectics means getting ourselves into the water, not only by becoming involved in decisions about water but by actively becoming mixed up in the solution that produces what water is and what it might be. Analytical hydrolectics means identifying the particular structure of water in any given situation. That water is different things to different people, depending on their modes of engagement, is the main principle behind this mode of analyzing problems and issues involving water. After discussing practical hydrolectics, I propose a rudimentary method for this kind of analysis.

Practical Hydrolectics

Practical hydrolectics reorients our response to water problems so that instead of striving to master a presumed water nature, we strive to change water's social nature. Something of this reorientation is suggested in the concept of integrated water resources management. Recall the second and third principles adopted at the 1992 International Conference on Water and the Environment (the Dublin Conference): "Water development and management should be based on a participatory approach, involving users, planners, and policy makers at all levels," and "Women play a central part in the provision, management, and safeguarding of water." However, that there may be fundamental differences between the *waters* of users, planners, and policy makers (including women in each of these roles) needs to be recognized. The modern-water bias of IWRM makes it difficult to put these excellent principles into practice. As Ken Conca has pointed out, "Not surprisingly an approach grounded in expert knowledge, scientific rationality, and increasingly bureaucratic organization has often reinforced a limited, hub-and-spoke notion of participation. Helpful information about uses, preferences, behavior, and effects flow in from society to expert centers; scientific truths as guides to social action flow out."[28]

Practising hydrolectics might be considered a means of allowing IWRM to live up to its participatory potential. However, it stresses *different modes* of participation. By its constitutional separation from people, modern water limits participation to the act of making decisions *about* water; thus, the participatory ideal becomes one of allowing everyone to have a hand in making these decisions. One might consider a different sort of participation, one that is forbidden by modern water's constitutional separation of water from people. Our bodily engagements with water – from diving to drinking to draining – constitute a kind of participation by which we make different waters rather than making decisions about water. The notion of getting into water can be taken literally as well as figuratively, in the sense of getting involved in the decision-making process. This more direct sense of participation, moreover, is a process in which we are constantly engaged. Practical hydrolectics is the business of recognizing and appreciating the different waters that get produced in these engagements. Once again, the examples given below are drawn from my own rather privileged experience and engagements with water in a wealthy, northern-temperate country. As examples, they are intended to be suggestive for readers in different places and circumstances.

In places where the trend of drinking branded bottled water threatens to undermine public water services, the simple act of drinking water from a public fountain effects the production of water as a public good rather than water as a commodity (see Figure 12.2). Drinking water from public fountains is one example of a practice that has the potential to influence or even change the constellation of things such as water identities, laws and regulations, administrative apparatus, and physical infrastructures that sustain historical instances of water. Even the simple act of drinking water from a public fountain therefore can be regarded as a way of practising hydrolectics, one that simultaneously (co)produces public water and a body public.

Apart from drinking it, our physical engagements with water can have the potential to change things. The strategy of getting bodies into the water (as well as getting water into bodies) can be an effective way to establish or restore public beaches. For several years in a row, a group of citizens in the city of Quebec has organized a day of swimming in the St. Lawrence River as a way of building political support for improved public access to the water in the city core. "Le grand splash," as this event is dubbed, garners media attention as well as the attention of City Hall by attracting local celebrities and large numbers of bathers to the cause.[29] A

FIGURE 12.5 Mass swim at Richardson beach, Kingston, Ontario, 22 July 2008 Photograph by Lynn Rees Lambert. Reproduced with permission of Lynn Rees Lambert and *Kingston This Week*.

similar event in which the author participated in the summer of 2008 drew attention to the need to restore an old public beach in the downtown area of Kingston, Ontario. The Mass Swim drew approximately three hundred people into the waters of Lake Ontario, making an argument for municipal investment in beach restoration (see Figure 12.5).[30] The need for such investment is supported by demands that all citizens, and not only those who can afford private summer cottages and vacations to exotic beaches, should have access to water for recreational purposes. By placing themselves in the water, people can physically produce public water spaces while simultaneously constituting themselves as a body politic with an interest in supporting public efforts to maintain high water-quality standards.

"People-Oriented Water Solutions" and "Local Water Management"

The principle that water is the product of peoples' involvement in the water process finds expression in new forms of water governance. For example, a new constellation of public participation, water infrastructure and regulatory apparatus seems to be forming in the wake of the experiment with urban water privatization in the 1990s. The shift toward privatizing water-service infrastructure may have had the salutary effect of breaking up some of the least responsive statist water regimes; however, after a decade or more of initial exuberance, it has became obvious, as discussed in Chapter 3, that privatization was no simple panacea: its widespread failure to provide adequate water services to the poor, to invest in new infrastructure and maintain existing infrastructure, to promote conservation, and even to make a profit has given rise to a rethinking of the simple dichotomy between statist and market water systems. In some jurisdictions, as Karen Bakker has shown, water services have been restructured and re-regulated so as to correct some of the more abject lapses of out-and-out privatization.[31] The combined shortcomings of both state and market-driven water systems have shown that the problem is neither quite the state nor quite the market but "a lack of democratic process in the public sector."[32] Thus, the "reclamation of public water" manifests as a process in which organizational structures respond to changing social and hydrological circumstances.

Balanyá and her colleagues have assembled a compendium of case studies where the citizens of urban regions in the global north and south are engaging with water in a way that patently bucks the modern trend of statist, technocratic water management.[33] The basic principle of these people-oriented "water solutions" is that local citizens are involved in all matters pertaining to water management.[34] While it may be too simplistic to characterize these solutions as a matter of substituting "community control" for privatization,[35] it is patent that when they are so involved, people identify water and assert its social nature in ways that differ markedly from modern water. The compendium shows

> how significant improvements in access to clean water and sanitation have been achieved by diverse forms of public water management. These people-centred public water solutions have occurred under a variety of socio-economic, cultural and political circumstances ... In [most of these cases], citizen and user participation in various forms is an essential factor behind the improvements in effectiveness, responsiveness and social achievements

of the water utility ... While it should not be considered a panacea to be implemented in every situation, and in some circumstances may not be feasible, participation and democratization in its multiple forms can be a powerful tool for positive change in most circumstances.[36]

In order to gain access to and control of water on their own terms, the players involved in these people-oriented water solutions have insisted on identifying water *in* their own terms, that is, as a public good, but one to which the state has no prior claim. What seems to be emerging at the progressive edge of urban water governance combines the rejection of state structures and rigid techniques with a renewed emphasis on democratic participation in decision making to produce a plethora of what might be called postmodern public waters.

The principle of local involvement in water governance is also an important lesson learned by international development agencies involved in water projects. "Local water management," writes resource economist David Brooks, "permits a democratizing decentralization of decision making and accountability. Well done, it empowers people (particularly the poor and otherwise disadvantaged) to take part in the decisions that define their own futures. And it encourages the integration of traditional knowledge with innovative science to promote fair and efficient supply management."[37]

The overriding principle of "local water management" as described by Brooks is that people "need to be engaged in the decisions affecting their lives."[38] Although this may seem an obvious prerequisite for success, it nevertheless poses a serious challenge to modern water and to the expert discourses by which modern water is sustained. Adhering to such a principle means paying attention to the particular hydrosocial relations and the different waters that inhere in specific social contexts. Further, it often means recognizing the legitimacy of, and adapting relatively simple techniques that embody, traditional or local knowledges of water. In many cases, these knowledges and techniques have been shown to be more successful and sustainable than complex, exogenous technologies, no matter how brilliant the latter may appear to their progenitors.[39]

The adaptation of traditional waters to twenty-first-century circumstances is discussed in Chapter 3 – for example, the tank system in southern India and other techniques of rainwater harvesting. These examples begin with understandings of, and relations with, water that are indigenous to particular peoples and places. Exogenous techniques may be adapted to such circumstances – indeed, they often must be so in order for traditional waters to meet today's needs – but in these cases, there is no imposition of modern water.

A kind of counter-example is provided in cases where inattentiveness to local waters has resulted in the failure of what otherwise appear to be ideally appropriate technologies. Among the many illustrations that may be cited are technologies for harvesting water from fog that were developed with funding from the International Development Research Centre and UNESCO in the 1980s. The construction of "fog catching" apparatus (nets) on the highlands and water conduits leading to communities in the rain shadow of the Andes mountains in Chile materialized an ingenious solution to the problem of water scarcity in these communities. Before being introduced to this novel source of water supply, these communities had relied almost entirely on very expensive water trucked over long distances. By the early 1990s, one installation of fog catchers was providing villagers with 11,000 litres of clean water daily, enough to furnish 33 litres for every person in the village at very low operating costs – this was more than twice the per capita supply that the community was paying to have delivered by truck.

Despite their technical elegance and initial functional success, the fog catchers in Chile ultimately failed, partly because of higher than expected costs and because they required more coordinated maintenance of the nets for harvesting fog and pipes for transporting water than the community was able to sustain. More significantly, they failed in part for a reason that their designers hadn't considered beforehand: the water from fog catchers was deemed inferior – or second class – to water supplied via so-called modern supply systems such as municipal distribution networks, or even trucks. The perceived inferiority of the water provided by the new technology made this technology unsustainable in this particular social context.[40] Although the fog-water may have been regarded by outside observers as far superior to more expensive truck-delivered water, its production could be sustained only within a set of social relations, attitudes, and expectations that were then foreign to the target community.

A further example of practical hydrolectics can be drawn from the concept of "water soft paths." Based on earlier work on energy by resource economist Amory Lovins, water soft paths describes an approach that is at the cutting edge of thought in the water management community; its main insight is to recognize that in most cases it is not water that people need but the various services that water may provide. Among other things, this approach attends to the potential efficiencies that can be realized by matching the quality of water supplies with the requirements of water demands and satisfying social ends by alternative means that may not require the use of water at all.[41] Due to the fluidity of the water process, these potential services are practically limitless and, as a result, water has come to be regarded as an essential resource for myriad purposes ranging from agriculture to industrial production to sewage disposal. Water-soft-path thinking is an approach that facilitates a freeing-up of this essential resourcefulness of water – a basis for realizing the potential inherent in the notion that people, for the most part, need services rather than water itself. "By looking at water as a bundle of services rather than a natural resource or commodity," writes David Brooks, "many more options can be conceived to satisfy demands."[42]

Analytical Hydrolectics

Analytical hydrolectics simply means applying a relational-dialectical approach to the study of water and people. This entire book could be considered an exercise in analytical hydrolectics. Here, however, we consider a kind of framework for the analysis of specific water issues in accordance with principles of hydrolectics. The value of such a framework is suggested by recognizing that the water "users, planners, and policy makers" invoked in the third principle enunciated at the Dublin Conference in 1992 each has a different relation to, and understanding of, water. The need to consider these different waters is implied in various calls to involve all "stakeholders," or "participants," in water decisions. For the most part, however, the very language of stakeholding and participation belies the homogenizing effects of the process: IWRM and related prescriptions for reforming water management generally call for the involvement of all stakeholders and participants but with the underlying notion that all have a stake in, and can participate in a discussion about, *the same thing.* By making such a presumption, these discourses have the effect of naturalizing modern water, which in turn disciplines the discussion and constrains the potential outcomes. Although it does perhaps *integrate* the interests of different stakeholders, IWRM is still water *resources management.* Stakeholders and participants are thereby consulted but not so as to elicit how different players may relate to water in ways other than as a resource whose natural disposition is to be managed.

A framework for analytical hydrolectics might be suggested in order to provide an approach to ensure that the different waters of different actors are considered when analyzing water issues and problems. This framework would involve two complementary moves: first, identifying the different actors involved; and second, associating different actors with different waters. The example below builds on a recent article dealing with the question of how to manage scarce water resources in Israel. Over half the freshwater used west of the Jordan River comes from groundwater, which makes this an extremely important source of water for Israelis. Although the social and hydrologic circumstances presented in this article are unique to modern Israel, the approach used here could be applied to different places, peoples, and waters.

In "Political Economy of Groundwater Exploitation: The Israeli Case," Eran Feitelson observes that despite possessing adequate knowledge, expertise, and power, the office of the state water commissioner nevertheless oversees a regime of unsustainable aquifer exploitation, allowing farmers to withdraw water during droughts at a rate that exceeds the rate of natural recharge. Here, all the "steps and measures" that would need to be put in place in order to regulate aquifers sustainably are known to the water commissioner, an agency that, moreover, "has more power than almost any other water manager in the world." The question, therefore, is, "Why have the best measures advanced not been used?"[43]

To analyze this problem, Feitelson identifies numerous actors in the "political game" of allocating water in aquifers. First there are the farmers (the "pumpers"), who use the water. Second are the water managers, namely the hydrologists and water resource experts who possess the knowledge and skills required to manage aquifers sustainably. Third are the decision makers who have ultimate responsibility for weighing various hydrological, ecological, political, and economic considerations relevant to the question of the level of abstractions that are to be permitted. Both the water managers and the decision makers are represented within the office of the state water commissioner. Fourth is the general electorate, whose interests in water are far more diffuse and attenuated than those of the water users themselves and who therefore have less political clout than the farmers who use the water directly. Feitelson points out that because decision makers are certain to lose the support of farmers if they cut back on abstractions, they are acting reasonably when they allow overpumping against the advice of water managers. Because the extent of abstractions is not of direct interest to most voters, organized groups of pumpers tend to be very effective. In fact, water managers are the only group likely to oppose pumpers' demands for greater abstractions, out of their concern for being held accountable for overdrafts:

The political game that ensues, therefore, is one in which users strain to increase abstractions, mainly by using political leverages, while water managers strive to constrain them, often using legal means or administrative authority. The conflict between these two parties is likely to come to a head in drought situations when both the shadow value [the value of water that would most likely pertain should there be a free market for water, that is, if water were not a subsidized commodity] of water and the threats to the aquifers rise.

The risk posed to aquifers from over-draft in droughts is a function of the attributes of the aquifer, the extremity of the drought and the extent of over-pumping. In most cases, it is unclear when would the deleterious implications of over-draft for nature and water quality be felt. In contrast, users feel any cut in allocations immediately and directly, and will attribute these losses to decision-makers. From a decision-makers' perspective any cut in allocations is thus a certain loss ... Therefore, they prefer to over-pump the aquifer, rather than to cut allocations to farmers, in contrast to the advice of water managers. In other words, the extensive abstraction from aquifers above the rate of recharge, especially in drought situations, is politically rational, and therefore should not be seen as an aberration or misconception by decision-makers.[44]

Feitelson writes from what is essentially a sustainable management perspective and with the overall purpose of suggesting "improvement in the management of the aquifers."[45] From this perspective, the problem is seen in terms of the inability of water managers to effect the kinds of controls that are deemed necessary to promote "sustainable development of aquifers."[46] His prescription is, in effect, to put more power into the hands of water managers. Because "a system that constrains and regulates abstraction is largely a function of the ability of professional or government agencies ... to push through [sustainable management] proposals," Feitelson argues it is necessary to vest more authority in these professionals and agencies.[47] He thus proposes to "overhaul the decision-making structures in the water-sector" specifically by giving the water commissioner authority to set water tariffs (prices) and to order alternative supplies (desalination and recycled wastewater) in addition to its authority for licensing abstractions from aquifers: "In essence, the Water Commissioner should be allowed to determine allocations, rates and the use of revenues, thereby allowing him to fund supply enhancement and quality assurance infrastructure."[48] Furthermore, it is proposed that all major policy decisions be approved by a revamped water council attached to the office of the water commissioner. Feitelson proposes that the water council, currently an advisory body, be enlarged and empowered to approve or reject important policy decisions:

> In this manner, the Water Council, which will represent all the affected parties as well as independent professionals representing different disciplines, will have real power to determine policy. Actually, this means that the [office of the water commissioner] will become the focal point of water politics. As this commission will be more representative than the current government structures, and will be more specialized than the forums in which politics take place today (the Knesset and Treasury) it can be hoped that this body will indeed strike the necessary balance between all the interests and considerations that have to be taken into account in determining abstractions from the aquifers.[49]

Feitelson's proposal comprises two basic moves: First, it concentrates authority for all matters pertaining to the management of aquifers in the office of the water commissioner; and second, it provides this office with the means to represent the various political interests involved or affected by water management decisions. This proposal, in other words, calls for expanding the power and scope of the central agency responsible for water management so that it embodies broader interests and concerns formerly beyond its remit; it thereby diversifies and expands the idea of water management and vests this new idea in a revamped water management structure and agency: "The conclusion that can be drawn from these findings is that the management of aquifers should not be considered mainly as a technical-hydrological issue. Rather, it should be analysed within the societal, institutional and political setting of water management and politics."[50]

In building on Feitelson's article to present an example of analytical hydrolectics, we should highlight this latter observation: managing aquifers (and water generally) should certainly "not be considered mainly as a technical-hydrological issue." But we can go further: The article provides a very good basis for illustrating the thoroughgoing social nature of water. By identifying the main players involved in this particular water issue, Feitelson already covers the first step in a rudimentary framework for hydrolectic analysis. It remains to take the second step – to identify the various waters of these different players.

Each of the four actors Feitelson identifies as being involved in this particular political game has a different understanding of water and promotes a different water discourse. First, there are the users (the farmers). The farmers are concerned with the water itself as a material substance; their aim is to get more of it. Second, there are the water managers. The managers are less interested in the water itself than in the hydroeconomic abstraction they make known as an "aquifer"; their concern is to manage the aquifer sustainably – that is, so as to ensure the long-term balance of outputs (abstractions) with inputs (recharge). Third, there are the decision makers. Although embodying the managerial view to some extent, the decision makers are more attentive to the broader political interests involved in making allocations of water; their role can be seen as mediating between users and managers. To the decision makers, water is an abstract resource to be allocated so as to procure the greatest social and political benefit. As Feitelson points out, they may decide to deplete the aquifers, a decision that is perfectly rational because it produces an immediate and certain political benefit. Fourth, there is the electorate, which figures in Feitelson's analysis only as having relatively little influence against that of the users-farmers. The relationship of such an aggregation of people to water is difficult to generalize. However, it might be taken to reflect and sustain the discourse of water as a national preoccupation in Israel. Something of the flavour of this discourse is suggested in the opening paragraph of Lonergan and Brooks' study of the role of water in the Israeli-Palestinian conflict:

> Both Biblical and modern Israel have been vitally concerned with water. The first explicit ecological reference in the Bible relates to the carrying capacity of pastures and water supplies that were coming under stress from the growth of the herds owned by Abraham and by his nephew Lot. One of the periodic droughts of the region starts the drama that leads to the Exodus story. More recently, the challenge to "make the desert bloom" brought Jewish settlers into what was first a Turkish colony and then a British protectorate. Later projects, such as the draining of the Hula wetlands and construction of the National Water Carrier, were hailed as symbols of the potential to live productively and comfortably in an arid region.[51]

Clearly, the four groups of players identified in Feitelson's article and the four corresponding views of water are interrelated in complex ways, a fact that is suggested by noting that the electorate comprises all the members of the three other groups. Nevertheless, analyzing the dynamics of water management in Israel from the perspective of hydrolectics draws attention not just to the different actors and different interests that are at play but to the fact that each involves a different view, understanding, and discourse of the "thing" in question. Farmers, water managers, decision makers, and the national electorate each consider water at a different level of abstraction, such that it *is* something different for each of them. In addition to the administrative and institutional reforms that may be necessary to promote "improved water management," it may therefore be desirable, or necessary, to procure some means of mutually recognizing and mediating between these heterogeneous discourses. In the particular instance examined here, perhaps the revamped water council – inasmuch as it represents all affected parties – could be given this function.

Here, the language and preconceptions entailed in the notion of bringing together various "stakeholders" with different "interests" in water would need to give way – as suggested above – to what Bruce Braun has described as "a sustained examination and appreciation" of the ways that a single aspect of nature (in Braun's case, a forest) may be invested with various meanings and values by different people, "an examination of our passionate attachments" to such things.[52] Furthermore, instead of presuming to identify a consensus among all such groups, Braun points out that a positive upshot of such an examination would be to move toward a co-construction: "If any consensus is to be reached, it will have to be something built rather than found, provisional rather than final, always open to examining its own constitutive exclusions."[53] This kind of practice might be suggested as the main role of a reorganized water council. In the course of fulfilling such a role, it might be surmised that the present overall aim of promoting "improved water management" might be abandoned to give way to a less singular formulation.

The management of aquifers in Israel is but one problem among a practically infinite number of water problems and issues. Each of these problems is unique, involving different actors and different waters. The basic aim of analytical hydrolectics is to identify these so as to gain a comprehensive appreciation of the full range of actors and waters involved with respect to the problem or issue in question.

This study began with the assertion that water is what we make of it, which was immediately followed by noting that there is nothing particularly new in such an assertion. Exactly the same can be said for hydrolectics. "Hydrolectics" might correctly be regarded as a catchy term describing what we always and continually do with water. We are always diving into water and coming up with words to describe the experience; we are constantly engaging with water in the process of eating, drinking, cooking, bathing, growing, soaking, feeding, drying, manufacturing, transporting, recreating ... and by virtue of these engagements, we are constantly producing new meanings, identities, discourses, and waters. We have a tendency to fix things, and as a result, we have an equal tendency to question, revise, reform, and revolutionize our fixations. Recognizing and working with these tendencies is the aim of hydrolectics.

Notes

Chapter 1: Fixing the Flow

1 Eliade, *Patterns in Comparative Religion*, 188-89; *Encyclopaedia Britannica*, 15th ed., 28:683, s.v. "Tissues and Fluids, Water"; Qur'an 21:30; Illich, *H₂O and the Waters of Forgetfulness*, 24; Thomas Mann is quoted in Chorley and Kates, "Introduction," 1; Melville, *Moby-Dick*, 94.

2 Tuan, *Hydrologic Cycle*, 20.

3 Genesis 1:1. Many North American cosmogonies also begin with a watery universe, from which the earth (Turtle Island) and its denizens came to be.

4 This moment in the history of science is discussed in Chapter 4.

5 I am indebted to Bruce Braun and Noel Castree for the term "social nature." Castree and Braun, *Social Nature*.

6 Sennett, *Flesh and Stone*, 139.

7 Tarnas, *Passion of the Western Mind*, xiii-xiv.

8 Collingwood, *The Idea of Nature*.

9 Foucault, *The Order of Things*.

10 Ibid., xxi-xxii.

11 The newspaper article to which I refer is Lyon, "Yemen Sleepwalks."

12 For example, Nuttle, "Taking Stock of Water Resources"; Nuttle, "Is Ecohydrology One Idea or Many?"; Falkenmark and Rockstrom, *Balancing Water*.

13 For example, Strang, *The Meaning of Water*.

14 For example, Bakker, *An Uncooperative Commodity*; Swyngedouw, *Social Power*.

15 Nietzsche is quoted in Gare, *Postmodernism and the Environmental Crisis*, 58.

16 Tuan, *Topophilia*, 7.

17 Foucault, "*Society Must Be Defended.*"

18 Some of the exceptions discussed below include Strang, *The Meaning*; Illich, *H₂O*; Goubert, *The Conquest*; Bachelard, *Water and Dreams*.

19 This rupture describes a radical cultural shift in the West over the past couple of generations. In general, faith in such tenets of modernism as rationalism, individualism, and science began to erode in the 1970s as critics drew attention to how "the achievements of modernity rested on race, class, and gender domination, colonialism and imperialism, anthropocentrism and destruction of nature." Jody Emel and Jennifer Wolch, "Witnessing the Animal Moment," 16. See Jameson, *Post-Modernism;* Harvey, *Condition of Postmodernity.*

20 Kaika, *City of Flows,* 12.

21 This concept of hegemony is drawn from the political thinker Antonio Gramsci, for whom "hegemony" referred to the intellectual and cultural domination by which a given social order is maintained.

22 Harvey, *Justice, Nature and the Geography of Difference,* 94.

23 Day and Quinn, *Water Diversion and Export,* 20.

24 Desbiens, "Political Geography."

25 The term "sclerosis," used in this context/sense, is from Harvey, *Spaces of Hope,* 185.

26 Walkem, "The Land Is Dry," 303-19.

27 The kayak is the traditional watercraft of the Inuit in northern Quebec; the canoe is that of the Cree.

28 Desbiens, "Political Geography," 112. See also Carlson, *Home Is the Hunter.*

29 Rodda and Ubertini, *The Basis of Civilization.*

30 Gregory, "(Post)Colonialism and the Production of Nature," 96-7. In several places, particularly in Chapter 3, we consider what might be described as a post-colonial approach to modern water, a topic that deserves a separate study. For the most part, the hegemony of modern water is treated below more in terms of the globalization of scientific discourse.

31 *Webster's Third New International Dictionary,* s.v. "water."

32 *Concise Oxford English Dictionary,* (Fifth Ed.). s.v. "water."

33 Marsh, J.H., ed., *The Canadian Encyclopedia,* s.v. "Water."

34 Wikipedia, s.v. "Water."

35 Giblett, *Postmodern Wetlands.*

36 The material flow is relatively obvious. That water constitutes an important part of our thoughts and dreams is perhaps revealed in the impossibility of proceeding very far along the stream of life without producing watery metaphors. For a good dive into the flow of water through the human psyche, try Bachelard, *Water and Dreams.*

37 Kalaora, "De l'eau sensible à OH2," 1.

38 Ward, *Reflected in Water.*

39 Strang, *The Meaning of Water,* 246.

40 Worster, *Rivers of Empire,* 5.

41 Hamlin, "'Waters' or 'Water'?" 321.

42 "Natural philosophers" refers to the thinkers and writers who made nature, or various aspects of nature, the focus of their work before the term "scientist" came into use in the nineteenth century. In an historical sense, it represents a very broad category. Natural philosophy is distinguished from modern "science" mainly in that its practitioners integrated, or did not make a distinction between, philosophy and the (scientific) study of nature.

43 Tuan, *Hydrologic Cycle.*

44 Horton, "Field, Scope, and Status."

45 Latour, *We Have Never Been Modern.*

Chapter 2: Relational Dialectics

1 Rescher, *Dialectics*, 46-60.

2 Marx, "The German Ideology," 163-65.

3 See, for example, Harvey, *Social Justice;* Harvey, *The Limits to Capital;* Harvey, *The Condition of Postmodernity;* Harvey, *Justice, Nature and the Geography of Difference;* Harvey, *Spaces of Hope;* Harvey, *Spaces of Capital.* For an overview and critical discussion and appraisal of Harvey's work see Castree and Gregory, *David Harvey.*

4 Neil Smith, *Uneven Development;* Swyngedouw, "Neither Global nor Local."

5 A complete study of this sort has yet to be written. Erik Swyngedouw (e.g., *Modernity and Hybridity* and *Social Power*) sets the stage in drawing from Henri Lefebvre as well as Marx, David Harvey, and Bruno Latour to elaborate a theory of "the production process of socionature" as it applies to water. The choice of Lefebvre as the theorist upon whose work a full elaboration of the "production of water" might be based is warranted by Lefebvre's supra-structural appreciation of the factors that enter into the production of space, namely in his triad of spatial practice (perceived space), representations of space (conceived space), and representational spaces (lived space). See Lefebvre, *The Production of Space,* chap. 1.

6 Among the well-known works in this field are Collingwood, *The Idea of Nature;* Collingwood, *The Idea of History;* Lovejoy, *The Great Chain of Being;* Weiner, *Dictionary of the History of Ideas.* A classic example of the history of geographical ideas is Glacken, *Traces on the Rhodian Shore.* Arguably falling into this category of the history of geographical ideas is Livingstone, *The Geographical Tradition.* As noted in Chapter 1, the history of ideas approach has largely been eclipsed by Foucauldian analysis as well as by historical materialism, both of which grew in popularity among Western scholars beginning in the 1970s.

7 Harvey, "Population, Resources," 271; Harvey, *Justice, Nature and the Geography of Difference,* 6-7; Harvey, *Spaces of Hope,* 15.

8 For a good introduction to process philosophy see Rescher, *Process Philosophy;* and Rescher, "The Promise of Process Philosophy."

9 Harvey, *Justice, Nature and the Geography of Difference,* 47.

10 Marx, "The Grundrisse," 231.

11 Whitehead, *A Philosopher Looks at Science,* 14.

12 Whitehead, *Process and Reality.*

13 Swyngedouw, *Social Power,* 15.

14 Wittfogel, *Oriental Despotism;* Worster, *Rivers of Empire.*

15 Roy and Lane, "Putting the Morphology Back," 111, 120.

16 This lyric is from a song titled "I Could Be Nothing," performed by Great Lake Swimmers; the song was written by band member Tony Dekker; it is on the CD titled *Bodies and Minds,* published by weewerk, 2005.

17 Ollman, *Dialectical Investigations.*

18 Whitehead, *Process and Reality,* 92.

19 Ollman, *Dialectical Investigations,* 29.

20 The equation of dialectical with binary thinking is no doubt reinforced by the traditional notion of a process of confrontation between thesis and antithesis. In any case, dialectics is often criticized for failing to transcend binary logic. Latour, *We Have Never Been Modern,* 57; Whatmore, "Hybrid Geographies," 25; Whatmore, "Introduction"; Braun, "Towards a New Earth," 199.

21 Whitehead, *Nature and Life*.

22 Harvey, "Space as a Keyword," 274.

23 Harvey's account of relational space, the account I have given here of a bicycle, and the idea of relationality generally are derived largely from Marx's illumination of how the constitutive relations of the products of human labour get hidden or made invisible in the production of commodities under capitalism – otherwise known as the "commodity fetish." (See Kaika, *City of Flows*, 29-31.) The philosophy of Alfred North Whitehead provides another source for these ideas: "The fundamental order of ideas is first a world of things in relation, then the space whose fundamental entities are defined by means of those relations and whose properties are deduced from the nature of these relations." Whitehead quoted in Harvey, *Justice, Nature and the Geography of Difference*, 256.

24 Harvey, *Social Justice*, 290.

25 Marx, "German Ideology," 163-75.

26 Harvey, *Social Justice*, 298-99.

27 Harvey, *Justice, Nature and the Geography of Difference*, 85. See also Palmer, *Descent into Discourse*.

28 Harvey's dialectical approach acknowledges the insights of recent developments in social theory while avoiding the enthusiasm that often arises upon recognition of the relations that these theories elucidate. Writing against what Palmer described as "the descent into discourse" by which discourse analysis "has consequently become more and more hegemonic in social and literary theory," Harvey gives full recognition to the insight that "texts (discourses) internalize everything there is and that meaningful things can be said by bringing deconstructionist tactics to bear both upon actual texts ... as well as a wide range of phenomena in which the semiotic moment has clear significance." At the same time, he cautions that it is too much "to insist that the whole world is nothing other than a text needing to be read and deconstructed." Harvey, *Justice, Nature and the Geography of Difference*, 85, 87. This caution might be regarded as an imperative of internal relations by which the locus of change is necessarily dispersed among the various moments that can be picked out of the social process, rather than fixed to any particular moment, be it discourse, material practice, ideas, beliefs, or social relations.

29 Ibid., 80.

30 Ibid., 92.

31 Ibid., 106, 107.

32 Braun, "Towards a New Earth," 213.

33 Hayles, "Searching for Common Ground."

34 Ingold, "Materials against Materiality," 14.

35 Linton, "Is the Hydrologic Cycle Sustainable?"

36 As discussed in Chapter 5, the idea of the circulation of water that persisted in Europe until approximately the nineteenth century held that subterranean physical forces caused water to rise in underground channels to reach areas of high altitude, from which springs and rivers were derived.

37 In Harvey's scheme presented above, such "realities" might best be located in material practices: "Material practices are the sensuous and experiential nexus – the point of bodily being in the world – from which all primary knowledge of the world ultimately derives." Harvey, *Justice, Nature and the Geography of Difference*, 79.

38 Castree, "Nature of Produced Nature," 21-22. Examples of such authors include Neil Smith, *Uneven Development;* Neil Smith, "The Production of Nature"; Castree, "Nature of Produced

Nature"; Castree, "Marxism and the Production of Nature"; Castree, "Socializing Nature"; Castree, *Nature;* Harvey, *Justice, Nature and the Geography of Difference;* Gandy, *Concrete and Clay;* Swyngedouw, "Modernity and Hybridity"; Swyngedouw, *Social Power.*

39 Swyngedouw, "The City as a Hybrid"; Castree, "Marxism and the Production of Nature."

40 See Castree, "Environmental Issues"; Castree, *Nature.*

41 Latour, *We Have Never Been Modern,* 64 et passim; Haraway, "A Manifesto for Cyborgs," 8; Biagioli, "Introduction."

42 Swyngedouw, *Social Power,* 14.

43 Swyngedouw, "The City as a Hybrid," 69-70.

44 Gandy, *Concrete and Clay,* 22.

45 Swyngedouw, "Modernity and Hybridity," 447.

46 Swyngedouw, "Neither Global nor Local"; Swyngedouw, *Social Power;* Swyngedouw, "Modernity and Hybridity."

47 Swyngedouw, *Social Power,* 28.

48 Ibid., 28.

49 Neil Smith, *Uneven Development;* Swyngedouw, *Social Power,* 15, 19.

50 Swyngedouw looks to Henri Lefebvre's dialectical investigation of *The Production of Space* for a means of resolving the apparent contradictions between materialism and discourse as constitutive elements of the socio-natural process. Swyngedouw, "Modernity and Hybridity," 447-49; Swyngedouw, *Social Power,* 20-24.

51 Ibid., 21.

52 Castree, "Environmental Issues," 204. For similar critiques see Swyngedouw, *Social Power,* 21; Swyngedouw, "Modernity and Hybridity," 447; Kirsch and Mitchell, "The Nature of Things"; Wainwright, "Politics of Nature."

53 Swyngedouw, *Social Power,* 15.

54 Swyngedouw, "Modernity and Hybridity," 448; Swyngedouw, "Metabolic Urbanization," 22.

55 Foster, *Marx's Ecology.*

56 Swyngedouw, "Metabolic Urbanization," 26.

57 Tarnas, *Passion of the Western Mind,* 248-71.

58 Ingold, *Perception of the Environment.*

59 Whitehead, *Process and Reality;* Harvey, *Justice, Nature and the Geography of Difference;* Harvey, *Spaces of Hope.*

60 Whitehead, *Adventures of Ideas,* 21.

61 Wittfogel, *Oriental Despotism.* A discussion of Wittfogel's work is found in Chapter 3.

62 "Thus nature is a structure of evolving processes. The reality is the process. It is nonsense to ask if the colour red is real. The colour red is ingredient to the process of realization." Whitehead, *Science and the Modern World,* 70.

63 Leiss, *The Domination of Nature,* xi.

64 Foucault, *The Order of Things.* "Scientific revolution" is a term that came into common use in the mid-twentieth century. Steven Shapin describes the contemporary philosophical sentiment thus: "What was said to be overwhelmingly wrong with existing natural philosophical traditions was that they proceeded not from the evidence of natural reality but from human textual authority. If one wished to secure truth about the natural world, one ought to consult not the authority of books but the authority of individual reason and the evidence of natural reality." Shapin, *The Scientific Revolution,* 68-69.

65 Merchant, *The Death of Nature.*

66 Whitehead, *Science and the Modern World,* 105.
67 Cronon, "Introduction," 24. This quotation is drawn from Cronon's introductory chapter of a collection of papers presented at a multidisciplinary conference contributing to the general intellectual project he describes as "nothing less than rethinking the meaning of nature in the modern world." The critical geographical work done (sometimes) under the banner of "social nature" draws from and contributes to this project. Bruce Braun and Noel Castree make reference to work produced by "anthropologists, cultural analysts, ethnographers of science, philosophers, and sociologists as part of what is now a genuinely interdisciplinary concern with how societies recraft the natural." Braun and Castree, "Preface," xii.

We must add geography to the list of academic disciplines Cronon cites here. In fact, the study of the engagement between people and nature has long been a central theme of academic geography (Glacken, *Traces on the Rhodian Shore;* Livingstone, *The Geographical Tradition*), if not the very "interface" at which geography has "stake[d] its disciplinary identity" (Whatmore, "Introduction," 165). Over the past fifteen or twenty years, the study of this interface and engagement has made a very significant contribution to new ways of understanding the concept of nature. Margaret Fitzsimmons, who brought the term "social nature" into academic discourse in 1989, asked us to "try to see Nature, like History, Geography, and Space, as a material, practical and conceptual reconstitution and reification of what are essentially social relationships." Fitzsimmons, "The Matter of Nature," 108. Since Fitzsimmons' intervention, geographers have applied a variety of theoretical and empirical approaches to elaborate on the sense in which nature is "indeed social through and through," to borrow a phrase from Bruce Braun and Noel Castree. Braun and Castree, "Foreword," xiii. Several of these approaches have already been mentioned in the discussion above. The list of relevant works is too large to mention here, but for useful summaries and collections of work at the socio-natural interface in geography, see Braun and Castree, *Remaking Reality;* Castree and Braun, *Social Nature;* Castree, *Nature.*

CHAPTER 3: INTIMATIONS OF MODERN WATER

1   Gleick's documentation of the water crisis is described in Chapter 10. For the old-new water paradigm see Gleick, "The Changing Water Paradigm."
2   Bakker, "Political Ecology."
3   Tvedt and Oestigaard, *A History of Water,* vol. 3, *The World of Water,* xiv.
4   Molle, Mollinga, and Meinzen-Dick, "Water, Politics and Development," 4.
5   Bakker, "Political Ecology," 41; *Uncooperative Commodity,* 12-13.
6   Bakker, *Uncooperative Commodity.*
7   Ibid., 33.
8   Balanyá et al., "Empowering Public Water," 247.
9   Despite the global disillusionment with dams described here, it has been speculated that they are making a comeback of sorts; the need for more dams is suggested by projected demands for hydroelectricity and for water storage facilities in the face of climate change. Economist.com, "Sina Aqua Non." Such a resurgence would provide interesting opportunities for studying the the social nature of waters in the 21st century.
10 World Commission on Dams, *Dams and Development,* 8.
11 McCully, *Silenced Rivers,* 2-3.
12 Wescoat and White, *Water for Life,* 3.

13 Goldsmith and Hildyard, *Social and Environmental Effects*.
14 Blackwelder in foreword to Goldsmith and Hildyard, *Social and Environmental Effects*.
15 DiFrancesco, "The Debate over Large Dams," 7.
16 R. Clarke, *Water;* Pearce, *The Dammed;* McCully, *Silenced Rivers;* World Commission on Dams, *Dams and Development;* Worster, *Rivers of Empire;* Reisner, *Cadillac Desert;* McCutcheon, *Electric Rivers;* Shiva, *Staying Alive*, 179-217; Shiva, *Water Wars*.
17 DiFrancesco, "The Debate over Large Dams," 7.
18 Gleick, "The Changing Water Paradigm"; Gleick, "The Changing Water Paradigm: A Look."
19 The term "water withdrawal" refers to the act of removing water from its "natural" source for human purposes. For many of these purposes (e.g., cooling thermal power generators, municipal water use, large portions of irrigation water), water that is withdrawn is returned to the basin of origin. Water withdrawal is contrasted with water "consumption," which refers to water that is "lost" to evaporation in, or embodied in, the process of producing agricultural or industrial products, and which is therefore transferred from – and is not available for reuse in – its original basin.
20 The most marked increase in water withdrawals occurred in the period from around 1940 to 1990. Gleick, "The Changing Water Paradigm," 10-12.
21 Gleick, "The Changing Water Paradigm," 6.
22 O'Neill, *Rivers by Design*.
23 Worster, "Water in the Age of Imperialism."
24 DiFrancesco, "The Debate over Large Dams."
25 World Commission on Dams, *Dams and Development*, 10; Grossman, *Watershed*. The privilege of being able to free American rivers, it should be stressed, is granted to some extent by virtue of the relocation of certain hydro-intensive industrial activities and hydro-electrical generation itself to other (notably poorer) countries (but also including northern Canada), where the arrival of the new water paradigm is less evident.
26 Gleick, "The Changing Water Paradigm."
27 Postel, *Last Oasis,* Part II.
28 Mitchell and Shrubsole, *Canadian Water Management*, 61-62.
29 Figuertes, Tortajada, and Rockstrom, "Conclusion," 229. "Integrated water resources management" (IWRM), a term that became prominent in the 1980s, is used by many water experts to describe an approach to managing water that takes into consideration all possible aspects of water values and uses for a particular region (usually a river basin), including ecological as well as socio-economic values and uses. IWRM is discussed more fully in Chapter 11.
30 Gleick, "The Changing Water Paradigm: A Look," 127.
31 Geertz, "The Wet and the Dry"; Worster, *Rivers of Empire;* Groenfeldt, "Building on Tradition."
32 Groenfeldt, "Building on Tradition," 117.
33 To cite only a few examples: for the *qanats,* see Wulff, "The Quanats of Iran"; Pazwash, "Iran's Modes of Modernization." For the *subak* system of Bali, see Geertz, "The Wet and the Dry"; note, however, that Lansing has pointed out that the more relevant locus of analysis for Balinese water management (or hydrosocial relations) is the water temple rather than the *subak.* Lansing, "Balinese 'Water Temples'"; Lansing, *Priests and Programmers.* For Sri Lankan tanks, see Goldsmith, "Traditional Agriculture." For tanks in Tamil Nadu, India, see Mosse, "Local Institutions and Power" and, particularly, Mosse, *The Rule*

of Water. For traditional water management and irrigation systems in pre-Hispanic Valley of Mexico, see Naranjo and Bobee, "Le cycle hydrologique." For the water tribunals of southern Spain, see Ward, *Reflected in Water,* chap. 2.

34 Geertz, "The Wet and the Dry," 29, 30.

35 Lansing, "Balinese 'Water Temples'"; Lansing, *Priests and Programmers.*

36 Lansing, *Priests and Programmers,* 44, 48, 49.

37 Mosse, "Local Institutions and Power," 146.

38 Bennett, "Anthropological Contributions," 37, 38-39.

39 Examples of research suggesting the benefits of drawing on traditional water techniques include, for Africa, Reij, Scoones, and Toulmin, *Sustaining the Soil;* Scoones, Reij, and Toulmin, "Sustaining the Soil"; for Asia, Groenfeldt, "Building on Tradition"; Pawluk, Sandor, and Tabor, "Role of Indigenous Soil"; for the variety and sustainability of success-ful adaptations of cultural practices to the hydrological conditions of semi-arid regions, Fairhead and Leach, *Misreading the African Landscape;* Swift, "Desertification"; Leach and Mearns, *The Lie of the Land;* and for examples of sustainable exploitation of wetlands and coastal areas, Berkes, *Sacred Ecology,* 72-75.

40 Goldsmith and Hildyard, *Social and Environmental Effects,* vol. 1, chap. 26; Groenfeldt, "Building on Tradition," 117; Scoones, Reij, and Toulmin, "Sustaining the Soil," 24.

41 Groenfeldt, "Building on Tradition," 114, 115. Especially when invoked in this context of lauding alternative modes of water management, traditional water management systems are usually characterized as socially equitable, locally autonomous, and ecologically stable. However, it is important to stress the uniqueness of each and to note the problems with making any of these generalizations. See Mosse, "Local Institutions and Power," 147; Mosse, *The Rule of Water.*

42 Leach and Mearns, *The Lie of the Land;* Stott and Sullivan, *Political Ecology;* Forsyth, *Critical Political Ecology;* Calder, *The Blue Revolution.*

43 Donahue and Johnston, "Conclusion," 339.

44 For example, see case studies from Zimbabwe: Derman, "Balancing the Waters"; Arizona: Whitely and Masayesva, "Use and Abuse of Aquifers"; the Sonora Desert of New Mexico: Bowden, *Killing the Hidden Waters;* northern Quebec: Ettenger, "'A River That Was'"; Palestine: Hassoun, "Water between Arabs and Israelis."

45 Said, "Yeats and Decolonization," 297.

46 Worster, "Water in the Age of Imperialism."

47 Worster, *Rivers of Empire.*

48 The popular (local) and broader (cosmopolitan and global) opposition to the large dams constructed and planned by the Indian state on the Narmada River (with initial support from the World Bank) offers a good example of this kind of resistance. See Morse and Berger, *Sardar Sarovar;* Baviskar, *In the Belly of the River;* Baviskar, "Written on the Body."

49 Shiva, *Staying Alive,* 67-77; Shiva, *Water Wars;* Mosse, "Local Institutions and Power"; Mosse, *The Rule of Water;* Arnold and Guha, "Introduction"; Hardiman, "Small-Dam Systems of the Sahyadris"; Gilmartin, "Scientific Empire and Imperial Science"; Gilmar-tin, "Models of the Hydraulic Environment"; Whitcombe, "Environmental Costs"; Seenivasan, *Neerkattis.*

50 Gregory, "Post-Colonialism," 612.

51 Worster refers to the formation of an "international fraternity of hydraulic engineers" who understood that "they were engaged in a mission of conquest that was going on in all the arid parts of the world – in India, Egypt, the Sahara, Australia." Worster, *Rivers of Empire,*

143-56. He describes how the experience gained by hydraulic engineers in colonial settings was applied in the western United States, which sent scores of hydrologists and engineers on field trips to India to learn the tricks of the trade.

52 Gregory, "(Post)Colonialism and the Production of Nature," 85; Gregory, *The Colonial Present*. Gregory's *The Colonial Present* is a post-colonial analysis of Western hegemony in Afghanistan, Palestine, and Iraq.

53 Millner, "Post-Colonialism," 669-70.

54 Whitcombe, "Environmental Costs," 237; Worster, *Rivers of Empire*, 143-56; "From the Punjab Irrigation Manual" (n.d.), quoted in Gilmartin, "Scientific Empire and Imperial Science," 1132.

55 Gregory, "(Post)Colonialism and the Production of Nature," 96-97.

56 A few examples from British India are worth noting. First, Hardiman's study of the withering of small-scale "community-based systems of irrigation" that had been common in many arid and piedmont regions of India as long ago as 700 BC. Focusing on the Baglan region, Hardiman argues that the colonial state was interested only in large-scale irrigation works that could be centrally controlled to maximize revenue production and shows how "the state, therefore neglected the small systems ... Without proper attention, the canal systems disintegrated quickly." Hardiman, "Small-Dam Systems of the Sahyadris," 189-90. Second, as noted above, the fate of the tank system of southern India has received a great deal of attention from historians and cultural ecologists. Much of this research is focused on the "crippling effect" of modernity, in the form of colonialism, on the traditional tank system. Contemporary struggles to maintain or restore the system are hampered by the legacy of colonial hydrological discourse, particularly insofar as it disregards or ignores the social bonds by which the tank system is constituted. As Seenivasan and Kumar note in *Vision for Village Tanks*, "The centralized administration introduced by the British colonial rule had almost wiped out the role of the community in conserving and developing [the tanks]" (5). And the problem has worsened under the post-colonial state: "Even in free and independent India it is [sic] continued to be even more retrogressive in keeping people and locals away in matters related to the tanks ... The decay of tanks represents the typical death of a village ecosystem ... The tank management systems which are directly related to the village social systems are breaking down across the state." Seenivasan and Kumar, *Vision for Village Tanks*, 8. Finally, Kamal's study of "the ruin of the ancient system of irrigation" by the British in what is now Bangladesh gives an account of how such ruin was brought about less by British technical ignorance than by the imposition of a foreign, "businessman's" model of hydrosocial relations. Kamal, "Living with Water," 201.

57 Mosse's study of the tank system in southern India, for example, points out that these have always been subject to uncertainty, political contestation, and vulnerability – even in pre-colonial times. He leaves no doubt, however, that the overall effect of the colonial era was to aggravate these problems and promote neglect of the traditional tank system. Mosse, "Local Institutions and Power."

58 Gilmartin, "Scientific Empire and Imperial Science"; Gilmartin, "Models of the Hydraulic Environment."

59 Gilmartin, "Models of the Hydraulic Environment," 226-27.

60 That is to say, they did not operate according to an idealized vision of community organization. Gilmartin does not raise the question of whether such instability and internal conflict might have been associated with the effects of colonization, or indeed with the

introduction of modern water in the basin. However, his remarks on the eventual break-down of these local systems would certainly suggest this to be the case. See Gilmartin, "Models of the Hydraulic Environment," 229-30; and below.

61 Gilmartin, "Models of the Hydraulic Environment," 226-7. Elsewhere Gilmartin offers an interesting study of how advances in river hydrology in colonial India – specifically the science of silt management and control – had the effect of breaking traditional social bonds with water in the sense that it freed canal management from a reliance on community organization. The engineer R.G. Kennedy made a series of studies late in the nineteenth century that yielded a reliable scientific formula for the slope and water velocity that could be maintained in an unlined canal channel without causing silting or scouring. This formula, in effect, allowed engineers to define mathematically what were known as "regime channels," or canals in which the rate of silting balances that of scouring during continued periods of operation. This had the effect of radically changing social relations with water, as it meant that "large-scale annual silt clearances should be viewed not as a normal part of canal operation, but as evidence in flaws in canal design. Annual silt clearances, in such a view, only perpetuated the problems that caused heavy silting. As one engineer put it in 1916: 'Mr. Kennedy's work was to explain scientifically that silt clearances were a hopeless task.' A new 'scientific' view of canal operation thus transformed the role of indigenous labor in canal maintenance ... Regular mobilization of labor for clearance was replaced by constant scientific monitoring, management, and remodeling. As one engineer commented: 'An irrigation system in its parts comprises a very delicate machine.' In the case of silt control, therefore, labor mobilized by 'communities' could be replaced, to a very large extent, by more sophisticated techniques of scientific adaptation to a changing environment." Gilmartin, "Scientific Empire and Imperial Science," 1137-38.

62 Worster, "Water in the Age of Imperialism," 13.

63 Ibid., 11.

64 Wittfogel, *Oriental Despotism;* Biswas, *History of Hydrology;* Norman Smith, *Man and Water;* Jackson, *Dams.*

65 Donahue and Johnston, "Conclusion," 339.

66 Describing the recent flourishing of academic interest in water history, the editors of this journal note, "This recent spate of research on water confirms that water history has developed into a vibrant historical subfield – one that incorporates and contributes to environmental history, urban history, and the history of technology and landscape" (Tempelhoff et al., "Where Has the Water," 4). Historical geography should be added to this list of subfields.

67 Tvedt and Jakobsson, *A History of Water,* vol. 1, *Water Control and River Biographies;* Coopey and Tvedt, *A History of Water,* vol. 2, *The Political Economy of Water;* Tvedt and Oestigaard, *A History of Water,* vol. 3, *The World of Water.*

68 Tvedt and Oestigaard, "Introduction," xiv.

69 Ibid.

70 Wittfogel, *Oriental Despotism.*

71 Wittfogel, "The Hydraulic Civilizations," 156.

72 For a summary of these implications see Worster, *Rivers of Empire,* 22-30.

73 Ibid., 22.

74 Ibid., 7.

75 Evenden, *Fish Versus Power,* 268.

76 Ibid., 1.
77 Ibid., 3.
78 Ibid., chap. 7.
79 Ibid., 241-42.
80 For example, see Castree, *Nature,* chap. 5.
81 This was the title of a series of six sessions at the 2007 meeting of the American Association of Geographers. Association of American Geographers 2007: 180, 203, 226, 247, 454, 475. In a similar vein, the 2008 meeting featured sessions on the theme of "the hydrosocial cycle." The hydrosocial cycle is discussed further in Chapter 12.
82 The Columbia River in northwestern North America is one of the most heavily dammed and regulated watercourses on earth and one of the world's greatest producers of hydro-electricity. But to environmental historian Richard White, "It is a mistake to read back the fate of the Columbia as a plan to denature it." R. White, *The Organic Machine,* 57. For White, the Columbia is an indivisible combination of human labour and the energy of falling water; it is an "organic machine," to put it in White's terms. The phrase "organic machine" thrusts together ideas that are usually held apart, often as opposites. Not unlike "socio-nature" (social nature) and "cyborg" (cybernetic organism), introduced in Chapter 2, the notion of an organic machine is slightly offensive to the modern – and the anti-modern – ear. And not unlike Donna Haraway, Bruno Latour, and the many other writers who use such terms, White intends to offend his audience slightly, just enough to promote new ways of thinking: "This is a book which seeks to blur boundaries, emphasize impurity, and find, paradoxically, along those blurred and dirty boundaries ways to better live with our dilemmas." Ibid., xi.
83 Crifasi, "Reflections in a Stock Pond," 626; see also Swyngedouw, "Modernity and Hybridity."
84 Urban and Rhoads, "Conceptions of Nature."
85 For example, Gleick, "The Human Right to Water"; Barlow and Clarke, *Blue Gold: The Battle.*
86 Swyngedouw, "The City as a Hybrid," 76, 80.
87 Gandy, "The Making of Metropolitan Nature," 1022; Gandy, "Cyborg Urbanization," 40-41.
88 Bakker, "Political Ecology"; Bakker, *Uncooperative Commodity;* Swyngedouw, *Social Power;* Swyngedouw, "Dispossessing H$_2$O"; Kaika, *City of Flows;* Budds, "Contested H$_2$O."
89 Budds, "Whose Scarcity?"
90 Goubert, *The Conquest of Water.*
91 Swyngedouw, "The City as a Hybrid," 79.
92 Gandy, "Editorial," 118.
93 Political ecology draws from political economy and (more recently) has employed post-structural insights on the social construction of nature to produce a diverse field of critical study of environmental issues. Bryant describes the field in terms of an "inquiry into the political sources, conditions, and ramifications of environmental change." Bryant, "Putting Politics First," 165. Originally focusing on ecological questions in Third World regions, research in political ecology is increasingly concerned with nature-society relations in all parts of the world. Among its central aims is to analyze questions of access to and control of resources and how these impinge on environmental sustainability and socio-economic livelihood. Watts, "Political Ecology," 257; see also R.J. Johnston et al., *Dictionary of Human Geography,* 590; Castree and Braun, "Construction of Nature," 11-12. To the

more foundational class-based and historical materialist analyses of how different people gain or lose access to resources have been added research on the way scientific and development discourses relate to these questions. "Recent observations in political ecology," notes Fiona Mackenzie, "demonstrate how struggles over material resources – the means of production – are simultaneously struggles over the symbolic meanings of these resources, over the discursive ways through which these contests are instigated and positions legitimated." Mackenzie, "Contested Ground," 697.

94 Shiva, *Water Wars,* ix.
95 See particularly Forsyth, *Critical Political Ecology.*
96 Stott and Sullivan, *Political Ecology.*
97 Swyngedouw, *Social Power,* 175.
98 B.R. Johnston, "Introduction to the Political Ecology."
99 Downing and Bakker, "Drought Discourse and Vulnerability," 13.
100 Blaikie et al., *At Risk,* 3. The critique of discourses of drought can be seen as one aspect of the social construction of resource scarcity. For over two decades, geographers and others have elaborated the social construction of the scarcity of resources (as well as generalized scarcity). For example, Harvey, "Population, Resources"; Sandbach, *Environment, Ideology and Policy;* Xenos, *Scarcity and Modernity;* Ross, "The Lonely Hour"; Watts, "Political Ecology," 262. This critical analysis of scarcity has been especially responsive to perceived Malthusian threats of imminent depletions of non-renewable resources beginning in the 1970s (see Chapters 8 and 9).
101 Bakker, "Privatizing Water, Producing Scarcity"; Kaika, "Constructing Scarcity"; Kaika, *City of Flows.* An example of the former is the temporary reduction or shutting down of municipal water services. An example of the latter is a government campaign describing a water shortage as a "natural" rather than a "distributional" problem.
102 Neil Smith, *Uneven Development,* 60.
103 Swyngedouw, *Social Power,* 47.
104 Ibid.
105 Haughton, "Private Profits – Public Drought"; Bakker, "Privatizing Water, Producing Scarcity"; Bakker, "Political Ecology"; Stehlik, Lawrence, and Gray, "Gender and Drought," 38; Nevarez, "Just Wait"; Kaika, "Constructing Scarcity"; Kaika, *City of Flows.*
106 Aguilera-Klink, Perez-Moriana, and Sanches-Garcia, "Social Construction of Scarcity"; B.R. Johnston, "Culture, Power."
107 Lansing, *Priests and Programmers,* 3.
108 B.R. Johnston, "Introduction to the Political Ecology."
109 Ibid., 74-85.
110 Ibid., 85.

CHAPTER 4: FROM PREMODERN WATERS TO MODERN WATER

1 Hamlin, "'Waters' or 'Water'?"; Goubert, *The Conquest of Water;* Illich, *H₂O and the Waters of Forgetfulness;* Kalaora, "De l'eau sensible à OH2."
2 Illich, *H₂O and the Waters of Forgetfulness.*
3 Cayley, *Ivan Illich in Conversation,* 246. Those familiar with Bachelard's meditation on *Water and Dreams* will recognize Illich's main source of inspiration for *H₂O and the Waters of Forgetfulness.*
4 Illich, *H₂O and the Waters of Forgetfulness,* 3-4.

5 Hamlin, "'Waters' or 'Water'?"

6 Ibid, 321.

7 Illich, *H₂O and the Waters of Forgetfulness*, 7.

8 For a good introduction to changing ideas of water through the ages beginning with Thales, see Ball, *Life's Matrix*.

9 Hamlin, "'Waters' or 'Water'?" 316.

10 This phrase is from Pliny's *Natural History*, Vol. 9: 379. For Pliny, water was the foremost element primarily because of its fecundity. Indeed, this "watery principle," as Hamlin puts it, can be described as a kind of universal fecundity. Among ancient philosophers, this principle is usually associated with Thales, but it has shown great persistence. The first century BC Roman engineer-architect Vitruvius, who devoted one of his ten books on architecture to water, described water as "the first principle of all things" and something that possesses a vital power of which "all things are composed." Vitruvius, *Ten Books on Architecture*, 96. The classicist R.G. Tanner suggests that water was understood as a vital principle and that this understanding was eventually imparted to the other elements: "Though the notion of one material substance as the originator of life later disappeared, it left a permanent legacy to philosophy. All these elements – water, fire and air – have the quality of being in constant motion, and all at times appear to initiate capricious spontaneous motion in the eyes of men unschooled in the laws of modern physics and chemistry. Such autonomous willed motion seemed to affirm the substance was alive, sentient and purposeful." Tanner, "Philosophical and Cultural Concepts," 30.

11 *Encyclopaedia Britannica*, 11th ed., s.v. "Empedocles."

12 Sambursky, *The Physical World of the Greeks*, 32.

13 Hamlin, "'Waters' or 'Water'?" 316, 317.

14 It is impossible to assign precise dates to changes in ideas even among relatively well-defined intellectual communities. At any given time in the history of Western thought, a wide variety of (often mutually contradictory) views have persisted, as they do today. For example, ancient, medieval, and Renaissance traditions of witchcraft, astrology, alchemy, and so on remained very much alive in some quarters well into the eighteenth century and, in some cases, beyond. Dear, *Revolutionizing the Sciences*, 29. Hence, the difficulty of signifying specific eras or historical periods of thought by terms such as "classical era," "modern era," "Scientific Revolution," and so on. But despite the uncertainties, it remains useful to apply such categories, albeit in a careful, limited sense. The historian of science Steven Shapin describes an uneasiness among some colleagues with the Scientific Revolution as a distinct historical category. Shapin, *The Scientific Revolution*, chap. 1. Nevertheless, that natural philosophers in the late sixteenth and seventeenth centuries clearly believed that "they were proposing some very new and very important changes in knowledge of natural reality and in the practices by which legitimate knowledge was to be secured" suggests that "it is possible to write about the Scientific Revolution unapologetically and in good faith." Ibid., 5. As I have found it useful to revert to these and other gross modes of periodization below, it needs to be borne in mind that these are always intended as approximations at best.

15 Ball, *Life's Matrix*, 133.

16 Fifth edition, s.v. "water."

17 Ibid., 145.

18 Part of the explanation for the power of Lavoisier's naming of water as a chemical compound can be attributed to the particular form of discourse that he initiated – a new chemical

nomenclature through which he and his colleagues articulated what has become known as the "chemical revolution." As Jan Golinsky points out, "On one reading, Lavoisier's radical challenge to the chemistry of his time appears to be embodied in a series of innovative factual claims, concerning the non-existence of phlogiston, ... the reinterpretation of combustion as a process of oxidation, the designation of water as a compound, and so on. But, as a complementary part of their program, Lavoisier and his allies also advanced a new nomenclature for chemical substances and a new model of chemical discourse as a 'demonstrative' process of reasoning akin to geometry" (*Making Natural Knowledge,* 117). Golinsky reports that a rival British chemist exclaimed, "We cannot speak the language of the new Nomenclature, without thinking as its authors do." Ibid., 118.

19 Laugier and Dumon, "D'Aristote à Mendeleev."

20 Hamlin, "'Waters' or 'Water'?" 315.

21 Although this painting gives the impression that Roman aqueducts were characterized by magnificent above-ground structures, some 80 to 90 percent of the combined length of all eleven aqueducts was in fact underground. Hodge, "Engineering Works," 67.

22 Ibid., 90.

23 Frontinus, *Stratagems and the Aqueducts,* 421.

24 Quoted in Pinto, *The Trevi Fountain,* 7.

25 Tanner, "Philosophical and Cultural Concepts," 30, 31. Tanner describes how the waters from two different sources were made by the Romans to flow for a length of seven kilometres in the same channel and then were separated again into two channels bearing the same proportions of the original supplies: "It is hard to see why the division was made in the same proportion as the previous mixture unless it was thought that the waters were living beings withdrawing from an embrace rather than an inert and passive divisible substance." Ibid., 32.

26 Frontinus, *Stratagems and the Aqueducts,* 357.

27 Tanner, "Philosophical and Cultural Concepts," 34.

28 Horton, "Field, Scope, and Status," 192.

29 Hodge, "Purity of Water," 96-97.

30 Natural history – "A former branch of knowledge embracing the study, description and classification of natural objects." *Webster's Third New International Dictionary,* s.v. "Natural history." Note the distinction from natural philosophy: "the study of nature in general," a branch of knowledge that evolved into modern natural science. Ibid, s.v. "Natural philosophy"

31 We might read this list in a manner similar to Michel Foucault's reading of a passage from "a certain Chinese encyclopedia," which famously set Foucault to pondering *The Order of Things.* Foucault, *The Order of Things,* xvi-xix. See also Gregory, *The Colonial Present,* 1-2. Like the Chinese dictionary, Pliny's natural history of water seems humorous to us, the effect of a very wide epistemological gulf between reader and writer.

32 Pliny, *Natural History,* Vol. 1: 141.

33 Ibid., Vol. 9: 391-95.

34 Hamlin, "'Waters' or 'Water'?" 318.

35 Biswas, *History of Hydrology,* 140.

36 Pliny, *Natural History,* Vol. 9: 353.

37 Vitruvius, *Ten Books on Architecture,* 130.

38 In fact, as discussed in Chapter 5, many modern hydrologists have credited Vitruvius as the first person to have a true understanding of the hydrologic cycle.

39 *Webster's Third New International Dictionary*, s.v. "Natural history," "Natural philosophy."
    Describing the seventeenth-century distinction between these two fields, Steven Shapin
    writes, "[Inquiries] into what sorts of things existed in nature and into the causal structure
    of the natural world were referred to, respectively, as 'natural history' and 'natural philoso-
    phy.'" Shapin, *The Scientific Revolution*, 6.
40 *Encyclopaedia Britannica*, 15th ed., s.v. "hydrology."
41 Hamlin, "'Waters' or 'Water'?" 317.
42 Two phases in this demise may be noted here. The period known as the Scientific Revolu-
    tion is associated with reforming natural history – "purifying it" by driving out its now
    apparently ridiculous assertions and testimonials. "The techniques of intellectual quality
    control recommended for a reformed natural history could be used to winnow out testi-
    monial wheat from chaff." Shapin, *The Scientific Revolution*, 138. Natural history (reformed)
    continued to thrive in the eighteenth and (to a lesser extent) nineteenth centuries, eventu-
    ally becoming "a former branch of knowledge" and "The Forgotten Science" in the
    twentieth century. Worster, *Nature's Economy*, 286.
43 Shapin, *The Scientific Revolution*, 88.
44 The epistemic recontextualization of the study of water – from waters in natural history
    to water in natural philosophy and natural science – is illustrated by taking note of the
    tendency among practitioners of modern hydrologic science to pick from among the clas-
    sic texts certain themes that are consistent with modern views and to reject others that
    seem to them irrelevant. For example, Vitruvius' observations of the gross behaviour of
    water have been championed as prescient by hydrologists interested in locating the origins
    of the concept of the hydrologic cycle, while his observations of the idiosyncratic qualities
    of particular spring waters and their effects on the local inhabitants have been completely
    ignored (see Chapter 5). Even when water is described in the universal (i.e., as water),
    classical writers tend to imbue it with forces, meanings, and significance that contrasts
    strongly with the stripped-down raw material that water has become in modern (scien-
    tific) discourse. Thus, along with the elements of fire and air, Seneca (55 BC-AD 39) ascribes
    to water a "principle of life," about which he was in agreement with Pliny and Cicero.
    Seneca, *Naturales Quaestiones*, 82-83. And Vitruvius – alluding to Thales' cosmogony of
    water as "the principle of all things" – justifies devoting an entire book on the subject of
    water (in his opus on architecture) with the assertion that "all things consist of the power
    of water." Vitruvius, *Ten Books on Architecture*, 86. These aspects of water that were so
    apparent to Seneca and Vitruvius have been expunged from the modern, hydrological
    exegesis of their writings.
45 Jankovic, "Meteors under Scrutiny," 8, 9.
46 Descartes' natural "mechanical" philosophy (particularly by way of its substitution of the
    machine for organic metaphors) is deemed by many to have had the effect of taking much
    of the wonder out of nature, hence, Weber's well-recited phrase. Shapin, *The Scientific
    Revolution*, 36. Carolyn Merchant's *The Death of Nature* can be read as an extended essay
    on this theme. See also Leiss, *The Domination of Nature*.
47 Hamlin, "'Waters' or 'Water'?" 319.
48 Eliade, *Patterns in Comparative Religion*, 188.
49 We are focusing our attention on the history of water in the Western world. The sanctity
    of water(s) certainly exists among people elsewhere, as with respect to the Ganges River,
    to cite a well-known example. Even in the West, modern water has not completely driven
    out hydrolatry, as anyone who visits the grotto at Lourdes must be aware.

50 Hamlin, "'Waters' or 'Water'?" 319.
51 Rattue, *The Living Stream*, 10. It is normal for us moderns to provide rational explanations for phenomena such as "hydrolatry." In the *Encyclopaedia of Religion and Ethics*, Hastings commences his article on "Water, Water Gods" as follows: "Since water is a first need of man in a primitive state of culture, it is little wonder that it is regarded as possessed of *mana* [spirit] and that it figures prominently in magico-religious cults." Hastings, *Encyclopaedia of Religion and Ethics*, 12:704.
52 Rattue, *The Living Stream*, 26. Eliade, *Patterns in Comparative Religion*, 199.
53 Rattue, 26.
54 Bord and Bord, *Sacred Waters*, 19.
55 Rattue, *The Living Stream*, 88.
56 Ibid., 97.
57 Goubert, *The Conquest of Water*, 3.
58 Ladourie, "Introduction," 2.
59 Hamlin, "'Waters' or 'Water'?" 319.
60 Rattue, *The Living Stream*, 130.
61 Ibid., 130.
62 Tozer, *History of Ancient Geography*, 267-72.
63 Schama, *Landscape and Memory*, 349.
64 Tanner, "Philosophical and Cultural Concepts," 34.
65 Glacken, *Traces on the Rhodian Shore*, 134-35.
66 Ibid., 135. Tacitus reports that, for whatever reason, "the motion of Piso, 'that nothing be changed,' was agreed to." Quoted in ibid., 135. In addition to highlighting the almost personal respect accorded to rivers in antiquity, we might conclude from this anecdote that the invocation of "nature" has provided a powerful political argument throughout the history of the West.
67 Hamlin, *A Science of Impurity*, 319-20.
68 Hamlin, "'Waters' or 'Water'?" 320.
69 Glacken, *Traces on the Rhodian Shore*, 82-88.
70 Rattue, *The Living Stream*.
71 Hamlin, "'Waters' or 'Water'?" 320.
72 Hamlin, *A Science of Impurity*, 319-20.
73 Ibid., 57.
74 Hamlin, "'Waters' or 'Water'?" 321.
75 Baker, *Quest for Pure Water*.
76 Hamlin, "'Waters' or 'Water'?" 321.
77 Another example is found in the attention to the variety of waters encountered by people involved in various industrial processes for which water quality was of the utmost importance. For example, Canada's Department of Mines conducted an inquiry into "the industrial waters of Canada" between 1934 and 1940. The study covered all industrial uses of water and was concerned about water quality "not because of environmental concerns but because dissolved or suspended materials in water could cause operating problems for manufacturers and resource extracting industries, thereby increasing production costs." Gossage, "Water in Canadian History," 104. As a matter of speculation, it would seem that the widespread adoption of more effective techniques to purify water – known as "water conditioning technologies" – after the Second World War has curtailed the industrialists' concerns with the qualities of different "waters," as these technologies produce a kind of

water that is useful for the purposes of most industrial applications. Permutit Company, *Water Conditioning Handbook.* The variety of waters encountered in the context of industrial applications can be traced far back into history. To cite one example, Agricola's (1494-1555) interest in waters stemmed from his concerns to promote mineralogy and mining. His written works, including *De re metallica,* published posthumously in 1556, revealed his fascination for classifying and describing water according to quality, taste, colour, provenance, and occurrence, such as in springs, wells, rain, snow melt, oceans, lakes, and swamps. In a letter written in 1546, he appended a dictionary of earth science and mining terms, translated from Latin to German, in which there are included "no less than 27 definitions of different waters." Pfannkuch, "Medieval Saint Barbara Worship," 46.

78 Gordon and Thompson, *Physiological Principles,* 24.
79 Ibid., 73.
80 Gandy, "The Paris Sewers," 33.
81 Goubert, *The Conquest of Water.*
82 Benedickson, *The Culture of Flushing,* 4, 12.
83 Goubert, *The Conquest of Water,* 24.
84 Dooge, "On the Study of Water," 23.
85 Kula, *Measures and Men,* 43.
86 Biswas, *History of Hydrology.*
87 Dooge, "Hydrology in Perspective," 62. In another article, Dooge describes the first documented reference to a rain gauge in a book known as the *Arthashastra,* authored by one Kautilya (also known as Chanakya), a Brahmin minister living in India in the fourth century BC. The subject of the book was politics and administration. Dooge points out that the land tax described in Kautilya's book was calibrated against the rain gauge. Dooge, "Waters of the Earth," 150-51. See also Dooge, "Background to Modern Hydrology."
88 Dooge, "Hydrology in Perspective," 63.
89 Kula, *Measures and Men.*
90 Crosby, *The Measure of Reality,* 17.
91 Collingwood points out that an earlier instance occurred among the followers of Pythagoras, who mixed mathematics and empiricism as early as the fifth century BC. Collingwood, *The Idea of Nature,* 54.
92 Dear, *Revolutionizing the Sciences,* 55; B.S. Hall, "The Didactic and the Elegant," 22; Leiss, *The Domination of Nature,* 91.
93 Whitehead, *A Philosopher Looks at Science,* 62-63; Evernden, *Social Creation of Nature,* 44.
94 Collingwood, *The Idea of Nature,* 103.
95 Ibid., 102.
96 Lefebvre, *The Production of Space,* 52.
97 Ibid., 285.
98 Illich, *H₂O and the Waters of Forgetfulness,* 24.
99 Collingwood, *The Idea of Nature,* 25.
100 Although chemistry treats water at the molecular scale (10^{-10} m), and hydraulics at the mesoscale (10^{-2} m), Dooge points out that "the scales of interest to the hydrologist are macroscales that are about 10^6 larger than the mesoscales of continuum mechanics and about 10^{12} larger than the microscales of physical chemistry." Dooge, "On the Study of Water," 40-41.
101 The Romans never calculated the amount of water serving their city. The legendary quantity of Rome's water supply is a modern preoccupation that seems to have been initiated when the American hydraulic engineer Clemens Herschel translated Frontinus' treatise *The*

Two Books on the Water Supply of the City of Rome in 1899. After estimating losses from leak-age and theft, Herschel calculated that the nine aqueducts serving the city in Frontinus' time delivered something like 38 million gallons daily to Rome's inhabitants. Based on an estimated population of 1 million at the time, this calculation produced a per capita figure of 38 gallons (144 litres) per day – a figure that exceeded that of every city in the world when Herschel's translation was published. Herschel's figure even suggests the Romans had more water per capita than is available to people in many cities today. See *Encyclopaedia Britannica*, 15th ed., 26:423. s.v. "Public Works"; also Biswas, *History of Hydrology*, chap. 5; and Norman Smith, *Man and Water*, chap. 7. Furthermore, it might be considered that because the water transported by aqueducts into Rome was free flowing – i.e., it was not closed by valves – an entirely different meaning, or standard, of "water availability per capita" pertains. Also, see Hodge for the difficulty of making such calculations and for the suggestion that Rome likely received less water than is commonly believed. Hodge, "Aqueducts," 47.

102 This has been interpreted by almost all modern hydrologists and most historians as a gross failure or oversight on the part of Roman science, "a startling ignorance of what is required," as Smith puts it, to cite but one example. Smith, *Man and Water*, 93. However, as Blackman and Hodge point out, although the Roman concept of rate "at very best was qualitative, certainly not quantitative," there was no perceived need for the scientific quantification of the flow of water: "What occupied them more was the administration of the distribution of water to users and of the associated licenses [sic]. By modern standards a bit rough and ready, but the *calices* [pipes of standard sizes] ensured a stable system which apportioned what all shared in a way which was perceived to be just." Blackman and Hodge, *Frontinus' Legacy*, 21, 24.

103 An earlier effort to produce a "scientific" understanding of the flow of water is associated with the work of Leonardo da Vinci (1452-1519), who stressed the importance of combin-ing measurements of wetted cross-sectional areas with the velocity of streamflow and who conducted experiments to measure the distribution of velocities in open channels. However, Leonardo appears not to have used this data to compute actual discharge values, and, in any case, his work had little immediate influence on the development of the hydrological sciences. For an account of Leonardo's hydrometric experimentation see Macagno, "Leo-nardo da Vinci."

104 Morello, "La question des eaux douces." Literally, "The Academy of Experiment," a sci-entific society founded in Florence in 1657 by Galileo's students.

105 Maffioli, *Out of Galileo*, 10. The law of continuity provides a good example of the kind of abstract thinking that was made possible by the application of mathematics to natural philosophy. "In order to formulate the laws governing the movement of natural objects," William Leiss points out, it was "necessary to disregard the sense-qualities of things ... and to posit the existence of a uniform substance common to all objects. Since this substance or matter is assumed to be everywhere the same, the differences between things can be reduced to simple quantifiable proportions ... As only relations of quantity are involved, the laws of motion could be set down entirely in mathematical and geometrical terms." Leiss, *The Domination of Nature*, 91.

106 Rouse and Ince, *History of Hydraulics*, 59-61.

107 Livingstone, *The Geographical Tradition*, 104.

108 For Wren in particular, see Biswas, "The Automatic Rain-Gauge"; Biswas, *History of Hy-drology*, 234-39. For others see Chapter 5.

109 Pierre Perrault was the brother of Claude Perrault, architect of the Louvre and – most significantly in the present discussion – author of a French translation of Vitruvius' *Ten*

Books of Architecture, published in 1673, the year before Pierre's book was published. Another Perrault brother, Charles, was controller of the Department of Public Works under Colbert and became well known as an author, particularly of fairy tales. *Encyclopaedia Britannica,* 11th ed., s.v. "Perrault."

110 Perrault, *On the Origin of Springs,* 144-45.

CHAPTER 5: THE HYDROLOGIC CYCLE(S)

1 Providing an exact number is impossible because many of the "hits" reported by Internet search engines such as Google are duplicates.

2 Linton, "Is the Hydrologic Cycle Sustainable?"

3 There is, moreover, a sense in which the hydrologic cycle, as a representation of the circulation of water, produces the very need to question its reality. As Timothy Mitchell has shown with respect to the development of abstract ideas such as "the economy," the notion of the reality of such things becomes a problem only by virtue of the gap that *appears* to open up between representational practices and the real things they purport to stand for. T. Mitchell, "Fixing the Economy." See also Latour, *Pandora's Hope.* By this way of thinking, the scientific hydrological practices that produce the concept of the hydrologic cycle raise the question of whether, and how well, the representation corresponds to the reality of water. And the more natural the representation becomes, that is, the more its familiarity obscures the practices by which it is produced, the more natural becomes the apparent need to question its reality.

4 Langbein and Hoyt, *Water Facts,* 3; Chorley and Kates, "Introduction," 3.

5 Tuan, *Hydrologic Cycle.*

6 Natural theology is a school of Christian theology that bases reasoning on observable ("natural") evidence rather than on the revealed Word of God in the Bible. A fuller description is given in the next section.

7 Tuan, *Hydrologic Cycle,* 6.

8 Like most research in the history of science, recent scholarship manifests a greater interest in micro-histories – studies of hydrological developments or applications of hydrological science in specific times and places. The research and most of the publications that make up the more conventional history of hydrology tend to be of an older vintage. A handful of books and collections of articles written mostly by hydrologists and published by or for scientific or engineering associations or government agencies make up the bedrock of this literature, along with works on the history of the related field of hydraulics. The only book available in English on the *general* history of modern hydrology is Biswas, *History of Hydrology,* though we might also include Landa and Ince, *The History of Hydrology,* in this category. Standard works on the history of hydraulics – which touch on various aspects of hydrologic history, include Rouse and Ince, *History of Hydraulics,* and Garbrecht, "Hydrologic and Hydraulic Concepts." A few articles that may be described as general histories of hydrology appeared in scientific journals in the 1930s in association with efforts to establish hydrology as a distinct scientific discipline in the United States, notably Meinzer, "History and Development"; and Baker and Horton, "Historical Development." In the 1960s and 1970s, articles began appearing in journals such as *Hydrological Sciences Bulletin, Water International, Water Resources Research,* and *Journal of Hydrology* that contributed to this literature, many of which are discussed below. Most of these and other

contemporary publications dealing with the history of hydrology sprang from work related to the International Hydrological Decade (1965-74), which was organized under the aegis of UNESCO and is discussed in greater detail in Chapter 8. Among its large volume of publications, the International Association of Hydrological Sciences has published at least one that deals with the history of hydrology in general; Rodda and Matalas, *Water for the Future*. Historical essays of various lengths on the origins of the discipline are usually found in the introductory chapters of hydrology textbooks and handbooks published in English. Not quite fitting into any of the above categories but nevertheless contributing to the literature of the general history of hydrology are F.D. Adams, "Origin of Springs"; F.D. Adams, *Birth and Development;* P.B. Jones et al., "Development of the Science"; Parizek, "The Hydrologic Cycle Concept"; Biswas, "The Hydrologic Cycle"; and various parts of National Research Council, *Opportunities*. In 2001, a conference was held in Dijon, France, on the history of hydrology (Colloque International OH2, 9-11 May 2001, "Origines et histoire de l'Hydrologie"). Several of the papers presented at this conference are discussed below. Finally, the eminent hydrologist James C. Dooge has been among the most prolific exponents of the history of the hydrological sciences, and I have drawn much from his work in compiling what I describe as the "conventional" history of the hydrologic cycle. Dooge's most recent work in this vein is found in Dooge, "Background to Modern Hydrology."

9 National Research Council, *Opportunities*, 38.

10 Golinsky, *Making Natural Knowledge*, 2.

11 Back, "Foreword," ix.

12 Chow, *Handbook of Applied Hydrology;* Biswas, *History of Hydrology*.

13 Perrault, *On the Origin of Springs*.

14 Biswas, *History of Hydrology*, 140.

15 F.D. Adams, *Birth and Development;* Perrault, *On the Origin of Springs*.

16 Dooge, "Concepts of the Hydrologic Cycle."

17 Ibid.

18 The more focused historical accounts of the question of the origin of springs include F.D. Adams, "Origin of Springs"; Meinzer, "History and Development"; Baker and Horton, "Historical Development"; F.D. Adams, *Birth and Development;* Meinzer, *Hydrology;* Krynine, "On the Antiquity of 'Sedimentation'"; Biswas, *History of Hydrology;* Nace, "General Evolution"; Nace, "The Hydrological Cycle"; Dooge, "Concepts of the Hydrologic Cycle."

19 F.D. Adams, *Birth and Development*, 426. Chapter 12, "The Origin of Springs," is based mainly on F.D. Adams earlier paper, "Origin of Springs"; Baker and Horton, "Historical Development"; and Meinzer, "History and Development." Baker and Horton, and Meinzer, in turn, cite a 1912 publication by Keilhack, whose (translated) title reads: "Groundwater and the Hydrology of Springs." Although I have not been able to find this work, Horton, in "Field, Scope, and Status," 190, notes that in it, Keilhack describes the history of the dispute over the origin of springs.

20 F.D. Adams, *Birth and Development*, 427.

21 Ibid., 429.

22 Ibid., 430. Given the doctrine of the transmutation of elements (see Chapter 4), this source of water was probably thought by Aristotle to be the earth itself.

23 The notable exception is Vitruvius, as discussed below.

24 Meinzer, "History and Development," 8. Meinzer was a groundwater specialist whose work was instrumental in establishing scientific hydrology as a bona fide scientific discipline in the United States in the 1930s. Meinzer's contribution to American hydrology is discussed in Chapter 6.

25 F.D. Adams, *Birth and Development*, 431.

26 Biswas, *History of Hydrology*, 71.

27 In his book *History of Hydrology*, Biswas devotes a brief chapter to the period from AD 200 to 1500. Biswas, *History of Hydrology*, 123-34.

28 F.D. Adams, *Birth and Development*, 432.

29 These include processes ranging from "vaporization and subsequent condensation, to rock pressure, to suction of the wind, to a vacuum produced by the flow of springs, to pressure exerted on the sea by the wind and waves, to 'the virtue of the heavens,' to capillary action, and to the curvature of the surface of the sea whereby the sea was believed to stand higher than the springs and hence to furnish the necessary head." Meinzer, *Hydrology*, 9.

30 F.D. Adams, *Birth and Development*, 433-40. The practice of alluding to the "fantastic" notions of Kircher and reproducing his illustrations of the subterranean hydrologic cycle was copied in subsequent historical treatments of hydrology. For example, Biswas, "The Hydrologic Cycle," 72-73; Biswas, *History of Hydrology*, 175-80; Hanor, "History of Thought," 85.

31 "Experiments to the Rescue" is a section heading in Biswas' history of the hydrologic cycle. Biswas, "The Hydrologic Cycle," 73.

32 F.D. Adams, *Birth and Development*, 448.

33 Ibid., 448-49.

34 Many examples are discussed below, but here I'll note that John Rodda, who was then director of the Hydrology and Water Resources Department of the World Meteorological Organization and subsequently served as president of the International Association of Hydrological Sciences, specified "the birth of scientific hydrology in 1674, following the publication of 'De l'origine des fontaines' by Pierre Perrault." Rodda, "Guessing or Assessing," 361-62.

35 Biswas, "The Hydrologic Cycle," 73; Meinzer, "History and Development," 11.

36 Biswas, *History of Hydrology*, 213; Meinzer, "History and Development," 12.

37 Parizek identifies these three as "the founding fathers of the science of hydrology." Parizek, "The Hydrologic Cycle Concept," 9. See also Meinzer, *Hydrology*, 14; P.B. Jones et al., "Development of the Science," 13; Biswas, "The Hydrologic Cycle," 73; Biswas, *History of Hydrology*, 207; Nace, "General Evolution," 46; Dooge, "Development of Hydrological Concepts," 280; Moser, "Hydrological Cycle," 416; Veissman and Lewis, *Introduction to Hydrology*, 4; Dooge, "Concepts of the Hydrologic Cycle," 6-7; Detay and Gaujous, "De la cosmologie," 6-7; Mather, "200 Years of British Hydrogeology," 1. (Although for Mather, it was Perrault, Halley, and Dalton who developed the concept of the hydrological cycle.)

38 Meinzer, "History and Development," 11

39 Goudie, "Hydrology," 256.

40 Dooge, "Development of Hydrological Concepts," 282.

41 Nace, "General Evolution," 46. The same point is made by Dooge: "Whatever one may think of the experiment it must be conceded that the result was of the right order of magnitude ... Thus Halley appears to have been the first to add an estimation of the evaporation part of the cycle to the previous procedure of comparing only rainfall and

runoff. Since his estimates were necessarily crude, *his main contribution was in the formulation of the concept of the complete cycle* rather than in the experimental verification of the water balance involved." Dooge, "Development of Hydrological Concepts," 282 (emphasis added).

42 Dooge, "Development of Hydrological Concepts," 288-90.

43 European Geosciences Union, *Awards and Medals*.

44 De Villiers' book *Water* won the Governor General's Award (Canada) for non-fiction in 1999. De Villiers, *Water*, 29.

45 That the sacred hydrologic cycle had ceased to retain adherents by the late nineteenth century is described below. Even today, however (and particularly with the most recent incarnation of the argument from design, known as "intelligent design"), the hydrologic cycle continues to provide an argument for the divine presence in the world. See, for example, the April 2002 issue of *Natural History Magazine* (published by the American Museum of Natural History), which features a special series of articles on intelligent design.

46 Glacken, *Traces on the Rhodian Shore;* Tuan, *Hydrologic Cycle*.

47 The roots of natural theology can be found notably in the Judeo-Christian *theologia naturalis* of the early Christian era and Middle Ages, the fundamental principle of which was "the belief that one can find in the creation the handiwork of a reasonable, loving, and beneficent creator." It may be opposed to another tradition in Judeo-Christian thought, that of "a *contemptus mundi*, a rejection literally of the earth as the dwelling place of man, a distaste for, and disinterest in, nature." Glacken, *Traces on the Rhodian Shore*, 162.

48 The works of Galileo, Descartes, and others are introduced in the previous chapter in the context of the discussion of the mathematization and the "disenchantment" of nature. Particularly through Descartes' deployment of machine metaphors against the organic metaphors used by his philosophical opponents, one effect of this work was to render a mechanistic view of creation by which "the universe of which the earth is a part is like a great machine and is to be understood in geometrical terms." In such a universe, "Nature owes its harmonies to an underlying mechanical order," an idea that threatened the long-standing teleological view that "the earth was a divinely designed environment." Glacken, *Traces on the Rhodian Share*, 391.

49 Merchant, *The Death of Nature*, 246.

50 Ibid., 244. John Ray described this impulse as originating in God and mediated by "the subordinate Ministry of some inferior Plastic Nature." This formulation allowed for the simultaneous presence and absence of the Creator in the material universe, thus avoiding the question of whether He has a hand in matters less seemly than one would wish to imagine: God is not immediately present in nature, directing its every activity, but has appointed, as it were, his "subordinate Ministry of some inferior Plastic Nature ... It is not decorous in respect of God, that he should ... set his own hand as it were to every work, and immediately do all the meanest and triflingst things himself drudgingly, without making use of any inferior or subordinate Ministers." Ray, *Wisdom of God*, 39.

51 Tuan, *Hydrologic Cycle*, 122.

52 Ibid., 149.

53 Kennedy, "Inventing the Earth," 23, 31-32; Hettner, *Surface Features*, 104.

54 Huxley, *Physiography*, 76.

55 Tuan, *Hydrologic Cycle*, 7.

56 Ray, *Wisdom of God*, 75.

57 Via a monograph published in 1698 by the mathematician and natural theologian John
   . Keill, who in turn had drawn notably from Halley's work on evaporation. Tuan, *Hydrologic Cycle*, 96.
58 Ibid., 10.
59 Ibid., 11.
60 Ibid., 6.
61 Ibid., 23.
62 Ibid., 144. Tuan's critique of the hydrologic cycle anticipates David Livingstone's approach
   to the geography of the production of scientific knowledge: "Place matters in the way
   scientific claims come to be regarded as true, in how theories are established and justified,
   in the means by which science exercises the power that it does in the world. There are
   always stories to be told of how scientific knowledge came to be made where and when it
   did. The appearance of universality that science enjoys, and its capacity to travel with
   remarkable efficiency across the surface of the earth, do not dissolve its local character."
   Livingstone, *Putting Science in Its Place*, 14. For an application of Livingstone's approach
   to the development of the hydrologic cycle concept, see Linton, "Is the Hydrologic Cycle
   Sustainable?"
63 Tuan, *Hydrologic Cycle*, 144. Italics added to emphasize the mixing of tenses.
64 Ibid., vii.
65 Ibid., vii. One example of other factors is the biblical renderings of "wilderness" and
   "deserts," inhospitable places bereft of water as well as God's mercy (e.g., Jeremiah 17:5-8).
66 Tuan, *Hydrologic Cycle*, 133.
67 Linton, "Is the Hydrologic Cycle Sustainable?"
68 Tuan, *Hydrologic Cycle*, 4.
69 Tuan, *Hydrologic Cycle*, 23-24. Tuan cites F.D. Adams, *Birth and Development;* Meinzer,
   "History and Development"; Baker and Horton, "Historical Development"; Parizek, "The
   Hydrologic Cycle Concept"; and Biswas, "The Hydrologic Cycle" as examples of the short
   summaries available. All these are cited or discussed in my section on the conventional
   history above.

CHAPTER 6: THE HORTONIAN HYDROLOGIC CYCLE

1 Golinsky, *Making Natural Knowledge*, ix.
2 For a discussion of the insights and the perils of constructionism, see Hayles, "Searching
   for Common Ground"; Hacking, *Social Construction;* Demeritt, "Being Constructive about
   Nature"; Demeritt, "What Is the 'Social Construction of Nature'?"
3 Demeritt, "Being Constructive about Nature," 35.
4 Golinsky, *Making Natural Knowledge*, 6. With this formulation, Golinsky is able to subsume
   under the term "constructivist history of science" the approaches of researchers who have
   disagreed with aspects of the construction metaphor, such as Ian Hacking (see Hacking,
   Social Construction) and Bruno Latour. Latour objects to the "social construction" of
   (scientific) knowledge, as it "implies that the initiative of action always comes from the
   human sphere, the world itself doing little more than offering a sort of playground for
   human ingenuity." Latour, *Pandora's Hope*, 114; see also Latour, *Politics of Nature*, chap. 1.
   For Latour, there is no question that scientists are involved in the production of facts and
   representations of nature, but in his view these are co-productions or co-constructions
   between human (scientists) and non-human (laboratory instruments, technologies, etc.)

actants. Thus: "Construction and fabrication ... have to be reconfigured totally, like all the other concepts that have been handed down to us, if we really wish to understand science in action." Latour, *Pandora's Hope*, 115.

5 Stichweh, "Sociology of Scientific Disciplines," 5.

6 Latour, *Science in Action*, 27.

7 Ibid., 43.

8 Huxley, *Physiography*, 76.

9 Horton, "Field, Scope, and Status."

10 Ibid., 190.

11 Davis described the "geographical cycle," or the "cycle of erosion," in terms of "the sequence in the developmental changes of land forms." Davis, "The Geographical Cycle," 254. Essentially, it comprised a sequence (now considered oversimplistic) by which convexities in the landscape were worn down by hydraulic processes to eventually form a flat plain, now at a lower elevation. The sequence could then be repeated by uplift of the terrain. For the disciplinary significance of Davis' geographical cycle and its contribution to intellectual history generally, see Kennedy, *Inventing the Earth*, 87-97.

12 Littlehales, "Inception and Development," 189.

13 Nemec, "International Aspects of Hydrology," 334. The name was changed to International Association of Hydrological Sciences in 1971, reflecting "a step towards breaking down the boundaries between hydrology and sciences in the diverse field of water resources development and management" and the abiding desire "to conserve the word 'science' in its title." Ibid.

14 Gordon and Thompson, *Physiological Principles*, 24.

15 "The purpose of the new Section is to contribute to the development of scientific research into geophysical phenomena associated with rivers, lakes, glaciers and underground water, which form the subject of Continental Hydrography, or Scientific Hydrology, the latter name perhaps being preferable in order to avoid possible confusion with other branches of geophysics" (author's translation). Section Internationale d'Hydrologie Scientifique, "Première Réunion plénière," 6-7.

16 This would have served the purpose of distinguishing hydrology from meteorology, a distinction that was particularly important in light of a suggestion, offered by the president of the International Union of Geodesy and Geophysics, that the Section on Hydrology be fused with the Section on Meteorology. Ibid., 9.

17 Kalinin, *Global Hydrology*, 7; L'vovich, *World Water Resources*.

18 National Research Council, *Opportunities*, 40.

19 Horton, "Field, Scope, and Status," 190, 192.

20 Ibid., 190-1, 202.

21 The Russian term "krugovorot vody v prirode" (literally "rotation of water in nature") is defined as "the continual process of circulation of water on the globe, taking place under the influence of solar radiation and the force of gravity." Robert North, personal email correspondence, 18 May, 2007. As pointed out by the Russian/Soviet hydrologist M. L'vovitch, it was on the basis of this "great process" that Russian hydrologists worked out calculations for regional and global water balances beginning in the late nineteenth century. L'vovitch, "World Water Balance," 402. Although this concept bears obvious similarities to that of the Hortonian hydrologic cycle, there are also differences, as discussed in Chapter 8.

22 Meyer, *The Elements of Hydrology*, 6.

23 Mead, *Hydrology*, 1.

24 National Research Council, *Opportunities*, 40.

25 Horton, "Field, Scope, and Status," 192.

26 Latour, *We Have Never Been Modern*.

27 Horton, "Field, Scope, and Status," 190.

28 A slightly more comprehensive version, and one that enables the equation to actually work out, is expressed as $R = P - E$ +/- *change in storage;* "this states that for any arbitrary volume and during any period of time, the difference between total input and output will be balanced by the change of water storage within the volume." UNESCO, *Scientific Framework*, 12.

29 L'vovitch argues that it is Brickner rather than Penck who deserves the credit: "That is why justice would demand the major formula of water balance to be named [the all-Russian] Brickner-Oppokov's equation." L'vovitch, "World Water Balance," 408.

30 UNESCO, *Scientific Framework*, 14.

31 Golinsky, *Making Natural Knowledge*, 33.

32 Horton, "Field, Scope, and Status," 199.

33 Marvin, "Status, Scope and Present-Day Problems," 54.

34 Marvin, "Status and Problems of Meteorology," 566-67.

35 Meinzer, "Formation of the Section of Hydrology," 228.

36 Meinzer, *Hydrology*, 3.

37 Wisler and Brater, *Hydrology*, 3. The meteorologists responded to hydrology's incursions in the mid-1930s by literally redrawing the hydrologic cycle to account for important atmospheric phenomena that they felt hydrologists had left out of the picture. See Thornwaite, "The Hydrologic Cycle Re-Examined"; Leighly, "Role of Atmospheric Circulation"; Jenkins, "Forests, Land and Sea." This episode amounted to something of a turf war between the two emerging sciences, with different versions of the hydrologic cycle serving as the means of claiming portions of the hydrosphere for their respective subdisciplines.

38 Goudie, "Hydrological Cycle," 256.

39 The geographic focus here is on the United States. However, partly as a result of the translation of American hydrology texts into other languages and partly as a result of the diffusion of the idea via other forms of literature, the hydrologic cycle has been taught as the central concept of hydrology in most other countries since the Second World War. The centrality of the hydrologic cycle to the teaching of hydrology generally is apparent in a comparative international study of hydrology textbooks published in 1974. UNESCO, *Textbooks on Hydrology*, vol. 1; UNESCO, *Textbooks on Hydrology*, vol. 2.

40 Recall that Horton's 1931 paper asserted the field and scope of hydrology in terms of "deal[ing] with the natural occurrence, distribution, and circulation of water on, in, and over the surface of the Earth." Horton, "Field, Scope, and Status," 190. In 1962, the Ad Hoc Panel on Hydrology of the US Federal Council for Science and Technology asserted: "Hydrology is the science that treats of the waters of the Earth, their occurrence, circulation, and distribution, their chemical and physical properties, and their reaction with their environment, including their relation to living things ... Hydrology is the scientific examination and appraisal of the whole continuum of the water cycle conceived as a circle made up of numerous arcs, some of which traverse the domains of other related earth sciences." United States Federal Council for Science and Technology Ad Hoc Panel on Hydrology,

Scientific Hydrology, 2-3. Sixty years after the publication of Horton's paper, a major stock-taking exercise by the US National Research Council identified hydrology in remarkably similar terms: "Hydrologic science deals with the occurrence, distribution, circulation, and properties of water on the earth ... Water moves through the earth system in an endless cycle that forms the framework of hydrologic science." National Research Council, *Opportunities,* ix.

41 Hydrology has been qualified as "an applied science" and "an observational science." American Society of Civil Engineers, *Hydrology Handbook,* viii; Maidment, *Handbook of Hydrology,* 1.1-1.2.

42 Horton, "Hydrologic Research," 527.

43 Klemes, "A Hydrological Perspective," 4. The quote continues rather caustically, "because ... most of them are themselves technologists in their hearts." For similar expressions of concern around this time, see Sutcliffe, "Introduction," xii; Dooge, "Looking for Hydrologic Laws"; Bras and Eagleson, "Hydrology"; Dooge, "Hydrology in Perspective"; Nash, "Hydrology and Hydrologists."

44 Bras and Eagleson, "Hydrology," 227.

45 National Research Council, *Opportunities,* vii.

46 Nash, "Hydrology and Hydrologists," 192.

47 The heightening of expressions of concern from within the discipline in the 1980s may be explained in part by the critique from outside the discipline, particularly from environmentalists "for alleged narrowness of outlook and for our inability, or at least our failure, to assess the consequences of our interferences in the watery environment." Ibid. In any case, as we know, the 1980s marked a period of erosion of popular faith in the "engineering" solution to water problems, as manifest in growing opposition to large dams around the world (see Chapter 3). That these developments outside the discipline would give rise to a heightened degree of introspective activity within is not surprising. In any case, the identity crisis has passed in recent years. The growing popularity of atmospheric and global hydrological processes in the context of concerns about climate change, and the popularity of hydrology in an era of increasing general interest in water issues, has invigorated many hydrologists with a sense of purpose that appears to have been lacking in the 1980s and early 1990s.

48 Dooge quoted in Klemes, "A Hydrological Perspective," 19.

49 National Research Council, *Opportunities,* xii.

50 Ibid., 35.

51 Saville, "Basic Principles of Water Behavior," 1. For contemporary illustrations of the conflation of – or at least the very close association between – the hydrologic cycle and its diagrammatic representation, see Thornwaite, "The Hydrologic Cycle Re-examined"; and Jenkins, "Forests, Land and Sea."

52 B.S. Hall, "The Didactic and the Elegant," 4.

53 Ibid., 21.

54 Horton, "Field, Scope, and Status," 190.

55 Ibid., 190.

56 Ibid., 192-93.

57 Meinzer, *Hydrology.*

58 Ibid., 1-2.

59 Chow, *Handbook of Applied Hydrology,* 1-2.

60 In 1940, geographer George Jenkins described the hydrologic cycle as a kind of composition: "The elements of the occurrence of water and its movements as related to the earth have been composed into a generalized scheme, called the 'hydrologic cycle.'" Jenkins, "Forests, Land and Sea," 309. For similar examples of the 'hydrologic cycle' in brackets or italics, see American Society of Civil Engineers, *Hydrology Handbook*, 1; Wisler and Brater, *Hydrology*, 3.

61 Following Meinzer's 1942 textbook, Wisler and Brater's *Hydrology* (1949) affirms that hydrology "is the science that treats of the various phases of the hydrologic cycle" and notes "the *hydrologic cycle* ... provides the groundwork upon which the science of hydrology is constructed." Wisler and Brater, *Hydrology*, 3 (emphasis in original). However, although dedicating the book to the memory of Robert Horton, "whose untiring efforts throughout more than a quarter of a century contributed immeasurably toward the development of the science of hydrology," the only specific contributions of his that they cite are with respect to the theory of infiltration capacity. Ibid., 5. Horton's major feat of issuing the "birth certificate" and identifying the hydrologic cycle as the midwife is invisible. They include a hydrologic cycle diagram that is evidently based on Horton's original but again neglect to acknowledge that it was, in essence, Horton's diagram that they reproduced. Ibid., 2. One – perhaps the only – reference to Horton as having identified hydrology as the science whose purpose is "to trace out and account for the phenomena of the hydrologic cycle" is found in National Research Council, *Opportunities*, 40.

62 H.E. Thomas, "Changes in Quantities," 544; Langbein and Hoyt, *Water Facts*, 3; Chorley and Kates, "Introduction," 3.

63 American Society of Civil Engineers, *Hydrology Handbook*, 1.

64 Saville, "Basic Principles of Water Behavior."

65 Saville acknowledges that "the most comprehensive diagrammatic representation of [the hydrologic cycle] was published by Horton in 1931" and points out that "a somewhat less comprehensive but perhaps more readily visualized diagram is presented herein." Saville, "Basic Principles of Water Behavior," 1. Like Horton, Saville stresses that although hydrological phenomena "may appear an obvious and simple cycle of natural events, especially as pictured in figure 1," they are in fact extremely complex. Ibid., 5.

66 Chow's 1964 *Handbook of Applied Hydrology* identifies three types of hydrologic cycle diagram: "Many diagrams have been designed to illustrate the hydrologic cycle; some are qualitative, some descriptive, and some quantitative." A variation of Horton's 1931 diagram is given as an example of a qualitative diagram. Chow, *Handbook of Applied Hydrology*, 1-3. Although Horton's authorship of the diagram is acknowledged, there is nothing to indicate that it was the first. A diagram much like Saville's is given as an example of a descriptive diagram.

67 Leighly, "Role of Atmospheric Circulation," 335; Jenkins, "Forests, Land and Sea," 309.

68 Most of the diagrams currently in circulation depict a continental land mass to the left of an ocean. This is consistent with the (1934) prototype. As a matter of speculation, such a proclivity suggests the American (as opposed to the European) origin of this way of representing the hydrologic cycle, as most American hydrologists of the era, including Horton (who was from New York State and whose diagram placed an ocean to the right side), were based in the eastern part of the country and would likely have visualized an ocean lying to the east (the right). However (thanks to Alfred Kalantar from the Department of Chemistry at the University of Alberta for pointing this out to me), the Western practice

of reading from left to right on the page would probably incline illustrators of the hydrologic cycle to represent the flow of water over the landscape in the same direction.

69 B.S. Hall, "The Didactic and the Elegant," 37.

70 De Villiers, "Water Works," 52.

71 F.R. Hall, "Contributions of Robert E. Horton"; National Research Council, *Opportunities*, 41; Reuss, "Hydrology," 275; Leopold, "The Alexandrian Equation," 27; Nace, "Development of Hydrology," 23.

72 Horton, "Erosional Development." For assessments of the significance of this paper and Horton's reputation among hydrologists, see National Research Council, *Opportunities*, 41; Dooge, "Looking for Hydrologic Laws," 52s; Kennedy, *Inventing the Earth*, 102-6.

73 Meinzer, "History and Development."

74 F.D. Adams, "Origin of Springs," 3.

75 Meinzer, "History and Development"; Baker and Horton, "Historical Development"; F.D. Adams, *Birth and Development*.

76 Meinzer, "History and Development."

77 Meinzer, *Hydrology*, 8.

78 F.D. Adams, *Birth and Development*, 446.

79 Biswas, *History of Hydrology*, 151.

80 Nace, "General Evolution"; see also Krynine, "On the Antiquity of 'Sedimentation'"; P.B. Jones et al., "Development of the Science"; Parizek, "The Hydrologic Cycle Concept"; Biswas, "The Hydrologic Cycle"; Larocque, "Translator's introduction"; Tixeront, "L'Hydrologie en France"; Dooge, "Concepts of the Hydrologic Cycle, Ancient and Modern"; among others.

81 UNESCO-WMO-IAHS, "Foreword," 11.

82 Ibid.

83 Meinzer, *Hydrology*, 11-13.

84 Biswas, *History of Hydrology*, 74.

85 Garbrecht, "Hydrologic and Hydraulic Concepts," 14.

86 *Encyclopaedia Britannica*, 15th ed., 20:770, s.v. "hydrosphere"; *Encyclopaedia Britannica*, 15th ed., 17:573, s.v. "history of earth sciences"; Edmunds, 2004, 194.

87 The anachronism with which the hydrologic cycle is inserted into earlier literature is also evident in recalling that none of the ancient and more modern writers to whom the hydrologic cycle concept is attributed ever actually discussed it as such. And yet, in order to make the claim that the origins of the modern science of hydrology may be associated with great thinkers of the past, it has sometimes been necessary to make it *appear* as though they did. Here I cite only a couple of the more reconstructive examples from the conventional history: Although they were interested in the "historical development of ideas regarding the origin of springs and groundwater" and not the hydrologic cycle per se, Baker and Horton illustrate the problem in their 1936 article "Historical Development of Ideas Regarding the Origin of Springs and Groundwater." After describing what the likes of Thales, Pythagoras, Aristotle, and Lucretius had to say about hydrological processes, Baker and Horton note: "The first engineer to describe the origin of springs was the Roman, Vitruvius." Baker and Horton, "Historical Development," 398. Their exegesis of Vitruvius' text proceeds as follows:

> Vitruvius was treating primarily of sources and quality of water-supplies. *Taking out and arranging in logical order what he wrote* on evaporation and condensation, precipitation and springs, we have the following:

[Evaporation and condensation] That vapor and mists and humidity come from the Earth, seems due to the reason that it contains burning heat, mighty currents of air, and a great quantity of water. So soon as the Earth, which has cooled off during the night, is struck by the rays of the Sun, and the winds begin to blow while it is yet quite dark, mists begin to rise upward from damp places ...

[Precipitation] Whenever the winds carry the vapor which rolls in masses from springs, rivers, and marshes and the sea it is brought together by the heat of the Sun, drawn off, and carried upward in the form of clouds. Then the clouds are supported by the current of air until they come to the mountains where they are broken up from the shock of collision and the gales, turn into water on account of their fullness and weight and in that form are dispersed upon the Earth (as rain) ...

[Springs] The valleys among the mountains receive the rain most abundantly, and on account of the thick woods, the snow is kept in them longer by the shade of trees and mountains. Afterwards, in melting, it filters throughout the fissures in the ground and thus reaches the very foot of the mountains, from which gushing springs come belching out." Ibid., 398 (emphasis added).

What Baker and Horton have done here is to reorganize Vitruvius' writing so as to make it fit a certain "logical order," specifically "evaporation and condensation," followed by "precipitation" and then "springs." Although presented in a different sequence, this account corresponds to what Horton had identified in his 1931 paper as the three phases of the hydrologic cycle, "namely rainfall, runoff, evaporation." Horton, "Field, Scope, and Status," 192. The reconstruction of Vitruvius' writings here and elsewhere (e.g., Biswas, *History of Hydrology*, 83-84) to produce a site onto which the scientific hydrologic cycle is projected into antiquity is ironic, given that the same writings have been used elsewhere to provide an example of premodern, heterogeneous, multiform waters; e.g., Hamlin, "'Waters' or 'Water'?" 317-18.

    Another example of forging the hydrological present by projecting elements of modern hydrology into the past is found in Dooge's "The Development of Hydrological Concepts in Britain and Ireland between 1674 and 1874." Here the work of Edmond Halley is repositioned with respect to the "hydrologic cycle": "It is probable that when writing [his 1687] paper Halley was aware of the comparison of rainfall and runoff made by Perrault and Mariotte and that *he attempted to close the cycle* by estimating the amount of evaporation and comparing this with the amount of rainfall and runoff. Since his estimates were necessarily crude, *his main contribution was in the formulation of the concept of the complete cycle* rather than in the experimental verification of the water balance involved." Dooge, "Development of Hydrological Concepts," 287 (emphasis added). To be sure, Halley's paper was taken as proof that evaporation provided a quantity of water sufficient to account for the source of springs and rivers. But to say that he "attempted to close" and formulated the concept of "the complete cycle" attributes something to Halley that wasn't quite there. For one thing, it isn't at all certain that Halley was aware of the work of Perrault and Mariotte when he wrote his paper in 1687. (This is shown by Dooge himself: ibid., 286-87.) In any case, it is obvious that Halley never mentioned Perrault or Mariotte in his 1687 paper. He could not then have been attempting to "close" or formulate the hydrologic cycle concept, as such a concept would have required acknowledgement of their work on precipitation and streamflow for its completion. Only from the perspective of the present –

when we can position the work of Perrault, Mariotte, and Halley in relation to the modern, scientific hydrologic cycle – can Halley be seen to have done such a thing. To speak of Halley as having "attempted to close the cycle" makes sense only if we presume that "the cycle" existed in Halley's time.

88 Latour, *Science in Action*, 27.
89 In addition to the many references to Perrault found in the historical introductions of hydrology textbooks, see, for example, F.D. Adams, *Birth and Development*, 448; Meinzer, *Hydrology*, 14; P.B. Jones et al., "Development of the Science"; Parizek, "The Hydrologic Cycle Concept," 9; Biswas, "The Hydrologic Cycle"; Larocque, "Translator's introduction"; Biswas, *History of Hydrology;* UNESCO-WMO-IAHS, "Foreword"; Nace, "General Evolution"; Nace, "The Hydrological Cycle"; Dooge, "Development of Hydrological Concepts"; Dooge, "The Waters of the Earth," 156; Dooge, "Concepts of the Hydrologic Cycle"; Carbonnel, "Introduction"; Dooge, "Background to Modern Hydrology," 7.
90 Larocque, "Translator's introduction," 1.
91 Perrault, *De l'origine des fontaines.*
92 Larocque, "Translator's introduction," 3.
93 F.D. Adams, *Birth and Development*, 448; Meinzer, "Introduction," 14.

CHAPTER 7: READING THE RESOURCE

1 Worster, *Rivers of Empire*. Worster's (1985) *Rivers of Empire* is the classic study of twentieth-century hydro-nationalism in the United States. For Spain, see Erik Swyngedouw's study of the modernization of the state thorugh hydrological engineering in the late-nineteenth and early twentieth-centuries. Swyngedouw, "Modernity and Hybridity." For (the Soviet hydrologist) Davidov's plan to divert Siberian rivers in order to (as he put it) "correct the mistakes of nature" in order to maximize the production of the Soviet economy, see Duke, "Seizing Favours," 3. For hydro-nationalism in late twentieth-century Quebec, see Desbiens, "Political Geography." For Nationalist China in the 1930s, see Pietz, "Controlling the Waters." For how "a mammoth project such as [the Three Gorges Project] indulges the nationalistic feelings of Chinese leaders" today, see Padovani, "The Chinese Way."
2 Reisner, "Unleash the Rivers."
3 Sachs, *Planet Dialectics.*
4 Worster, "Water in the Age of Imperialism," 11.
5 UNESCO, *Textbooks.*
6 Scott, *Seeing Like a State.* There are interesting parallels between the fixation of water in the hydrologic cycle and the fixation of economic transactions in "the economy" around this time. Timothy Mitchell has shown "how the modern understanding of 'the economy' as the totality of the relations of production, distribution and consumption of goods and services within a given country or region ... came into being between the 1930s and 1950s as the field of operation for new powers of planning, regulation, statistical enumeration and representation." T. Mitchell, "Fixing the Economy," 82, 91. "Fixing the economy" as "a new discursive object" was partially an effect of the nascent practice of econometrics begun in the 1930s as an "attempt to create a mathematical representation of the entire economic process as a self-contained and dynamic mechanism." Ibid., 91. Thus, "the economy" was discursively produced in the process of measuring relations of exchange over a predefined space. As what might be considered a representation of the economy of water, the hydrologic cycle has enjoyed a history similar to that of the economy: via

Horton, hydrological science established a mathematically structured representation of the water process as a self-contained and dynamic mechanism, one that – like the economy – was readily taken up by the state as a discursive object and eventually became a fixture in popular discourse.

7 McGee, "Water as a Resource," 522-23. McGee has been identified as "the chief theorist of the conservation movement." Hays, *Conservation,* 102. He "emerged as the prophet of the new world which conscious purpose, science, and human reason could create out of the chaos of a laissez-faire economy where short-run individual interest provided no thought for the morrow. McGee was a key figure in disseminating the expanding concepts of the conservation movement." Ibid., 124. A self-made scientist and author of numerous articles in fields ranging from anthropology to hydrology, McGee was instrumental in establishing the National Geographic Society, the American Anthropological Society (both of which he served as president), and the Geological Society of America. In 1909, he was head of the Inland Waterways Commission, which had been established, at McGee's own urging, by Roosevelt in 1907. The IWC epitomized the conservation movement as it applied to water, promoting the view of rivers as integrated hydrological systems providing multiple benefits to society, the rational exploitation of which required coordinated, centralized control of entire river basins. Ibid., 102-9.

8 McGee, "The Beginning of Agriculture," 372-73.

9 McGee, "Our Great River."

10 Mumford, *Technics and Civilization.*

11 Among the Canadian economic and political establishment, this resourcefulness became "the most striking feature" of the landscape, as averred by T.C. Keefer, president of the Royal Society of Canada in 1899 in describing the potential for hydroelectrical generation: "An examination of any good map of our broad Dominion, reveals, as its most striking feature, an extraordinary wealth and remarkable uninterrupted succession of lakes and rivers ... an almost continuous distribution of lakes; lakelets and rivers ... and many possessing facilities for the storage of their flood waters. In many places the outlet from the lake or the connection between a chain of lakes is a narrow cleft in a rock where an inexpensive dam will hold back the water supplied by the winter's accumulation of snow." Quoted in Gossage, "Water in Canadian History," 98-99.

12 Nelles, *The Politics of Development,* 215-75.

13 McGee, "Principles of Water-Power Development"; Leighton, "Water Power in the United States"; Newell, *Water Resources.*

14 Hays, *Conservation,* 2.

15 Ibid., 262.

16 Ibid., 271.

17 Scott, *Seeing Like a State,* 11.

18 Ibid., 4.

19 Worster, *Rivers of Empire,* 135.

20 Worster provides an illustration of how this program made water legible to the state, and how this in turn was tied in to an even more ambitious program of social engineering: The purpose of the water survey, he argues, was "to identify all lands in the public domain that would be suitable for reservoirs, ditches, canals, or irrigated agriculture ... Powell had a corps of young engineering-school graduates learning the art of stream gauging under Frederick Newell's direction ... That was the hydrographic branch of the survey. Another group traveled by horse and rail, looking for reservoir sites while compiling a series of

topographic maps of every western river basin, beginning with those in Colorado, Nevada, Montana, and New Mexico. First the grand map of the West, Powell decided, with statistics on precisely how much water was running through the land. Then could come the grander scheme, the mapping of a new society." Ibid., 135.

21 Nace, "Development of Hydrology," 22.

22 Hays, *Conservation*, 6.

23 Newell, *Water Resources*, 149.

24 McGee, "Principles of Water-Power Development," 822.

25 Reuss, "Hydrology," 275. The text referred to was a compendium of course notes compiled by Mead in 1904, which served as the basis for his 1919 text on hydrology. Mead, *Hydrology*.

26 McGee, "Water as a Resource," 522.

27 Ibid., 523. McGee's devotion to the quantification of things included his very person. He and his friend and colleague John Wesley Powell made an extraordinary wager that involved having scientists at the Smithsonian Institution in Washington remove the brains from their skulls and weigh them – posthumously, of course. Powell won; his brain can still be found on a shelf at the Smithsonian Institution, where it "rests in a preservative." Miller, *John Wesley Powell*, 114; The whereabouts of McGee's brain remains an unknown quantity.

28 McGee, "Water as a Resource," 532.

29 Langbein and Hoyt, *Water Facts*.

30 To quantify water resources, McGee proposed the "kilostere," which equals one thousand cubic metres. The phrase "our stock of water" is from McGee, "Water as a Resource." Here McGee provides data on flows: "The World Supply," in terms of mean annual global precipitation, is reported as 125 billion kilosteres. "The National Supply" totals six billion kilosteres of annual precipitation, which he breaks down regionally between the "more humid two-fifths of the country east of the ninety-fifth meridian," the "semi-arid fifth ... between the ninety-fifth and hundred and third meridians" and "the western two-fifths of the country, including our arid lands." The total volume of annual national precipitation is reported as being disposed of in the following manner: "Of the total rainfall, over half is evaporated; about a third flows into the sea; the remaining sixth is either consumed or absorbed." McGee, "Water as a Resource," 40-41. It isn't clear where these data come from; they are the same as those found in the report of the National Conservation Commission published in the same year, in which McGee played a leading role. National Conservation Commission, *Report*, 21-23; 39-45. Possibly, the figures he reports for the so-called world supply are derived from Brickner's 1905 calculation of the global water balance, which is discussed in Chapter 8. The data for the "national supply" appear to come from a variety of US federal sources, including the Bureau of Reclamation, the Hydrographic Branch of the US Geological Survey, and the US Weather Bureau.

31 McGee, "Water as a Resource," 532.

32 Rogers, *America's Water*, 50. The Boulder (later renamed the Hoover) Dam, begun in 1928, was one exception.

33 Zimmermann, *World Resources and Industries*, 542-58.

34 Ibid., 542.

35 Zimmermann, *World Resources and Industries*, rev. ed., 571-72.

36 Ibid., 572.

37 National Resources Committee, *Drainage Basin Problems and Programs*, 7.

38 Hays, *Conservation*, 91.

39 National Resources Board, *Report on National Planning*, 255.

40 Reuss, "Development of American Water"; O'Neill, *Rivers by Design*.
41 National Resources Board, *Report on National Planning*, 260-63.
42 Ibid., 292.
43 McGee, "Water as a Resource," 521-34.
44 National Resources Committee, *Deficiencies in Basic Hydrologic Data*.
45 Ibid., v.
46 Ibid., 1-2.
47 Saberwal, "Science and the Desiccationist Discourse," 311.
48 Worster, *Rivers of Empire*, 194.
49 Major water projects of this era involving federal government agencies included the Tennessee Valley Authority (created in 1933); dams on the Colorado River, including the Hoover Dam (completed in 1936), Parker Dam (begun in 1934), Imperial Dam (completed in 1940) Davis Dam (completed in 1951), and Glen Canyon Dam (begun in 1956); the All-American Canal, which brings Colorado River water to the Imperial Valley in California (completed in 1942); the multipurpose Central Valley Project in California (begun in 1935); and projects on the Columbia River, including the Bonneville Dam (completed in 1937), the Grand Coulee Dam (completed in 1942), and Chief Joseph Dam (completed in 1955). Meanwhile, flood control had become an imperative for the federal state, especially after the immense losses suffered by periodic flooding in the Mississippi and other river basins. In 1936, controlling floods was deemed "in the interest of the general welfare" and thus became a federal responsibility under the Flood Control Act passed by Congress in that year. O'Neill, *Rivers by Design*, 165. With this responsibility, the federal government's control of the nation's water was enhanced across the country, and particularly on the Mississippi River.
50 Reisner, *Cadillac Desert*, 172.
51 Worster, *Rivers of Empire*.
52 Ibid., 279.
53 Reisner, *Cadillac Desert*, 168.
54 Worster, *Rivers of Empire*, 59.
55 Chorley and Kates, "Introduction," 3.

Chapter 8: Culmination

1 Varenius, *Cosmography and Geography*, 142; Tuan, *Hydrologic Cycle*, 99. Varenius seems to have given up before trying: "For finding out the accurate and true quantity of water and land, first we ought to know both the whole Superficies of the water, as also its depth in divers parts of the Sea: also the subterraneous heaps of water ought to be examined. All which, feeling that we cannot find out by any method, therefore we cannot find out the accurate quantity of water." Varenius, *Cosmography and Geography*, 142. Despite this admission, he suggests that it might be possible to compute the quantity of "subterraneous heaps of water" on the estimation that the superficies of the water is half that of the earth and the profundity is a quarter or half a mile. However, admitting too many uncertainties, he doesn't attempt to follow through with a calculation.
2 UNESCO, *Scientific Framework*, 14. This, it must be admitted, is an exception to the general observation made earlier that the proponents of the sacred hydrologic cycle were little interested in quantifying flows of water. Perhaps such an exception can be granted for Keill, as he was, after all, a mathematician.

3 Littlehales, "Status, Scope, and Problems," 69-70.

4 Linton, "Global Hydrology."

5 Nace, "Water of the World"; Chow, "Water as a World Resource"; Shiklomanov and Rodda, *World Water Resources;* Rodda, "Whither World Water?" And see the series titled *The World's Water: The Biennial Report on the Freshwater Resources,* with Peter Gleick as lead author, published every other year by Island Press.

6 McGee, "Water as a Resource," 523.

7 Ibid., 524.

8 Rees, *Natural Resources,* 17.

9 In a companion piece to McGee's article, M.O. Leighton, chief hydrographer of the United States Geological Survey, argued for the need to develop American hydropower potential, warning that "the menace to American industrial leadership is already on the horizon" with the construction of hydro facilities in other countries. Future world power, he points out, would depend to a large extent on exploitation of hydro potential: "If we are to place ourselves in an economic position in which we may finally prevail as a world power, our point of view regarding fuel and water power must be changed, and our policies must be altered so that we may wisely prepare to use the vast power resource which we are now neglecting." Leighton, "Water Power in the United States," 81.

10 McGee also employed a water balance approach, particularly to represent the water supply of the United States, as mentioned in Chapter 7.

11 L'vovich, *World Water Resources,* 14.

12 Ibid., 15-17.

13 Some sources have Brickner spelled "Bruckner." In any case, the consensus among the international hydrological community is that Brickner was the first to calculate a global water balance. See UNESCO, *Scientific Framework,* 14; L'vovitch, "World Water Balance," 402; Baumgartner and Reichel, "Preliminary Results," 585.

14 Baumgartner and Reichel, "Preliminary Results," 585.

15 Bassin, "History and Philosophy."

16 L'vovich, *World Water Resources,* v.

17 Ibid., v.

18 Kalinin, *Global Hydrology;* Korzun and USSR Committee for the International Hydrological Decade, *World Water Balance;* L'vovich, *World Water Resources;* Nace, "Editor's Preface," iv; Dooge, "Waters of the Earth," 156-57; Rogers, "Fresh Water," 260; Rogers, "Hydrology and Water Quality," 234.

19 Van Hylckama, *The Water Balance of the Earth;* Barry, "The World Hydrological Cycle," 17.

20 An exception is found in the work of Van Hylckama, the first American effort to calculate the water balance for the earth's continental areas. Van Hylckama, *The Water Balance of the Earth.*

21 Nace, "Editor's Preface," iv.

22 Nace, "Water of the World"; Nace, "Water Resources"; Nace, "World Water Inventory"; Nace, "World Hydrology"; Nace, "Development of Hydrology"; Nace, "Editor's Preface," iv.

23 Along with fellow American Walter Langbein and L.J. Tison, the German head of the IASH, Nace put a proposal for the IHD forward in 1962 in his capacity as special advisor to the Ad Hoc Panel on Hydrology of the Federal Council of Science and Technology. United States Federal Council for Science and Technology Ad Hoc Panel on Hydrology, *Scientific Hydrology,* 24-37; see also Nace, "Water Resources," 554. Nace's summary of the

program's purpose is worth citing: "The purpose of the proposed international program is to strengthen the whole science, to broaden the base of world water facts, and to advance understanding of hydrologic processes, all leading to improved ability to bend the forces of nature to the benefit of man." United States Federal Council for Science and Technology Ad Hoc Panel on Hydrology, *Scientific Hydrology*, 27.

24 Section Internationale d'Hydrologie Scientifique, "Première Réunion plénière."

25 American Geophysical Union, *International Union of Geodesy and Geophysics (IUGG)*.

26 Livingstone, *Putting Science in Its Place*, 171-77; see also Latour, *Science in Action*, 224-25; Golinsky, *Making Natural Knowledge*, 172-82.

27 Nemec, "International Aspects of Hydrology," 335.

28 Dooge, "Waters of the Earth," 154.

29 Chow, "Water as a World Resource," 3.

30 Nace, "Water Resources," 555.

31 L'vovich, *World Water Resources*, 23.

32 Korzun and USSR Committee for the International Hydrological Decade, *World Water Balance*.

33 Dooge, "Waters of the Earth," 156; Maidment, *Handbook of Hydrology*, 1.3-1.6; J.A.A. Jones, *Global Hydrology*, chap. 2.

34 Nace, "World Water Inventory"; Kalinin, *Global Hydrology*; Baumgartner and Reichel, "Preliminary Results"; and L'vovich, *World Water Resources*.

35 Houghton, "Introduction to Techniques of Management," 2.

36 Rodda, "Foreword," xi.

37 Conca, *Governing Water*, 132, 135-37.

38 Ibid., 131. In addition to these factors, there is another possible explanation for the popularity of global water in the 1990s: the geopolitical sea change that had occurred during the years intervening between Houghton's and Rodda's observations might have had an influence on how global water circulated. The collapse of the Soviet Union and the end of the Cold War had had the effect of freeing a number of Soviet hydrologists – who practised hydrology in the tradition and with the methods of L'vovich and his contemporaries – to work in the West for the United Nations and other organizations. As a matter of speculation, these geopolitical changes might also have had the effect of freeing the inhibitions of scientists working in places such as the United States and England who might formerly have been disinclined to reject Soviet methods.

39 W.L. Thomas et al., *Man's Role*.

40 Rogers, "Hydrology and Water Quality," 234-37; L'vovich et al., "Use and Transformation."

41 Livingstone, *Putting Science in Its Place*, 89.

42 Rockstrom, "Managing Rain for the Future," 70.

Chapter 9: The Constitution of Modern Water

1 Latour, *We Have Never Been Modern*.

2 At the same time, the concept of hybridity is compatible with relational dialectics, as both approaches are relational in their philosophical stance. Castree, "Geographies of Nature"; Castree, "Environmental Issues"; Castree, *Nature*. Both hybridity and relational dialectics rest on the fundamental idea that things occur by virtue of their relations and processes rather than in and of themselves.

3 Worster, *Rivers of Empire,* 19.
4 Whatmore, "Hybrid Geographies," 27; Whatmore, "Introduction."
5 Latour, *We Have Never Been Modern,* 6.
6 Ibid., 8.
7 Ibid., 32. There is nothing particularly novel about this observation. A common theme in social theory is to distinguish "the modern secular and rational social universe" by its disembeddedness from nature and from the lifeworld. Lansing, *Priests and Programmers,* 8. This characteristic of modernity has been stressed by many others, writing from different perspectives, and it may be taken as something of a critique of Latour that he does not acknowledge this as well as he might. Here are some examples: First, Clarence Glacken's history of ideas of nature devotes considerable attention to how, beginning in the seventeenth century, increasing awareness of the damages inflicted by human agency in the physical environment led to the sense that "man was going his own separate way" from nature, and how this marked a departure from the prevalent notion of humankind and the changes wrought by humankind in the physical environment as irreducibly part of nature. Glacken, *Traces on the Rhodian Shore,* particularly chapters 3, 7, 10, and 14. Second, Neil Smith's Marxian-inspired analysis of bourgeois representations of nature, which he calls "the ideology of nature," that has made it appear, as Noel Castree and Bruce Braun have pointed out, "resolutely external to society." Neil Smith, *Uneven Development;* Castree and Braun, "Construction of Nature," 7. Third, environmental historians, to make a rather gross but nonetheless valid generalization, have shown how nature is far more a product of human history – not only in a physical sense but also in a conceptual and discursive sense – than has been acknowledged. To quote William Cronon: "'Nature' is not nearly so natural as it seems. Instead, it is a profoundly human construction. This is not to say the nonhuman world is somehow unreal or a mere figment of our imaginations – far from it. But the way we describe and understand that world is so entangled with our own values and assumptions that the two can never be fully separated. What we mean when we use the word 'nature' says as much about ourselves as about the things we label with that word." Cronon, "Introduction," 25. Finally, drawing largely from the work of Michel Foucault, poststructuralist accounts of the construction and the effects of discourses of nature have drawn attention to how "discursive relations infuse our relation with nature at every turn, including even at the micro-level of knowledge and practice." Castree and Braun, "Construction of Nature," 17. Latour may perhaps be distinguished from all of these in illustrating that/how the world refuses to behave in the manner decreed by the Modern Constitution, as evidenced in the proliferation of hybrids.
8 Scarce, *Fishy Business,* 3.
9 Latour, *Politics of Nature,* 43.
10 Swyngedouw, *Social Power,* 14.
11 Latour, *We Have Never Been Modern,* 42.
12 Swyngedouw, *Social Power,* 14.
13 Latour, *We Have Never Been Modern,* 31.
14 Ibid., 78. Latour's reasoning, it must be admitted, is perhaps too clever in places: the notion of mediating between things that do not come into play until the event of mediation suggests a kind of performativity that he doesn't take up fully, at least in *We Have Never Been Modern.*
15 Ibid., 31.

16 Ibid., 79.

17 This silence has already been mentioned in Chapter 2 as a source of difficulty, especially in reconciling these ideas to the concerns of political economy. Moreover, Latour invites such criticism with his constitutional metaphor, as well as with his frequent invocation of things political, as in "The Parliament of Things" (ibid., 142-45) and "The Politics of Nature" (Latour, *Politics of Nature*).

18 Latour, *We Have Never Been Modern*, 32.

19 Ibid.

20 Ibid., 34.

21 Ibid., 49, 50-51.

22 The contemporaneousness of the first stirrings of the Modern Constitution and modern water is relevant here. Latour locates the historical origins of the Modern Constitution in the mid-seventeenth century, specifically in the outcome of an intellectual dispute between Robert Boyle (a key figure of the Scientific Revolution) and Thomas Hobbes (a political philosopher who made an important contribution to the idea of the modern state). In locating the origin of the Modern Constitution in this dispute, Latour draws from a study by science studies scholars Steven Shapin and Simon Schaffer. Shapin and Schaffer, *Leviathan and the Air-Pump*. This dispute occurred about a generation before the proto-hydrologists (such as Perrault and Mariotte) began the work of purifying modern water.

23 Latour, *We Have Never Been Modern*, 198-99.

24 Ibid., 7.

25 Ibid., 32.

26 Langbein and Hoyt, *Water Facts*.

27 Ibid., 3-4.

28 Ibid., 86-87.

29 Heidegger, *The Question Concerning Technology*, 21.

30 Ibid., 21, 23.

31 Ibid., 27.

32 Zimmerman, *Heidegger's Confrontation with Modernity*, 203.

33 Nash, "Hydrology and Hydrologists," 192.

34 Linton, *Beneath the Surface*. 31-35; McCutheon, *Electric Dams*.

35 Latour, *We Have Never Been Modern*, 31.

36 Ibid., 32.

37 Horton, "Field, Scope, and Status," 192.

38 Chorley and Kates, "Introduction," 3.

39 Horton, "Field, Scope, and Status," 190.

40 Newell's *Water Resources: Present and Future Uses* gives a particularly good illustration of this attitude. His chapter on evaporation begins as follows: "A force is at work day and night, summer and winter, steadily robbing water from lakes, streams, trees, animals, and all objects that contain it ... Man's ability to use water in all of its varied forms and applications is confined largely to that portion of it which is left after evaporation has taken its full share." Newell, *Water Resources*, 65.

41 Klemes, "A Hydrological Perspective," 20.

42 Wisler and Brater, *Hydrology*, 3.

43 Ibid., 4.

44 H.E. Thomas, "Changes in Quantities," 556. As noted above, a little over a decade later, it was estimated that fully "ten percent of the national wealth of the United States is found

in capital structures designed to alter the hydrologic cycle: to collect, divert and store about a quarter of the available surface water, distribute it where needed, cleanse it, carry it away and return it to the natural system." Chorley and Kates, "Introduction," 3.

45 H.E. Thomas, "Changes in Quantities," 544.

46 Ibid.

47 By the mid-1970s, "man" had begun to find his situation in the hydrologic cycle rather uncomfortable. He therefore began to define the relation in a slightly different way. Instead of merely improving it to his advantage, he now saw himself as impacting it and influencing it. See Pereira, "The Influence of Man." The shift was perceptible by the latter part of the International Hydrological Decade. "Man's Influence on the Hydrological Cycle" was the title chosen by the Coordinating Committee of the IHD for a study published in 1973. UNESCO/FAO Working Group, *Man's Influence.* A key symposium organized in 1974 by the International Association of Hydrological Sciences, UNESCO, and the World Meteorological Organization was titled "Effects of Man on the Interface of the Hydrological Cycle with the Physical Environment." Fournier, "Preface," v. And see also Dooge, Costin, and Finkel, "Man's Influence." Eventually, the situation became so unbearable that "man" even changed his own identity and became "humans," or "human society." Describing the growing "human impact on the hydrologic cycle," the National Research Council in 1991 suggested that "it was the dramatic color photographs of the earth in space ... that crystallized active interest in the interconnectedness of nature and in the changes being wrought by humans. This realization has found its way into contemporary views of the interactive role of man in the hydrologic cycle." National Research Council, *Opportunities,* 45. The main point, however, is that throughout it all, the basic integrity of "man"/ "humanity"/"human society" and the basic waterproofness of the hydrologic cycle have remained fundamentally intact.

48 Gleick, "Water in Crisis: Paths," 574.

49 C.E. Hunt, *Thirsty Planet,* 1.

CHAPTER 10: MODERN WATER IN CRISIS

1 As a measure of the cultural embeddedness of the "water crisis," the most recent edition of *The Fontana Dictionary of Modern Thought* includes an entry under this term, which reads in part: "Water is a fundamental resource that is integral to all environmental and societal processes ... Demand for water is burgeoning, and fresh water resources are unevenly distributed ... there is fear that, as competition for adequate water resources intensifies, 'water wars' may result ... There is a disturbing trend towards the use of force in water-related disputes. Some nations are willing to use water supply systems as targets and tools of war. There remains little international legal agreement about coping with conflicts over resource disparities or environmental damage." Goudie, "Water Crisis," 916.

2 Gleick, "The Changing Water Paradigm."

3 For example, Postel, *Last Oasis.*

4 Exceptions include Stott and Sullivan, *Political Ecology;* Biro, "Wet Dreams"; Conca, *Governing Water.* There is a long tradition of critique of the methods used to construct the idea of resource scarcity in the face of population growth, from which this chapter draws heavily. The origins of this critique are found in Marx's response to Malthus. Using the logical-empirical approach (and drawing heavily on demographic data provided by Humboldt), Malthus had posited the inevitability of poverty and scarcity and used this position

to justify a politically conservative policy. Marx responded with the argument that poverty is not the inevitable outcome of natural processes (as Malthus had argued) but, rather, was the product of the capitalist law of accumulation. As described by David Harvey: "[Marx] replaces the inevitability of the 'pressure of population on the means of subsistence' ... by an historically specific and necessary pressure of labor supply on the means of employment produced internally within the capitalist mode of production. Marx's distinctive method permitted this reformulation of the population-resources problem, and put him in a position from which he could envisage a transformation of society that would eliminate poverty and misery rather than accept its inevitability." Harvey, "Population, Resources," 269.

5 The social dimensions of some of these other environmental issues are discussed in Chapter 3, particularly in the section dealing with the political ecology of water. See Forsyth, *Critical Political Ecology*, for a good summary of critical assessments of these issues in the political ecology tradition. For critical sociological analyses of global environmental issues, see Taylor and Buttel, "How Do We Know"; Buttel and Taylor, "Environmental Sociology"; Wynne, "Scientific Knowledge"; and Yearley, *Sociology, Environmentalism, Globalization*.

6 This idea comes mainly from Latour. Environmental crises, he points out, "are never presented in the form of 'crises of nature.' They appear rather as *crises of objectivity*, as if the new objects that we produce collectively have not managed to fit into the Procrustean bed of the two-house politics, as if the 'smooth' objects of tradition were henceforth contrasted with 'fuzzy' or tangled objects that the militant movements disperse in their wake. We need this incongruous metaphor to emphasize to what extent the crisis bears on *all* objects, not just on those on which the label 'natural' has been conferred – this label is as contentious, moreover, as those of *appellations d'origine controlee*. [The environmental movement] thus does not reveal itself owing to a crisis of ecological objects, but through a generalized constitutional crisis that bears upon *all objects*." Latour, *Politics of Nature*, 20.

7 Geoffrey Lean quoted in Yearley, *Sociology, Environmentalism, Globalization*, 50.

8 Figuertes, Tortajada, and Rockstrom, "Conclusion," 229.

9 *Encyclopaedia Britannica*, 15th ed., 17:580, s.v. "earth sciences: hydrologic sciences."

10 Postel, *Pillar of Sand*, 3.

11 Latour, *Politics of Nature*, 18-19.

12 Ibid., 3.

13 Brown, Gardner, and Halweil, "Beyond Malthus," 16.

14 Douglas, *Purity and Danger*.

15 Hamlin, *A Science of Impurity*.

16 Furon, *The Problem of Water*, 3.

17 McGee, "Water as a Resource," 523.

18 Quoted in Zimmermann, *World Resources and Industries*, 103. For a contemporary assessment of potential water shortages affecting the entire country see R.M. Brown, "Man and Water Supply."

19 Meigs, "Water Problems," 346. According to a federal government climatologist, about half of the US west was under contract to rain makers ("pluviculturists") in the early 1950s, at charges ranging from one to ten cents an acre. Ibid., 358. For modern faith in pluviculture in the United States and abroad, see Langewiesche, "Stealing Weather."

20 Nikolaieff, "Editor's introduction," 9.

21 Langbein and Hoyt, *Water Facts*, ix, 3.

22 United States Federal Council for Science and Technology Ad Hoc Panel on Hydrology, *Scientific Hydrology*, 24.

23 Wright, *The Coming Water Famine*, 15.
24 Moss, *The Water Crisis*, 276. For similar expressions of concern, see Nikolaieff, "Editor's introduction," 9.
25 There were important exceptions, and it could be argued that these hinted at a critique of modern water. The geographer Gilbert White, for example, stressed a more flexible approach to predicting future demands for water, one that admitted of alternative possibilities depending on prospective management strategies. In denying the proclamation of a "water crisis," White recognized that the United States was in a "situation [in which] there are major opportunities to improve public management of water by making it less rigid and more responsive to accelerating technological change. The methods of allocating resources can be fashioned so as to increase efficiency by national standards as well as to be more sensitive to human needs for spiritual and aesthetic expression. The most menacing aspect of the prophecy that the United States is running out of water is that it may become self-fulfilling. For the view that water will be in short supply and requires augmentation sets in motion forces that in time worsen the situation. It is more likely that human welfare in the United States will be impaired through degradation of water quality or through inept management than from a physical scarcity of water." G.F. White, *Strategies of American Water Management*, 2. And most insightfully, White concludes: "A vast arena for water management of other types remains available to man on the North American continent." Ibid., 5. White responded in a similar fashion to the rather dire warnings of a generalized water crisis predicted a decade later in the *Global 2000 Report to the President*. A relevant section of this response is worth citing in light of the present critique of the global water crisis: "To learn how adequate the water resource is for a present or future society it is necessary to look into how the physically available *waters* are being used and what physical, social, economic and political constraints apply to further use. In some places the physical limits are severe, and changes in technology, as with central pivot irrigation, can alter the estimates drastically. More frequently, the crucial factors are social, economic and political. The capacity of a society to manage demand, to assure continuity in management policy, and to place values on environmental effects and the maintenance of the quality of surface and ground *waters* is more likely to influence the course of development." G.F. White, "Water Resource Adequacy," 263 (emphasis added to highlight White's use of the term "waters").
26 Bocking, *Canada's Water*, 1.
27 Bradley, "Some Population Limits," 24.
28 Langbein and Hoyt, *Water Facts*, 36-37, chap. 4.
29 Because of their particular expertise, hydrologists may be expected to naturalize modern hydrosocial relations. Thus, their projections of future water demand have tended to be extrapolations of current (contemporary) use. The hegemony of modern water has perhaps led to a presumption that hydrologists are the most appropriate, qualified experts to make this sort of projection. Consideration of water as a hydrosocial phenomenon, by contrast, is likely to stress other forms of expertise and give rise to very different sorts of projections.
30 Raymond Nace quoted in Moss, *The Water Crisis*, 41.
31 Ibid., 36; Joseph L. Myler, a popular American environmental writer and author of the memorably titled article "The Dirty Animal – Man," also draws heavily from Nace's expert commentary to present a picture of water in terms of "the crisis on the horizon." Myler, "The Dirty Animal – Man," 118. Here, the same fixed relation between water use and standard of living is unequivocal: "The high standard of living in the United States and other affluent nations of the modern world depends on fresh water – lots of it." Ibid., 117.

32 The evidence is found in statistics showing reduced water use intensity in several wealthy, industrialized countries (particularly the United States) since the early 1980s. See Gleick, "The Changing Water Paradigm."

33 Fry, *Water*, 11.

34 For now, the economic costs of large-scale international water diversions alone would appear to make them unfeasible. Smaller-scale water transfers are more likely, but one cannot rule out the return of proposals for international diversions. For a recent assessment of the water export threat from a Canadian perspective, see Pentland and Hurley, "Thirsty Neighbours." It might be added that the idea of importing and exporting water (especially in bulk) is absurd unless we cannot see beyond modern water. As with all water issues and problems, what really counts is less the water itself than the way it is brought into relation with people. The notion of bulk trade in water usually rests on the mistaken notion that water problems are *problems with water,* and that, as such, they can be resolved with – or by adding more – water.

35 "America's water crisis" persists in the 21st century, albeit now reflecting 21st-century American concerns such as climate change and growing demands for biofuel production. For example, see Glennon, *Unquenchable: America's Water Crisis and What to Do about It.*

36 Furon, *The Problem of Water.*

37 Meadows et al., *The Limits to Growth,* 62-63.

38 Cited in R. Clarke, *Water,* 19.

39 It is instructive to contrast two monumental studies of the human-nature relationship: *Man's Role in Changing the Face of the Earth* (W.L. Thomas et al.) and *The Earth as Transformed by Human Action* (Turner et al.). In the introduction to *The Earth as Transformed by Human Action,* Kates, Turner, and Clark distinguish their more recent study from the earlier one by emphasizing "transformations of the biosphere at a global scale," noting that "the capability now exists ... to estimate their trajectories over the past several centuries with some degree of confidence; to identify some of the broad, direct processes underlying them; and to understand the interactions of these processes at the regional scale." The epistemological foundation of contemporary global environmental concern is also noted: "In formulating the study we have drawn upon three significant, indeed profound, developments of science – developments that have gained prominence since the *Man's Role* effort: 1) new ways to conceptualize the unity of the biosphere, symbolized by the wide currency of the term itself; 2) new ways and collective efforts to acquire data and analyze their detail and complexity; and 3) reassessment of some of the avenues that link social behavior with environmental transformations." Kates, Turner, and Clark, "The Great Transformation," 2.

40 Quoted in Sachs, *Planet Dialectics,* 43.

41 Falkenmark, "Massive Water Scarcity," 114.

42 Falkenmark, "Approaching the Ultimate Constraint." See also Biswas, "Deafness to Global Water Crisis."

43 Falkenmark, "Massive Water Scarcity," 112; Falkenmark, "Approaching the Ultimate Constraint."

44 Falkenmark, "Massive Water Scarcity," 118.

45 Falkenmark, "Fresh Water."

46 Ibid., 192.

47 Falkenmark, "Stockholm Water Symposium," 216.

48 Postel, *Last Oasis,* 7-8; Saeijs and van Berkel, "Global Water Crisis."

49 Leslie, "Running Dry."
50 In the article "Environmental Refugees: The Origins of a Construct," Patricia Saunders traces the history of the concept of "environmental refugees" through thirty "key documents" that "emerged from a wide-ranging examination of the literature." Saunders, "Environmental Refugees," 219. I have applied a similar method to trace the genealogy of the water crisis.
51 Postel, *Last Oasis.*
52 Gleick, *Water in Crisis.*
53 Postel, *Last Oasis.*
54 Ibid., 18.
55 Ibid., 166.
56 Ibid., 27.
57 Ibid., 28.
58 Information on the total volumes of saltwater and freshwater on earth is from Nace, as cited in van der Leeden, Troise, and Todd, *The Water Encyclopedia.* Nace's work drew from Soviet sources, especially L'vovich. Postel's data on the amount of renewable fresh water flowing through the global hydrologic cycle is from L'vovich, *World Water Resources.* The data on global runoff are from the Institute of Geography, National Academy of Sciences of the Soviet Union, as published in World Resources Institute, *World Resources 1992-1993.* And the "14,000 cubic kilometers as a relatively stable source of supply" is from L'vovich, *World Water Resources* (Postel, *Last Oasis,* 194). Other measures of global water in *Last Oasis* are taken indirectly from Russian and Soviet hydrologists via other secondary sources, for example, Todd, *The Water Encyclopedia;* van der Leeden, *Water Resources;* Spidel and Agnew, "World Water Budget"; L'vovich et al., "Use and Transformation"; van der Leeden, Troise, and Todd, *The Water Encyclopedia.*
59 Postel and others attribute the origin of these indices to Falkenmark, "Massive Water Scarcity." Falkenmark, in turn, relies on L'vovich, *World Water Resources,* for the regional water balance data with which to construct them. Falkenmark, "Massive Water Scarcity."
60 Postel, *Last Oasis,* 28-29.
61 Ibid., 29.
62 For example, the World Water Council points out that the Falkenmark indicator does not account for the temporal variability in water availability or for water use. Cosgrove and Rijsberman, *World Water Vision.* They use instead what is called "the criticality ratio of withdrawals for human use to renewable resources." A critical situation is considered to apply when withdrawals reach 20 percent in basins with highly variable runoff and 60 percent in temperate zone basins. "High water stress" is considered to apply in temperate basins when withdrawals reach 40 percent of renewable water resources. Ibid., 26. There are two implications of applying this method. First, by adding water use to the equation, the number of countries suffering from water stress is increased. Second, by using the figure of 40 percent as the threshold of "high water stress," the indicator is weighted in a fashion that defines more temperate countries as suffering from high water stress. Because the more temperate regions tend to have denser human populations, the World Water Council can conclude: "Under the business as usual scenario, by 2025 about 4 billion people – half the world's population – will live in countries with high water stress." Ibid., 28. The overall effect is to expand the occurrence of "water stress" to areas in temperate regions that otherwise would not be considered. The accompanying map projection and colour coding gives a strong impression that water stress affects, or could affect, places

such as the United States. The political implications of such a projection in terms of garnering support for the use of economic instruments, including privatization of water services, in wealthy markets might be suggested.

63 Postel, *Last Oasis,* 29.

64 Gleick, *Water in Crisis.*

65 In Gilbert White's foreword to the publication, it is described as "a solid, discriminating array of data on the spatial and temporal distribution of the world's fresh water resources. The tables selected for this collection encompass more completely than any other effort to date the basic statistical information available." *Water in Crisis* is cited with great frequency in the water literature of the 1990s and since. See, for example, among numerous others, Patrick, "Is Water Our Next Crisis?"; Ohlsson, *Hydropolitics;* de Villiers, *Water;* Buckley, "The Water Crisis"; UNESCO, "Water for People."

66 These reports have been published every other year from 1998 to the present.

67 Gleick, "An Introduction to Global Fresh Water Issues," 3.

68 Nace, "Editor's Preface," iv.

69 United Nations Commission on Sustainable Development, *Comprehensive Assessment.*

70 De Villiers, *Water,* 30, 364.

71 Shiklomanov's tables depicting global water resources and global water balances are based respectively on data "collected by Soviet scientists" and "based on research conducted by Soviet scientists." Shiklomanov, "World Fresh Water Resources," 13, 15. The data Shiklomanov used in 1993 were not new – they were from sources originally published in the 1960s and 1970s, during the International Hydrological Decade, as described in Chapter 8.

72 Gleick, "Water in the 21st Century," 105.

73 According to Shiklomanov's method, "The values of water use in various large regions of the earth are determined by three main factors: the level of economic development, population, and the geophysical (especially climatic) particularities of the territory in question." Shiklomanov, "World Fresh Water Resources," 19. Estimates for future water demands are based on extrapolation from past trends but also take into account expected improvements in water-use efficiency in irrigation processes. Overall, however: "In considering the dynamics of water use throughout the world, we should note that a continual increase in water withdrawals during this century has been characteristic of all regions, the largest growth occurring in the 1950s and 1960s. A significant increase in water requirements over 1980 levels is also expected through the turn of the century, with the largest increases expected to occur in South America and Africa (95 percent and 70 percent). Decreases are possible in many major industrialized countries." Ibid.

74 Gleick, *Water in Crisis,* 106.

75 Gleick, *The World's Water 2000-2001,* 39.

76 Ibid., 58.

77 Gleick was still describing "water in crisis" in 1998. Gleick, "Water in Crisis." But he has abjured use of the term in all of *The World's Water* series so far. Many, many other writers have not.

78 Gleick, *The World's Water 2000-2001,* 58.

CHAPTER 11: SUSTAINING MODERN WATER

1 Petrella, "Blue Gold of the 21st Century."

2 Cosgrove and Rijsberman, *World Water Vision,* 68.

3 To date, meetings of the World Water Forum have been held in Morocco in 1997, The Hague in 2000, Japan in 2003, Mexico in 2006, and Istanbul in 2009.

4 Conca, *Governing Water,* 1-2.

5 Ibid., 2.

6 Ibid., 4.

7 Cosgrove and Rijsberman, *World Water Vision,* xii.

8 Mahmoud A. Abu-Zeid, Egyptian minister of Water Resources and president of the World Water Council, in foreword to Cosgrove and Rijsberman, *World Water Vision,* iv.

9 Ibid., iv.

10 Gleick, "The Changing Water Paradigm"; Postel, *Last Oasis.* There's no question that it all adds up. As noted above, over the past twenty years or more, certain countries have experienced increased population and economic growth while reducing their total annual volume of water withdrawals. In the United States, to cite the best-documented example, total annual water withdrawals rose by a factor of ten between 1900 and 1980, and then fell by 10 percent between 1980 and 2000. Considering the increase in population from 1980 to 2000, per capita water withdrawals in the United States actually fell by 25 percent from their peak in 1980. Gleick, *The World's Water 2004-2005,* 313. At the same time, since the late 1970s, the economic productivity of water – the ratio of annual economic output (GDP) to annual water withdrawals – has risen substantially in all sectors of the US economy. Ibid., 317, 320. Similar patterns of reduced water use combined with increases in population and economic growth are evident in many parts of the world. Wolff and Gleick, "The Soft Path for Water," 23. The point, however, is to recognize that efficiency is less an object in itself than a by-product of changing hydrosocial relations.

11 The reference is to Francis Fukayama's essay "The End of History," in which he proclaimed the post-Cold War culmination of history in the triumph of capitalism over socialism. Fukuyama, *End of History.*

12 Frederick Newell used the term "hydro-economics" to describe a program for rationalizing the use of water in the early conservation era. For Newell, the term conveyed "the conception of [water's] efficient employment, of utility or of thrift." Newell, *Water Resources,* 31. Evidently, the fundamental idea hasn't changed much over the past century or so, as both old and new versions of hydro-economics are powered by modern water. However, in contrast with the ideal of the *World Water Vision,* Newell's is a more traditional hydro-economics, wherein water was regarded as a national, public good rather than a global commodity.

13 Conca, *Governing Water,* 140; Young, Dooge, and Rodda, *Global Water Resource Issues,* 26-41; Rodda, "Whither World Water?"; Gleick, "The Changing Water Paradigm: A Look."

14 Quoted in Gleick et al., "Globalization," 37.

15 In *Last Oasis,* the obvious – one might say the *natural* – alternative to engineering new water supplies is the procurement of market-based solutions to promote greater water-use efficiency. "Moving toward more efficient, ecologically sound, sustainable patterns of water use," Postel argues, "requires major changes in the way water is valued, allocated, and managed. Appropriate pricing, the creation of markets for buying and selling water, and other economic inducements for wise water use hardly exist in most places. They have a central role to play in the transition to an era of scarcity." Postel, *Last Oasis,* 166. And Gleick, in *Water in Crisis,* signals a transition from common resource to economic good by emphasizing, "water is a scarce resource in many parts of the world, yet it is still often

treated as a free good, available at no charge to whoever can pump it from the ground or remove it from a river or lake ... Until water is priced at its true value, there will be few incentives for wise and efficient water use." Gleick, *Water in Crisis*, 110.

16 Conca, *Governing Water*, 126.
17 Global Water Partnership, "Towards Water Security," 13.
18 Biswas, "Integrated Water Resources Management," 3.
19 Ibid., 6.
20 Quoted in Gleick et al., "Globalization," 38.
21 Quoted in Conca, *Governing Water*, 141.
22 Molle, Mollinga, and Meinzen-Dick, "Water, Politics and Development," 3-4.
23 Conca, *Governing Water*, 127-28.
24 Barlow, *Blue Gold: The Global Water Crisis*, 9.
25 Lamers, "IRN at Ten."
26 Conca, *Governing Water*, 4.

CHAPTER 12: HYDROLECTICS

1 Whitehead, *Nature and Life*, 42.
2 "Neutral stuff" is the term used by Erich Zimmermann to describe material aspects of nature prior to their appraisal as "resources" by people. Zimmermann, *World Resources and Industries*, rev. ed., 11-15.
3 A 2003 survey had found that of 740 First Nations community water systems assessed in Canada, "about 29 percent (218) posed a potential high risk that could negatively impact water quality." Canada, Indian and Northern Affairs Canada, *National Assessment*, i-ii.
4 Linton, *Beneath the Surface*, 108.
5 Canada accounts for about 7 to 9 percent of the total (global) supply of runoff. About 20 percent of the world's water stored in lakes is found in Canada, and over 20 percent of the world's wetlands are found in this country. Pearse, Bertrand, and MacLaren, *Currents of Change*. Much of Canada's water occurs in northern regions where there are relatively few people, and Canadians are by no means immune from serious pressures on water resources, particularly in southern parts of the country. Bakker, *Eau Canada*. Nevertheless, when set against the less than 0.5 percent of the world's human population – as is our modern habit – Canada's water resources must be deemed extremely abundant.
6 One useful way of thinking about such problems is to distinguish between first and second order scarcities. Ohlsson and Turton, *The Turning of a Screw*. First order scarcity describes a shortage of naturally occurring water as, say, measured in terms of water availability (streamflow) per capita. Second order scarcity can be considered the incapacity of a society to either manage first order scarcity or to translate natural water supplies into useful water services. Brooks, "Another Path Not Taken," 53-54. The water problems in Canada's north offers a prime example of second order scarcity in the midst of water plenty.
7 The difficulty entailed in transferring this knowledge and expertise from places where these technologies were developed to northern communities is manifest. A study conducted by the federal Department of Indian and Northern Affairs in 2003 found that only about 10 percent of the operators working in First Nations water and wastewater treatment facilities were trained to meet industry certification requirements. Canada, Indian and Northern Affairs Canada, *National Assessment*, 9. The importance of embedding water technologies in the fabric of local culture is the main theme of Brooks, *Water*.

8 Swain, *Report of the Expert Panel,* 15-16.
9 For a federal government assessment of the nature and scope of the problem, as well as recommendations on how to address it, see Canada, Indian and Northern Affairs Canada, *National Assessment.* In the spring of 2003, the federal government announced $600 million in new funding targeted toward improving the quality of water and wastewater treatment in First Nations communities across Canada. Ibid.
10 Falkenmark, *Towards Hydrosolidarity,* 5.
11 Brooks, *Water,* xi-xii.
12 For example, Brichieri-Colombi, "Hydrocentricity."
13 Gibbs, "Valuing Water," 77. Gibbs has produced a study of the social nature of waters at different phases of the hydrologic cycle: "River water, rain water, bore water, all have different values, and at a more subtle level, in-channel flow and overland flow; soaking rain, light steady rain and follow-up rain; Mungerannie bore water and South Galway bore water, all have different values ... This level of description of water in the Lake Eyre Basin reveals connections between water, place, temporality, scale, change, difference and the intimate locally specific knowledge of water and its effects held by people in the Basin." Ibid.
14 Hayles, "Searching for Common Ground," 56.
15 The interrelationship between the circulation of water and the circulation of capital and power is a theme that has been developed in studies of the political ecology of water. The rudiments of the concept of the "hydrosocial cycle" have emerged from the development of this theme, especially in the work of Erik Swyngedouw, Karen Bakker, Maria Kaika, and Jessica Budds (see the bibliography for references to their work), among others.
16 Swyngedouw, *Social Power,* 2.
17 Ibid., 4.
18 Linton, *Beneath the Surface.*
19 Hanson, "As Pure as Snow."
20 Tvedt and Jakobsson, *A History of Water,* vol. 1, *Water Control and River Biographies,* xiv.
21 Horton, "Field, Scope, and Status," 192.
22 Vörösmarty et al., "Humans Transforming the Global Water System," 509, 514.
23 Ibid., 514.
24 Maidment, *Handbook of Hydrology,* 3.
25 Swyngedouw, *Social Power,* 4.
26 For an analysis along these lines of the dereliction of public drinking fountains on university campuses in Canada, see Girard and Shaker, "Bottled Up or Tapped Out."
27 Budds and Linton, "Water, Science, Humans."
28 Conca, *Governing Water,* 158.
29 Lalonde, "A Beach Plan Worth Diving Into."
30 Press, "Group Seeks Beach Cleanup"; Switzer, "Swimmers Head Back"; Hsieh, "Hundreds Take Back the Beach."
31 Bakker, "Political Ecology."
32 D. Hall, "Introduction," 20.
33 Balanyá et al., *Reclaiming Public Water.*
34 Balanyá et al., "Empowering Public Water," 247.
35 Bakker, "The Ambiguity," 247.
36 Ibid., 247, 249, 254.
37 Brooks, *Water,* 5-6.
38 Ibid., 7.

39   Ibid., 17.
40   Ibid., 10-12.
41   Lovins, *Soft Energy Paths;* Brooks, "Another Path"; Brooks, "Beyond Greater Efficiency"; Brandes and Brooks, "The Soft Path"; Gleick, "Global Freshwater Resources."
42   Brooks, "Another Path," 11.
43   Feitelson, "Political Economy," 418, 413.
44   Ibid., 415.
45   Ibid., 415, 420.
46   Ibid., 413.
47   Ibid., 414.
48   Ibid., 420.
49   Ibid., 420-21. "Affected parties" include farmers, urban water users, industry, the Nature and Parks authority, the tourism, health, and environmental ministries, as well as environmental NGOs. Ibid., 422n13.
50   Ibid., 421.
51   Lonergan and Brooks, *Watershed,* 3.
52   Braun, *The Intemperate Rainforest,* 4-5.
53   Ibid., 6.

Bibliography

Adams, F.D. *The Birth and Development of the Geological Sciences*. Baltimore: Williams and Wilkins, 1938.

–. "The Origin of Springs and Rivers: An Historical Review." *Fennia* 50, 1 (1928): 3-18.

Adams, R. "Historic Patterns of Mesopotamian Irrigation Agriculture." In *Irrigation's Impact on Society*, ed. T.E. Downing and M. Gibson, 1-6. Tucson: University of Arizona Press, 1974.

Adams, W.M. "Irrigation, Erosion and Famine: Visions of Environmental Change in Marakwet, Kenya." In *The Lie of the Land: Challenging Received Wisdom on the African Environment*, ed. M. Leach and R. Mearns, 155-67. Oxford: International African Institute in association with James Currey and Heinemann, 1996.

Aguilera-Klink, F., Eduardo Perez-Moriana, and Juan Sanches-Garcia. "The Social Construction of Scarcity: The Case of Water in Tenerife (Canary Islands)." *Ecological Economics* 34, 2 (2000): 233-45.

Aldhous, P. "The World's Forgotten Crisis." *Nature* 422 (2003): 251.

Allan, J.A. "IWRM/IWRAM: A New Sanctioned Discourse?" London: Occasional Paper 50, School of Oriental and African Studies, King's College, University of London, 2003.

Allardyce, G. "'The Vexed Question of Sawdust': River Pollution in 19th Century New Brunswick." *Dalhousie Review* 52 (1972): 177-90.

American Geophysical Union (AGU). *International Union of Geodesy and Geophysics (IUGG)*. http://www.agu.org/iugg/internat.html.

American Society of Civil Engineers, Hydrology Committee. *Hydrology Handbook*. New York: American Society of Civil Engineers, 1949.

Arnold, D., and R. Guha. "Introduction: Themes and Issues in the Environmental History of South Asia." In *Nature, Culture and Imperialism: Essays on the Environmental History of South Asia*, ed. D. Arnold and R. Guha, 1-20. Delhi: Oxford University Press, 1995.

Association of American Geographers. *2007 Annual Meeting Program – April 17-21, San Francisco, California*. Washington: The Association of American Geographers, 2007.

Bachelard, G. *Water and Dreams: An Essay on the Imagination of Water*. Dallas: Dallas Institute of Humanities and Culture, 1983.

Back, W. "Foreword." In *The History of Hydrology*, ed. E.R. Landa and S. Ince, ix-x. Washington, DC: American Geophysical Union, 1987.

Baker, M.N. *The Quest for Pure Water: The History of Water Purification from the Earliest Records to the Twentieth Century*. New York: American Water Works Association, 1949.

Baker, M.N., and R.E. Horton. "Historical Development of Ideas Regarding the Origin of Springs and Groundwater." *Transactions, American Geophysical Union* 17 (1936): 395-400.

Bakker, K., ed. "The Ambiguity of Community: Debating Alternatives to Private-Sector Provision of Urban Water Supply." *Water Alternatives* 1, 2 (2008): 236-52.

–. *Eau Canada: The Future of Canada's Water*. Vancouver: UBC Press, 2007.

–. "A Political Ecology of Water Privatization." *Studies in Political Economy* 70 (Spring 2003): 35-58.

–. "Privatizing Water, Producing Scarcity: The Yorkshire Drought of 1995." *Economic Geography* 76, 1 (1999): 4-27.

–. *An Uncooperative Commodity: Privatizing Water in England and Wales*. Oxford: Oxford University Press, 2003.

Balanyá, B., B. Brennan, O. Hoedman, S. Kishimoto, and P. Terhorst. "Empowering Public Water – Ways Forward." In *Reclaiming Public Water: Achievements, Struggles and Visions from around the World*, ed. Belén Balanyá, Brid Brennan, Olivier Hoedman, Satoko Kishimoto, and Philipp Terhorst, 247-74. Amsterdam: Transnational Institute and Corporate Europe Observatory, 2005.

–, eds. *Reclaiming Public Water: Achievements, Struggles and Visions from around the World*. Amsterdam: Transnational Institute and Corporate Europe Observatory, 2005.

Ball, P. *Life's Matrix: A Biography of Water*. Berkeley: University of California Press, 2001.

Barlow, M. *Blue Gold: The Global Water Crisis and the Commodification of the World's Water Supply*. Rev. ed. San Francisco: International Forum on Globalization, 2001.

Barlow, M., and T. Clarke. *Blue Gold: The Battle against Corporate Theft of the World's Water*. Toronto: Stoddart, 2002.

Barry, R.G. "The World Hydrological Cycle." In *Introduction to Physical Hydrology*, ed. R.J. Chorley, 8-26. London: Methuen, 1969.

Bassin, M. "History and Philosophy of Geography." *Progress in Human Geography* 21, 4 (1997): 563-72.

Baumgartner, A., and E. Reichel. "Preliminary Results of New Investigations of World's Water Balance." In *World Water Balance: Proceedings of the Reading Symposium, July 1970*, ed. IASH-UNESCO-WMO, 580-92. Gentbrugge, Belgium: IASH-UNESCO-WMO, 1972.

Baviskar, A. *In the Belly of the River: Tribal Conflicts over Development in the Narmada Valley*. Delhi: Oxford University Press, 1995.

–. "Written on the Body, Written on the Land: Violence and Environmental Struggles in Central India." In *Violent Environments*, ed. N. Peluso and M. Watts, 354-79. Ithaca, NY: Cornell University Press, 2001.

Beardmore, N. *Manual of Hydrology*. London: Waterlow, 1862.

Bedford, D. "The Great Salt Lake as America's Aral Sea?" *Environment Magazine*, September/October 2009. http://www.environmentmagazine.org.

Bell, M. "The Water Decade Valedictory, New Delhi 1990: Where Pre- and Post-Modernism Meet." *Area* 24, 1 (1992): 82-89.

Benedickson, J. *The Culture of Flushing: A Social and Legal History of Sewage.* Vancouver: UBC Press, 2007.

Bennett, J.W. "Anthropological Contributions to the Cultural Ecology and Management of Water Resources." In *Man and Water*, ed. L.D. James, 34-81. Lexington: University Press of Kentucky, 1974.

Berkes, F. *Sacred Ecology: Traditional Ecological Knowledge and Resource Management.* Philadelphia: Taylor and Francis, 1999.

Beven, K. "Some Reflections on the Future of Hydrology." In *Water for the Future: Hydrology in Perspective (Proceedings of the Rome Symposium, April 1987)*, ed. J.C. Rodda and N.C. Matalas, 393-403. IASH publication no. 164. Wallingford, UK: International Association of Hydrological Sciences, 1987.

Bhaskar, R. "Dialectics." In *A Dictionary of Marxist Thought*, ed. T. Bottomore, L. Harris, V.G. Kiernan, and R. Miliband, 143-50. Oxford and Malden, MA: Blackwell, 1999.

Biagioli, M. "Introduction." In *The Science Studies Reader*, ed. M. Biagioli, xi-xviii. New York: Routledge, 1999.

Biro, A. "Wet Dreams: Ideology and the Debates over Canadian Water Exports." *Capitalism, Nature, Socialism* 13, 4 (2002): 29-50.

Biswas, A. K. "The Automatic Rain-Gauge of Christopher Wren, F.R.S." *Notes and Records of the Royal Society of London* 22 (1967): 94-104.

—. "Deafness to Global Water Crisis: Causes and Risks." *Ambio* 27, 6 (1998): 492-93.

—. *History of Hydrology.* Amsterdam: North-Holland Publishing and American Elsevier Publishing, 1970.

—. "The Hydrologic Cycle." *Civil Engineering: ASCE* 35, 4 (1965): 70-74.

—. "Integrated Water Resources Management: A Reassessment." *Water International* 29, 2 (2004): 248-56.

Blackman, D.R., and A.T. Hodge. *Frontinus' Legacy: Essays on Frontinus' 'de aquis urbis Romae.'* Ann Arbor: University of Michigan Press, 2001.

Blackwelder, B. Foreword. In *The Social and Environmental Effects of Large Dams* by E. Goldsmith and N. Hildyard. Bordeaux, France: European Ecological Action Group.

Blaikie, P., T. Cannon, I. Davis, and B. Wisner. *At Risk: Natural Hazards, People's Vulnerability, and Disasters.* New York: Routledge, 1994.

Boberg, J. *Liquid Assets: How Demographic Changes and Water Management Policies Affect Freshwater Resources.* Santa Monica: Rand Corporation, 2005.

Bocking, R.C. *Canada's Water: For Sale?* Toronto: James Lewis and Samuel, 1972.

Bolling, D.M. *How to Save a River: A Handbook for Citizen Action.* Washington, DC: Island Press, 1994.

Bond, P. *Valuing Water beyond "Just Price It": Costs and Benefits of Water for People and Nature.* Paper presented at the Water for People and Nature conference, organized by the Council of Canadians, Vancouver, 7 July 2001.

Bord, J., and C. Bord. *Sacred Waters: Holy Wells and Water Lore in Britain and Ireland.* London: Granada, 1985.

Bowden, C. *Killing the Hidden Waters.* Austin: University of Texas Press, 1977.

Bradley, C.C. "Some Population Limits." In *The Water Crisis*, ed. G.A. Nikolaieff, 23-24. New York: H.W. Wilson, 1967.

Bradnock, R.W., and P.L. Saunders. "Sea-Level Rise, Subsidence and Submergence: The Political Ecology of Environmental Change in the Bengal Delta." In *Political Ecology: Science, Myth and Power,* ed. P. Stott and S. Sullivan, 66-90. London: Arnold, 2000.

Brandes, O.M., and D.B. Brooks. "The Soft Path for Water in a Nutshell." Ottawa and Victoria: Friends of the Earth Canada and the POLIS Project on Ecological Governance, 2005.

Bras, R., and P.S. Eagleson. "Hydrology, the Forgotten Earth Science." *Eos* 68, 16 (1987): 227.

Braun, B. "Environmental Issues: Writing a More-Than-Human Urban Geography." *Progress in Human Geography* 29, 5 (2005): 635-50.

–. *The Intemperate Rainforest: Nature, Culture, and Power on Canada's West Coast.* Minneapolis: University of Minnesota Press, 2002.

–. "Nature and Culture: On the Career of a False Problem." In *A Companion to Cultural Geography,* ed. James S. Duncan, Nuala. C. Johnson, and Richard. H. Schein, 151-79. Malden, MA: Blackwell, 2004.

–. "Towards a New Earth and a New Humanity: Nature, Ontology, Politics." In *David Harvey: A Critical Reader,* ed. Noel Castree and Derek Gregory, 191-222. Malden, MA: Blackwell, 2006.

Braun, B., and N. Castree. "Foreword." In *Remaking Reality: Nature at the Millennium,* ed. Bruce Braun and Noel Castree, xi-xii. London: Routledge, 1998.

–. "Preface." In *Social Nature: Theory, Practice and Politics,* ed. Noel Castree and Bruce Braun. Malden, MA: Blackwell, 2001.

–, eds. *Remaking Reality: Nature at the Millennium.* London: Routledge, 1998.

Brichieri-Colombi, J.S. "Hydrocentricity: A Limited Approach to Achieving Food and Water Security." *Water International* 29, 3 (2004): 318-28.

Brooks, D.B. "Another Path Not Taken: A Methodological Exploration of Water Soft Paths for Canada and Elsewhere." Unpublished Report submitted to Environment Canada. Ottawa, 2003.

–. "Beyond Greater Efficiency: The Concept of Water Soft Paths." *Canadian Water Resources Journal* 30, 1 (2005): 83-92.

–. *Water: Local-Level Management.* Ottawa: International Development Research Institute, 2002.

Brown, L.R., G. Gardner, and B. Halweil. "Beyond Malthus: Sixteen Dimensions of the Population Problem." In *Worldwatch Paper #163.* Worldwatch Institute, 1998.

Brown, R.M. "Man and Water Supply." *Journal of Geography* 32, 6 (1933): 250-53.

Bryant, R.L. "Putting Politics First: The Political Ecology of Sustainable Development." *Global Ecology and Biogeography Letters* 1 (1991): 164-66.

Buckley, R. *The Water Crisis: A Matter of Life and Death.* Cheltenham, UK: Understanding Global Issues, 2001.

Budds, J. "Contested H$_2$O: Science, Policy and Politics in Water Resources Management in Chile." *Geoforum* 40, 3 (2009): 418-30.

–. "Whose Scarcity? The Hydrosocial Cycle and the Changing Waterscape of La Ligua River Basin, Chile." In *Contentious Geographies: Environment, Meaning, Scale,* ed. M. Goodman, M. Boykoff, and K. Evered, 59-68. Aldershot, UK: Ashgate, 2008.

Budds, J., and J. Linton. 2009. "Water, Science, Humans: Advancing the Hydrosocial Cycle." Call for papers, issued prior to the annual meeting of the Association of American Geographers, Las Vegas, March 2009.

Buttel, F., and P. Taylor. "Environmental Sociology and Global Environmental Change." In *Social Theory and the Global Environment*, ed. M. Redclift and T. Benton, 228-55. New York: Routledge, 1994.

Calder, I.R. *The Blue Revolution: Land Use and Integrated Water Resources Management.* 2nd ed. London: Earthscan, 2005.

Canada. Indian and Northern Affairs Canada. "First Nations Water Top Priority for Government of Canada." Press release, 14 May 2003.

–. *National Assessment of Water and Wastewater Systems in First Nations Communities: Summary Report.* Ottawa: Indian and Northern Affairs Canada, 2003.

Carbonnel, J.P. "Introduction." In Perrault, P. *De l'origine des fontaines (Textes fondateurs de l'hydrologie N. 2).* 1674. Reprint, Asnières, France: Association Internationale des Sciences Hydrologiques, Comité National Français des Sciences Hydrologiques, 2001.

Carlson, H.M. *Home Is the Hunter: The James Bay Cree and Their Land.* Vancouver: UBC Press, 2008.

Castree, N. "Birds, Mice and Geography: Marxisms and Dialectics." *Transactions of the Institute of British Geography,* n.s., 21 (1996): 342-62.

–. "Environmental Issues: Relational Ontologies and Hybrid Politics." *Progress in Human Geography* 27, 2 (2003): 203-11.

–. "Geographies of Nature in the Making." In *Handbook of Cultural Geography,* ed. K. Anderson, M. Domosh, S. Pile, and N. Thrift, 168-83. London: Sage Publications, 2002.

–. "Marxism and the Production of Nature." *Capital and Class* 72 (2000): 5-36.

–. *Nature.* London: Routledge, 2005.

–. "The Nature of Produced Nature: Materiality and Knowledge Construction in Marxism." *Antipode* 27, 1 (1995): 12-48.

–. "Socializing Nature: Theory, Practice and Politics." In *Social Nature: Theory, Practice and Politics,* ed. Noel Castree and Bruce Braun, 1-21. Malden, MA: Blackwell, 2001.

Castree, N., and B. Braun. "The Construction of Nature and the Nature of Construction." In *Remaking Reality: Nature at the Millennium,* ed. Bruce Braun and Noel Castree, 3-42. London: Routledge, 1998.

–, eds. *Social Nature: Theory, Practice and Politics.* Malden, MA: Blackwell, 2001.

Castree, N., and D. Gregory, eds. *David Harvey: A Critical Reader.* Malden, MA: Blackwell, 2006.

Cayley, D. *Ivan Illich in Conversation.* Concorde, ON: House of Anansi, 1992.

Chorley, R.J., ed. *Water, Earth, and Man: A Synthesis of Hydrology, Geomorphology, and Socio-Economic Geography.* London: Methuen, 1969.

Chorley, R.J., and R.W. Kates. "Introduction." In *Water, Earth, and Man: A Synthesis of Hydrology, Geomorphology, and Socio-Economic Geography,* ed. R.J. Chorley, 1-7. London: Methuen, 1969.

Chow, V.T. *Handbook of Applied Hydrology: A Compendium of Water-Resources Technology.* New York: McGraw-Hill, 1964.

–. "Water as a World Resource." *Water International* 4, 1 (1979): 3-24.

Chow, V.T., D.R. Maidment, and L.W. Mays. *Applied Hydrology.* New York: McGraw-Hill, 1988.

Clarke, R. *Water: The International Crisis.* London: Earthscan, 1991.

Clarke, R., and J. King. *The Water Atlas.* New York: New Press, 2004.

Clarke, T. *Inside the Bottle: Exposing the Bottled Water Industry.* Rev. ed. Ottawa: Polaris Institute, 2007.

Collingwood, R.G. *The Idea of History.* Oxford: Clarendon Press, 1946.
–. *The Idea of Nature.* Oxford: Oxford University Press, 1945.
Conable, B. "Address to the World Resources Institute." Washington, DC: The World Bank, 1987. Quoted in A. Escobar. "Construction Nature." *Futures* 28, 4 (1996): 325-343.
Conca, K. *Governing Water: Contentious Transnational Politics and Global Institution Building.* Cambridge, MA: MIT Press, 2006.
Coopey, R., and T. Tvedt, eds. *A History of Water.* Vol. 2, *The Political Economy of Water.* London: I.B. Tauris, 2006.
Cosgrove, D. "An Elemental Division: Water Control and Engineered Landscape." In *Water, Engineering and Landscape: Water Control and Landscape Transformation in the Modern Period,* ed. D. Cosgrove and G. Petts, 1-11. London: Bellhaven Press, 1990.
Cosgrove, D., and G. Petts. "Preface." In *Water, Engineering and Landscape: Water Control and Landscape Transformation in the Modern Period,* eds. D. Cosgrove and G. Petts, i-xv. London: Bellhaven Press, 1990.
–, eds. *Water, Engineering and Landscape: Water Control and Landscape Transformation in the Modern Period.* London: Bellhaven Press, 1990.
Cosgrove, W.J., and F.R. Rijsberman. *World Water Vision: Making Water Everybody's Business.* London: Earthscan, 2000.
Crifasi, R.R. "Reflections in a Stock Pond: Are Anthropogenetically Derived Freshwater Ecosystems Natural, Artificial, or Something Else?" *Environmental Management* 6, 5 (2005): 625-39.
Cronon, W. *Changes in the Land: Indians, Colonists, and the Ecology of New England.* New York: Hill and Wang, 1983.
–. "Introduction: In Search of Nature." In *Uncommon Ground: Toward Reinventing Nature,* ed. William Cronon, 23-56. New York: W.W. Norton, 1995.
–. *Nature's Metropolis: Chicago and the New West.* New York: W.W. Norton, 1991.
–. "Toward a Conclusion." In *Uncommon Ground: Toward Reinventing Nature,* ed. William Cronon, 447-59. New York: W.W. Norton, 1995.
–, ed. *Uncommon Ground: Toward Reinventing Nature.* New York: W.W. Norton, 1995.
Crosby, A.W. *The Measure of Reality: Quantification and Western Society, 1250-1600.* Cambridge: Cambridge University Press, 1997.
Dalby, S. *Environmental Security.* Minneapolis: University of Minnesota Press, 2002.
Daly, H., and J.B. Cobb Jr.. *For the Common Good: Redirecting the Economy toward Community, the Environment and a Sustainable Future.* Boston: Beacon Press, 1989.
Davis, W.M. "The Geographical Cycle." In *Geographical Essays (an unabridged republication of the 1909 edition),* 249-78. Reprint, New York: Dover Publications, 1954.
Day, J.C., and F. Quinn. *Water Diversion and Export: Learning from Canadian Experience.* Waterloo, ON: Department of Geography, University of Waterloo, 1992.
De Villiers, M. *Water: The Fate of Our Most Precious Resource.* Toronto: Stoddart,1999.
–. "Water Works." *Canadian Geographic* 120, 4 (May-June 2000): 50-58.
Dear, P. *Discipline and Experience: The Mathematical Way in the Scientific Revolution.* Chicago: University of Chicago Press, 1995.
–. *The Intelligibility of Nature: How Science Makes Sense of the World.* Chicago: Chicago University Press, 2006.
–. *Revolutionizing the Sciences: European Knowledge and Its Ambition, 1500-1700.* Princeton, NJ: University of Princeton Press, 2001.

Demeritt, D. "Being Constructive about Nature." In *Social Nature: Theory, Practice and Politics*, ed. Noel Castree and Bruce Braun, 22-40. Malden, MA: Blackwell, 2001.

–. "The Construction of Global Warming and the Politics of Science." *Annals of the Association of American Geographers* 91, 2 (2001): 307-37.

–. "Ecology, Objectivity and Critique in Writings on Nature and Human Societies." *Journal of Historical Geography* 20, 1 (1994): 22-37.

–. "Hybrid Geographies, Relational Ontologies and Situated Knowledges." Contribution to a review symposium on Sarah Whatmore's *Hybrid Geographies. Antipode* 37, 4 (2005): 818-23.

–. "What Is the 'Social Construction of Nature'? A Typology and Sympathetic Critique." *Progress in Human Geography* 26, 6 (2002): 767-90.

Derman, B. "Balancing the Waters: Development and Hydropolitics in Contemporary Zimbabwe." In *Water, Culture, and Power: Local Struggles in a Global Context*, ed. J.M. Donahue and B.R. Johnston, 73-93. Washington, DC: Island Press, 1998.

Desbiens, C. "A Political Geography of Hydro Development in Quebec." *Canadian Geographer* 48, 2 (2004): 101-18.

Descartes, R. *The Essential Descartes*. Edited and with an introduction by Margaret D. Wilson. New York: New American Library, 1969.

Detay, M., and D. Gaujous. "De la cosmologie au cycle de l'eau." In *Colloque International OH₂, Dijon, 9-11 mai 2001*. Dijon: Université de Bourgogne, 2003.

Diamond, J. *Guns, Germs and Steel: The Fates of Human Societies*. New York: W.W. Norton, 1999.

DiFrancesco, K. "The Debate over Large Dams." In *Global Perspectives on Large Dams: Evaluating the State of Large Dam Construction and Decommissioning across the World*. Report on a conference held 3-5 November 2006 at the Yale School of Forestry and Environmental Studies. Yale School of Forestry and Environmental Studies Report #13, ed. K. DiFrancesco and K. Woodruff, 7-14. New Haven, CT: Yale School of Forestry and Environmental Studies, 2008.

Donahue, J.M., and B.R. Johnston. "Conclusion." In *Water, Culture, and Power: Local Struggles in a Global Context*, ed. J.M. Donahue and B.R. Johnston, 339-46. Washington, DC: Island Press, 1998.

Dooge, J.C. "Background to Modern Hydrology." In *The Basis of Civilization: Water Science?* ed. J.C. Rodda and L. Ubertini, 3-12. Wallingford, UK: International Association of Hydrologic Sciences Press, 2004.

–. "Concepts of the Hydrologic Cycle, Ancient and Modern." In *Colloque International OH₂, Dijon, 9-11 mai 2001*. Dijon: Université de Bourgogne, 2003.

–. "The Development of Hydrological Concepts in Britain and Ireland between 1674 and 1874." *Hydrological Sciences Bulletin* 19, 9 (1974): 279-302.

–. "Hydrology in Perspective." *Hydrological Sciences Journal* 33, 1/2 (1988): 61-85.

–. "Looking for Hydrologic Laws." *Water Resources Research* 22, 9 (1986): 46s-58s.

–. "On the Study of Water." *Hydrological Sciences Journal* 28, 1 (1983): 23-48.

–. "The Waters of the Earth." *Hydrological Sciences Journal* 29, 2 (1984): 149-76.

Dooge, J.C., A.B. Costin, and L.H.J. Finkel. "Man's Influence on the Hydrological Cycle." *Irrigation and Drainage Paper No. 17*. Rome: FAO, 1973.

Douglas, M. *Purity and Danger: An Analysis of Concepts of Pollution and Taboo*. New York: Praeger, 1966.

Downing, T.E., and K. Bakker. "Drought Discourse and Vulnerability." In *Hazards and Disasters: A Series of Definitive Major Works,* ed. D.A. Wilhite, 213-30. London: Routledge, 1997.

Downing, T.E., and M. Gibson, eds. *Irrigation's Impact on Society.* Tucson: University of Arizona Press, 1974.

Duke, D.F. "Seizing Favours from Nature: The Rise and Fall of Siberian River Diversion." In *A History of Water.* Vol. 1, *Water Control and River Biographies,* ed. T. Tvedt and E. Jakobsson, 3-34. London: I.B. Tauris, 2006.

Eagleson, P.S. *Dynamic Hydrology.* New York: McGraw-Hill, 1970.

Economist.com. "Sina Aqua Non: Dams are Making a Comeback" *The Economist,* 21 March 2009. http://www.economist.com.

Edmunds, Wm. "Bath Thermal Waters: 400 Years in the History of Geochemistry and Hydrogeology." In *200 Years of British Hydrogeology,* ed. J.D. Mather, 193-99. London: The Geological Society, 2004.

Endter-Wada, J., J. Kurtzman, S.P. Keenan, R.K. Kjelgren, and C.M.U. Neale. "Situational Waste in Landscape Watering: Residential and Business Water Use in an Urban Utah Community." *Journal of the American Water Resources Association* 44, 4 (2008): 902-20.

Eliade, M. *Patterns in Comparative Religion.* London: Sheed and Ward, 1958.

Emel, J., and J. Wolch. "Witnessing the Animal Moment." In *Animal Geographies: Place, Politics, and Identity in the Nature-Culture Borderlands,* ed. Jennifer Wolch and Jody Emel, 1-24. London: Verso, 1998.

Environment Canada. *Water – Here, There and Everywhere.* Freshwater Series A-2. Ottawa: Minister of Supply and Services, 1992.

Escobar, A. "After Nature: Steps to an Antiessentialist Political Ecology." *Current Anthropology* 40, 1 (1999): 1-16.

–. "Constructing Nature: Elements for a Poststructural Political Ecology." In *Liberation Ecologies: Environment, Development, Social Movements,* ed. R. Peet and M. Watts, 46-68. London: Routledge, 1996.

–. "Culture Sits in Places: Reflections on Globalism and Subaltern Strategies of Localization." *Political Geography* 20 (2001): 139-74.

–. *Encountering Development: The Making and Unmaking of the Third World.* Princeton, NJ: Princeton University Press, 1995.

Ettenger, K. "'A River That Was Once So Strong and Deep': Local Reflections on the Eastmain Diversion, James Bay Hydroelectric Project." In *Water, Culture, and Power: Local Struggles in a Global Context,* ed. J.M. Donahue and B.R. Johnston, 47-71. Washington, DC: Island Press, 1998.

European Geosciences Union. "Awards and Medals: John Dalton." http://www.egu.eu/awards-medals/award/john_dalton/patron/15.html.

Evenden, M. *Fish versus Power: An Environmental History of the Fraser River.* Cambridge: Cambridge University Press, 2004.

Evernden, N. *The Social Creation of Nature.* Baltimore: Johns Hopkins University Press, 1992.

Fairhead, J., and M. Leach. *Misreading the African Landscape.* Cambridge: University of Cambridge Press, 1996.

Falkenmark, M. "Approaching the Ultimate Constraint: Water Shortage in the Third World." In *Resources and Population: National, Institutional and Demographic Dimensions of Development,* ed. B. Colombo, P. Demeny, and M.F. Perutz, 71-81. Oxford: Clarendon Press, 1996.

–. "Fresh Water: Time for a New Approach." *Ambio* 15, 4 (1986): 192-200.

–. "The Massive Water Scarcity Now Threatening Africa: Why Isn't It Being Addressed?" *Ambio* 18, 2 (1989): 112-18.

–. "Stockholm Water Symposium 1995: Human Dimensions of the Water Crisis." *Ambio* 25, 3 (1996): 216.

–. *Towards Hydrosolidarity: Ample Opportunities for Human Ingenuity (Fifteen-Year Message from the Stockholm Water Symposia).* Stockholm: Stockholm International Water Institute, 2005.

Falkenmark, M., and J. Rockstrom. *Balancing Water for Humans and Nature: A New Approach to Ecohydrology.* London: Earthscan, 2004.

Fasso, C. "Birth of Hydraulics during the Renaissance Period." In *Hydraulics and Hydraulic Research: A Historical Review,* ed. G. Garbrecht, 55-79. Rotterdam: A.A. Balkema, 1987.

Feitelson, E. "Political Economy of Groundwater Exploitation: The Israeli Case." *Water Resources Development* 21, 3 (2005): 413-23.

Ferrari-Comeau, L., and E.L. Chalecki. "International Relations Theory and the Commodification of Canadian Water." Paper presented at the annual meeting of the International Studies Association, New Orleans, 24-27 March 2002.

Ferroukhi, L., and S. Chokkakula. "Indigenous Knowledge of Water Management." In *Reaching the Unreached: Challenges for the 21st Century – Selected papers of the 22nd WEDC Conference, New Delhi, India, 1996,* ed. J. Pickford, 91-95. Loughborough, UK: Intermediate Technology Publications in association with The Water, Engineering and Development Centre, 1996.

Figuertes, C.M., C. Tortajada, and J. Rockstrom. "Conclusion: The Way Forward." In *Rethinking Water Management: Innovative Approaches to Contemporary Issues,* ed. C.M. Figuertes, C. Tortajada, and J. Rockstrom, 228-36. London: Earthscan, 2003.

Finnegan, W. "Letter from Bolivia: Leasing the Rain." *New Yorker,* 8 April 2002, 43-53.

Fishman, C. "Message in a Bottle." *Fast Company,* 19 December 2007. http://www.fastcompany.com.

Fitzsimmons, M. "Engaging Ecologies." In *Envisioning Human Geographies,* ed. P. Cloke, P. Crang, and M. Goodwin, 30-47. New York: Arnold Publishers, 2004.

–. "The Matter of Nature." *Antipode* 21, 2 (1989): 106-20.

Forsyth, T. *Critical Political Ecology: The Politics of Environmental Science.* London: Routledge, 2003.

Foster, J.B. *Marx's Ecology: Materialism and Nature.* New York: Monthly Review Press, 2000.

Foucault, M. *The Order of Things: An Archaeology of the Human Sciences.* London: Routledge, 2002.

–. *"Society Must Be Defended": Lectures at the College de France, 1975-76.* New York: Picador, 2003.

Fournier, F. "Preface." Paper presented at the International Symposium on the Effects of Man on the Interface of the Hydrological Cycle with the Physical Environment, Paris. IASH publication no. 113. Washington: IASH-UNESCO-WMO, 1974.

Francis, Chief K. "First They Came and Took Our Trees, Now They Want Our Water Too." In *Water Export,* ed. J.E. Windsor, 93-102. Cambridge, ON: Canadian Water Resources Association, 1992.

Frontinus, Sextus Julius. *Stratagems and the Aqueducts of Rome,* trans. Charles Bennett. Cambridge, MA: Harvard University Press, 1961.

Fry, A. *Water: Facts and Trends.* Geneva: World Business Council on Sustainable Development, 2005.

Fukuyama, F. *The End of History and the Last Man.* New York: Free Press; Toronto: Maxwell Macmillan Canada, 1992.

Furon, R. *The Problem of Water: A World Study,* trans. Paul Barnes, London: Faber and Faber, 1967.

Gandy, M. *Concrete and Clay: Reworking Nature in New York City.* Cambridge, MA: MIT Press, 2002.

–. "Cyborg Urbanization: Complexity and Monstrosity in the Contemporary City." *International Journal of Urban and Regional Research* 29, 1 (2005): 26-49.

–. "Editorial: Water and Landscape." *Landscape Research* 31, 2 (2006): 117-19.

–. "The Making of Metropolitan Nature: A Reply to Heiman, Lake and Mitchell." *Antipode* 35, 5 (2003): 1022-28.

–. "The Paris Sewers and the Rationalization of Urban Space." *Transactions of the Institute of British Geographers* 24, 1 (1999): 23-44.

Garbrecht, G., ed. *Hydraulics and Hydraulic Research: A Historical Review.* Rotterdam: A.A. Balkema, 1987.

–. "Hydrologic and Hydraulic Concepts in Antiquity." In *Hydraulics and Hydraulic Research: A Historical Review,* ed. G. Garbrecht, 1-22. Rotterdam: A.A. Balkema, 1987.

Gare, A.E. *Postmodernism and the Environmental Crisis.* London: Routledge, 1995.

Geertz, C. "Afterword." In *Senses of Place,* ed. S. Feld and K.H. Basso, 259-62. Santa Fe, NM: School of American Research Press, 1996.

–. "The Wet and the Dry: Traditional Irrigation in Bali and Morocco." *Human Ecology* 1, 1 (1972): 23-39.

Gibbs, L.M. "Valuing Water: Variability and the Lake Eyre Basin, Central Australia." *Australian Geographer* 37, 1 (2006): 73-85.

Giblett, R. *Postmodern Wetlands: Culture, History, Ecology.* Edinburgh: Edinburgh University Press, 1996.

Gilmartin, D. "Models of the Hydraulic Environment: Colonial Irrigation, State Power and Community in the Indus Basin." In *Nature, Culture and Imperialism: Essays on the Environmental History of South Asia,* ed. D. Arnold and R. Guha, 210-36. Delhi: Oxford University Press, 1995.

–. "Scientific Empire and Imperial Science: Colonialism and Irrigation Technology in the Indus Basin." *Journal of Asian Studies* 53, 4 (1994): 1127-49.

Girard, R., and E. Shaker. "Bottled Up or Tapped Out: Where Have All the Water Fountains Gone?" *Academic Matters: The Journal of Higher Education,* February 2009, http://www.academicmatters.ca.

Glacken, C.J. *Traces on the Rhodian Shore: Nature and Culture in Western Thought from Ancient Times to the End of the Eighteenth Century.* Berkeley: University of California Press, 1967.

Gleick, P.H. "The Changing Water Paradigm." In *The World's Water 1998-1999,* ed. P.H. Gleick, 5-37. Washington, DC: Island Press, 1998.

–. "The Changing Water Paradigm: A Look at Twenty-First Century Water Resources Development." *Water International* 25, 1 (2000): 127-38.

–. "Global Freshwater Resources: Soft Path Solutions for the 21st Century." *Science* 302 (28 November 2003): 1524-28.

–. "The Human Right to Water." *Water Policy* 1, 5 (1999): 487-502.

–. "An Introduction to Global Fresh Water Issues." In *Water in Crisis: A Guide to the World's Freshwater Resources*, ed. Peter H. Gleick, 3-12. New York: Oxford University Press, 1993.

–, ed. *Water in Crisis: A Guide to the World's Freshwater Resources*. New York: Oxford University Press, 1993.

–. "Water in Crisis: Paths to Sustainable Water Use." *Ecological Applications* 8, 3 (1998): 571-79.

–. "Water in the 21st Century." In *Water in Crisis: A Guide to the World's Freshwater Resources*, ed. Peter H. Gleick, 105-13. New York: Oxford University Press, 1993.

–. *The World's Water 1998-1999: The Biennial Report on Freshwater Resources*. Washington, DC: Island Press, 1998.

–. *The World's Water 2000-2001: The Biennial Report on Freshwater Resources*. Washington, DC: Island Press, 2000.

–. *The World's Water 2004-2005: The Biennial Report on Freshwater Resources*. Washington, DC: Island Press, 2004.

Gleick, P.H., H. Cooley, D. Katz, and E. Lee. *The World's Water 2006-2007: The Biennial Report on Freshwater Resources*. Washington, DC: Island Press, 2006.

Gleick, P.H., G. Wolff, E.L. Chalecki, and R. Reyes. "Globalization and International Trade of Water." In *The World's Water 2002-2003: The Biennial Report on Freshwater Resources*, 33-56. Washington, DC: Island Press, 2002.

Glennon, R. *Unquenchable: America's Water Crisis and What to Do about It*. Washington, DC: Island Press, 2009.

Global Water Partnership. *Towards Water Security: A Framework for Action*. Stockholm: Global Water Partnership, 2000.

Goldsmith, E. "Traditional Agriculture in Sri Lanka." *The Ecologist* 12, 5 (1982).

Goldsmith, E., and N. Hildyard. *The Social and Environmental Effects of Large Dams*. Bordeaux: European Ecological Action Group, 1984.

Golinsky, J. *Making Natural Knowledge: Constructivism and the History of Science*. Cambridge: Cambridge University Press, 1998.

Gordon, R.G., and F.G. Thompson. *The Physiological Principles of Hydrology*. London: Jonathan Cape, 1930.

Gossage, P. "Water in Canadian History: An Overview." *Inquiry on Federal Water Policy Research Paper #11*. Ottawa: Environment Canada, 1985.

Goubert, J.-P. *The Conquest of Water: The Advent of Health in the Industrial Age*. Princeton, NJ: Princeton University Press, 1986.

Goudie, A. "Hydrological Cycle." In *The Dictionary of Physical Geography*, ed. D.S.G. Thomas and A. Goudie, 254-56. Oxford: Blackwell, 2000.

–. "Hydrology." In *The Dictionary of Physical Geography*, ed. D.S.G. Thomas and A. Goudie, 256-57. Oxford: Blackwell, 2000.

–. "Water Crisis." In *The New Fontana Dictionary of Modern Thought*, ed. A. Bullock and S. Trombley, 916. London: HarperCollins, 2000.

Gram-Hanssen, K. "Objectivity in the Description of Nature: Between Social Construction and Essentialism." In *Between Monsters, Goddesses and Cyborgs: Feminist Confrontations with Science, Medicine and Cyberspace,* ed. N. Lykke and R. Braidotti, 88-102. London: Zed Books, 1996.

Gregory, D. *The Colonial Present: Afghanistan, Palestine, Iraq.* Malden, MA: Blackwell, 2004.

–. "Dialectic(s)." In *The Dictionary of Human Geography,* ed. R.J. Johnston, Derek Gregory, Geraldine Pratt, and Michael Watts, 172-73. Malden, MA: Blackwell, 2000.

–. "Post-Colonialism." In *The Dictionary of Human Geography,* ed. R.J. Johnston, Derek Gregory, Geraldine Pratt, and Michael Watts, 612-15. Malden, MA: Blackwell, 2000.

–. "(Post)Colonialism and the Production of Nature." In *Social Nature: Theory, Practice and Politics,* ed. Noel Castree and Bruce Braun, 84-111. Malden, MA: Blackwell, 2001.

Groenfeldt, D. "Building on Tradition: Indigenous Irrigation Knowledge and Sustainable Development in Asia." *Agriculture and Human Values* 8 (1991): 114-20.

Grossman, E. *Watershed: The Undamming of America.* New York: Counterpoint, 2002.

Hacking, I. *The Social Construction of What?* Cambridge, MA: Harvard University Press, 1999.

Hall, B.S. "The Didactic and the Elegant: Some Thoughts on Scientific and Technological Illustrations in the Middle Ages and Renaissance." In *Picturing Knowledge: Historical and Philosophical Problems Concerning the Use of Art in Science,* ed. B.S. Baigrie, 3-39. Toronto: University of Toronto Press, 1996.

Hall, D. "Introduction." In *Reclaiming Public Water: Achievements, Struggles and Visions from around the World,* ed. Belén Balanyá, Brid Brennan, Olivier Hoedman, Satoko Kishimoto, and Philipp Terhorst, 15-24. Amsterdam: Transnational Institute and Corporate Europe Observatory, 2005.

Hall, F.R. "Contributions of Robert E. Horton." In *History of Geophysics.* Vol. 3, *The History of Hydrology,* ed. E.R. Landa and S. Ince, 113-17. Washington, DC: American Geophysical Union, 1987.

Hamlin, C. *A Science of Impurity: Water Analysis in Nineteenth Century Britain.* Bristol: Adam Hilger, 1990.

–. "'Waters' or 'Water'? Master Narratives in Water History and Their Implications for Contemporary Water Policy." *Water Policy* 2 (2000): 313-25.

Hanor, J.S. "History of Thought on the Origin of Subsurface Sedimentary Brines." In *History of Geophysics.* Vol. 3, *The History of Hydrology,* ed. E.R. Landa and S. Ince, 81-91. Washington, DC: American Geophysical Union, 1987.

Hansen, J.E. "As Pure as Snow." *Science Briefs.* NASA Goddard Institute for Space Studies, December 2003. http://www.giss.nasa.gov/research/briefs/hansen_10.

Haraway, D. "A Manifesto for Cyborgs: Science, Technology, and Socialist Feminism in the 1980s." In *The Haraway Reader,* 7-45. New York: Routledge, 2004.

–. "Situated Knowledges." In *The Science Studies Reader,* ed. M. Biagioli, 172-88. New York: Routledge, 1999.

Hardiman, D. "Small-Dam Systems of the Sahyadris." In *Nature, Culture and Imperialism: Essays on the Environmental History of South Asia,* ed. D. Arnold and R. Guha, 185-209. Delhi: Oxford University Press, 1995.

Hartley, D. *Water in England.* London: Macdonald and Jane's, 1964.

Harvey, D. *The Condition of Postmodernity.* Oxford: Oxford University Press, 1989.

–. *Justice, Nature and the Geography of Difference.* Oxford: Blackwell, 1996.

–. *The Limits to Capital.* Oxford: Oxford University Press, 1982.

–. "The Nature of Environment: The Dialectics of Social and Environmental Change." *Socialist Register* 30 (1993): 1-51.

–. "Population, Resources, and the Ideology of Science." *Economic Geography* 50 (1974): 256-77.

–. *Social Justice and the City.* London: Edward Arnold, 1973.

–. "Space as a Keyword." In *David Harvey: A Critical Reader,* ed. Noel Castree and Derek Gregory, 270-93. Malden, MA: Blackwell, 2006.

–. *Spaces of Capital: Toward a Critical Geography.* New York: Routledge, 2001.

–. *Spaces of Hope.* Berkeley and Los Angeles: University of California Press, 2000.

Hassoun, R. "Water between Arabs and Israelis: Researching Twice-Promised Resources." In *Water, Culture, and Power: Local Struggles in a Global Context,* ed. J.M. Donahue and B.R. Johnston, 313-38. Washington, DC: Island Press, 1998.

Hastings, J., ed. *Encyclopaedia of Religion and Ethics.* Vol. 12. Edinburgh: T. and T. Clark, 1921.

Haughton, G. "Private Profits – Public Drought: The Creation of a Crisis in Water Management for West Yorkshire." *Transactions of the Institute of British Geographers* 23, 4 (1998): 419-33.

Hayles, N.K. "Searching for Common Ground." In *Reinventing Nature? Responses to Postmodern Deconstruction,* ed. M.E. Soule and G. Lease, 47-63. Washington, DC: Island Press, 1995.

Hays, S.P. *Beauty, Health and Permanence: Environmental Politics in the United States, 1955-1985.* Cambridge: Cambridge University Press, 1987.

–. *Conservation and the Gospel of Efficiency: The Progressive Conservation Movement, 1890-1920.* Cambridge, MA: Harvard University Press, 1959.

Heidegger, M. *The Question Concerning Technology and Other Essays.* New York: Harper and Row, 1977.

Helms, D. "The Early Soil Survey: Engine for the Soil Conservation Movement." In *Sustaining the Global Farm: Selected Papers from the 10th International Soil Conservation Meeting, May 1999,* ed. D.E. Stott, R.H. Mohtar, and G.C. Steinhart. 2001. http://www.tucson.ars.ag.gov.

Hettner, A. *The Surface Features of the Land: Problems and Methods of Geomorphology,* trans. Philip Tilley. New York: Hafner, 1972.

Hewitt, K., ed. *Interpretations of Calamity.* Boston: Allen and Unwin, 1983.

Hodge, A. Trevor. " Aqueducts." In *Handbook of Ancient Water Technology,* ed. O. Wikander, 39-66. Leiden, Germany: Brill, 2000.

–. "Engineering Works." In *Handbook of Ancient Water Technology,* ed. O. Wikander, 67-94. Leiden, Germany: Brill, 2000.

–. "Purity of Water." In *Handbook of Ancient Water Technology,* ed. O. Wikander, 95-99. Leiden, Germany: Brill, 2000.

Hölderlin, F. *Poems and Fragments.* Bilingual edition with preface, introduction, and notes, trans. Michael Hamburger. London: Routledge and Kegan Paul, 1961.

Horton, R.E. "Erosional Development of Streams and Their Drainage Basins: Hydrophysical Approach to Quantitative Morphology." *Geological Society of America Bulletin* 56 (1945): 275-370.

–. "The Field, Scope, and Status of the Science of Hydrology." *Transactions, American Geophysical Union* 12 (1931): 189-202.

–. "Hydrologic Research." *Science* 86, 2241 (1937): 527-30.

Houghton, J.T. "Introduction to Techniques of Management and Analysis: Atmospheric Processes." In *Variations in the Global Water Budget,* ed. A. Street-Perrot, M. Beran, and R. Ratcliffe, 1-4. Dordrecht: D. Reidel, 1983.

Hsieh, L.H. Tiffany. "Hundreds Take Back the Beach." *Kingston This Week,* 26 July 2008.

Hubbard, P., R. Kitchen, B. Bartley, and D. Fuller. *Thinking Geographically: Space, Theory and Contemporary Human Geography.* London: Continuum, 2002.

Hunt, C.E. *Thirsty Planet: Strategies for Sustainable Water Management.* London: Zed Books, 2004.

Hunt, E., and R.C. Hunt. "Irrigation, Conflict and Politics: A Mexican Case." In *Irrigation's Impact on Society,* ed. T.E. Downing and M. Gibson, 129-57. Tucson: University of Arizona Press, 1974.

Huxley, T.H. *Physiography: An Introduction to the Study of Nature.* 1877. Reprint of 1st ed., New York: D. Appleton, 1907.

Illich, I. *H₂O and the Waters of Forgetfulness.* Dallas: Dallas Institute of Humanities and Culture, 1985.

–. *Shadow Work.* London: Marion Boyars, 1981.

Ingold, T. "Materials against Materiality." *Archaeological Dialogues* 14, 1 (2007): 1-16.

–. *The Perception of the Environment.* London: Routledge, 2000.

Jackson, D.C., ed. *Dams.* Aldershot, UK: Ashgate, 1997.

James, P.E. *All Possible Worlds: A History of Geographical Ideas.* Indianapolis: Odyssey Press, 1972.

Jameson, F. *Post-Modernism, or, The Cultural Logic of Late Capitalism.* Durham, NC: Duke University Press, 1991.

Jankovic, V. "Meteors under Scrutiny: Private, Public and Professional Weather in Britain, 1660-1800." PhD diss., University of Notre Dame, 1998.

Jenkins, G. "Forests, Land and Sea in Rainmaking." *Journal of Geography* 34, 8 (1940): 309-14.

Jiaqu, C. "The New Stage of Development of Hydrology: Water Resources Hydrology." In *Water for the Future: Hydrology in Perspective (Proceedings of the Rome Symposium, April 1987),* ed. J.C. Rodda and N.C. Matalas, 17-26. IASH publication no. 164. Wallingford, UK: International Association of Hydrological Sciences, 1987.

Johnson, President Lyndon B. "Welcoming Address by the President of the United States." In *Water for Peace Conference Proceedings.* Vol. 1, 10-12. Washington, DC: Water for Peace Conference, 1967.

Johnston, B.R. "Culture, Power, and the Hydrological Cycle: Creating and Responding to Scarcity on St. Thomas, Virgin Islands." In *Water, Culture, and Power: Local Struggles in a Global Context,* ed. J.M. Donahue and B.R. Johnston, 285-312. Washington, DC: Island Press, 1998.

–. "An Introduction to the Political Ecology of Water." *Capitalism, Nature, Socialism* 14, 3 (2003): 73-90.

Johnston, R.J., Derek Gregory, Geraldine Pratt, and Michael Watts. *The Dictionary of Human Geography.* Malden, MA: Blackwell, 2000.

Jones, J.A.A. *Global Hydrology: Processes, Resources and Environmental Management.* Essex: Longman, 1997.

Jones, P.B., G.D. Walker, R.W. Harden, and L.L. McDaniels. "The Development of the Science of Hydrology." Circular No. 63-03. Dallas: Texas Water Commission, 1963.

Kaika, M. *City of Flows: Modernity, Nature, and the City.* London: Routledge, 2005.

–. "Constructing Scarcity and Sensationalizing Water Politics: 170 Days That Shook Athens." *Antipode* 35, 5 (2003): 919-54.

Kalaora, B. "De l'eau sensible à OH₂." In *Colloque International OH₂, Dijon, 9-11 mai 2001.* Dijon: Université de Bourgogne, 2003.

Kalinin, G.P. *Global Hydrology.* 1968, trans. N. Kaner. Jerusalem: Israel Program for Scientific Translation, 1971.

Kamal, A. "Living with Water: Bangladesh since Ancient Times." In *A History of Water.* Vol. 1, *Water Control and River Biographies,* ed. T. Tevdt and E. Jakobsson, 194-213. London: I.B. Tauris, 2006.

Kates, R.W., B.L. Turner, and W.C. Clark. "The Great Transformation." In *The Earth as Transformed by Human Action: Global and Regional Changes in the Biosphere over the Past 300 Years,* ed. B.L. Turner, W.C. Clark, R.W. Kates, J.F. Richards, J.T. Mathews, and W.B. Meyer, 1-17. Cambridge: Cambridge University Press with Clark University, 1990.

Kazmann, R.G. *Modern Hydrology.* New York: Harper and Row, 1965.

–. *Modern Hydrology.* 2nd ed. New York: Harper and Row, 1972.

Kennedy, B.A. *Inventing the Earth: Ideas on Landscape Development since 1740.* Malden, MA: Blackwell, 2006.

Kirsch, S., and D. Mitchell. "The Nature of Things: Dead Labor, Nonhuman Actors, and the Persistence of Marxism." *Antipode* 36, 4 (2004): 687-705.

Klemes, V. "A Hydrological Perspective." *Journal of Hydrology* 100 (1988): 3-28.

Knight, D. "Illustrating Chemistry." In *Picturing Knowledge: Historical and Philosophical Problems Concerning the Use of Art in Science,* ed. B.S. Baigrie, 134-63. Toronto: University of Toronto Press, 1996.

Korzun, V.I., and USSR Committee for the International Hydrological Decade. *World Water Balance and Water Resources of the Earth.* Studies and Reports in Hydrology, No. 25. Paris: UNESCO, 1978.

Kreutzwiser, R.D. "Water Resource Management: Canadian Perspectives and the Great Lakes Water Levels Issue." In *Resource and Environmental Management in Canada: Addressing Uncertainty,* ed. B. Mitchell, 259-85. Don Mills, ON: Oxford University Press, 1995.

Krynine, P.D. "On the Antiquity of 'Sedimentation' and Hydrology." *Bulletin of the Geological Society of America* 71 (1960): 1721-26.

Kuhn, T.S. *The Structure of Scientific Revolutions.* 2nd ed. Chicago: University of Chicago Press, 1970.

Kula, W. *Measures and Men.* Princeton, NJ: Princeton University Press, 1986.

Ladourie, E.L.R. "Introduction." In *The Conquest of Water: The Advent of Health in the Industrial Age,* by Jean-Pierre Goubert, 1-17. Princeton, NJ: Princeton University Press, 1989.

Lalonde, M. "A Beach Plan Worth Diving Into." *Montreal Gazette,* 1 July 2008.

Lamers, O. "IRN at Ten." *World Rivers Review* 11, 4 (September 1996). http://www.internationalrivers.org.

Landa, E.R., and S. Ince, eds. 1987. *History of Geophysics.* Vol. 3, *The History of Hydrology.* Washington, DC: American Geophysical Union, 1987.

Langbein, W.B., and W.G. Hoyt. *Water Facts for the Nation's Future.* New York: Ronald Press, 1959.

Langewiesche, W. "Stealing Weather." *Vanity Fair,* May 2008, 172-85.

Lansing, J.S. "Balinese 'Water Temples' and the Management of Irrigation." *American Anthropologist* 89 (1987): 326-41.

–. *Priests and Programmers: Technologies of Power in the Engineered Landscape of Bali.* Princeton, NJ: Princeton University Press, 1991.

Larocque, A. Translator's introduction to *On the Origin of Springs,* by Pierre Perrault, 1-16. New York: Hafner, 1967.

Latour, B. *Pandora's Hope: Essays on the Reality of Science Studies.* Cambridge, MA: Harvard University Press, 1999.

–. *Politics of Nature: How to Bring Sciences into Democracy.* Cambridge, MA: Harvard University Press, 2004.

–. *Science in Action: How to Follow Scientists and Engineers through Society.* Milton Keynes, UK: Open University Press, 1987.

–. *We Have Never Been Modern.* Cambridge, MA: Harvard University Press, 1993.

Laugier, A., and A. Dumon. "D'Aristote à Mendeleev: Plus de 2000 ans de symbolisme pour représenter la matière et ses transformations." *L'Actualité chimique* 240 (March 2001): 38-50.

Leach, M., and R. Mearns. "Environmental Change and Policy: Challenging Received Wisdom in Africa." In *The Lie of the Land: Challenging Received Wisdom on the African Environment,* ed. M. Leach and R. Mearns, 1-33. Oxford: International African Institute in association with James Currey and Heinemann, 1996.

–, eds. *The Lie of the Land: Challenging Received Wisdom on the African Environment.* Oxford: International African Institute in association with James Currey and Heinemann, 1996.

Lefebvre, H. *The Production of Space.* 1972. Oxford: Blackwell, 1991.

Leighly, J. "The Role of Atmospheric Circulation in the Hydrologic Cycle of the Lands." *The Geographical Review* 28 (1938): 334-35.

Leighton, M.O. "Water Power in the United States." *Annals of the American Academy of Political and Social Science* 33 (1909): 535-65.

Leiss, W. *The Domination of Nature.* Boston: Beacon Press, 1974.

Leopold, A. *A Sand County Almanac.* New York: Ballantine Books, 1966.

Leopold, L.B. "The Alexandrian Equation." In *The History of Hydrology,* ed. E.R. Landa and S. Ince, 27-29. Washington, DC: American Geophysical Union, 1987.

Leslie, J. "Running Dry: What Happens When the World No Longer Has Enough Freshwater?" *Harper's Magazine,* July 2000, 37-52.

Levins, R., and R. Lewontin. *The Dialectical Biologist.* Cambridge, MA: Harvard University Press, 1985.

Linton, J. *Beneath the Surface: The State of Water in Canada.* Ottawa: Canadian Wildlife Federation, 1997.

–. "Global Hydrology and the Construction of a Water Crisis." *Great Lakes Geographer* 11, 2 (2004): 1-13.

–. "The Hydrosocial Cycle and the Politics of Drinking." Paper presented at the annual meeting of the Association of American Geographers, Las Vegas, 22-27 March 2009.

–. "Is the Hydrologic Cycle Sustainable? A Historical-Geographical Critique of a Modern Concept." *Annals of the Association of American Geographers* 98, 3 (2008): 630-49.

–. "The Social Nature of Natural Resources: The Case of Water." *Reconstruction: Studies in Contemporary Culture* 6, 3 (2006). http://reconstruction.eserver.org/063/contents.shtml.

–. "Variations on the Meaning of Water." Paper presented at the annual meeting of the Canadian Association of Geographers, Toronto, 1 June 2002.

Lipsey, R.G., Gordon R. Sparks, and Peter O. Steiner. *Economics.* 2nd ed. New York: Harper and Row, 1976.

Littlehales, G.W. "The Inception and Development of the International Section of Hydrology." *Transactions, American Geophysical Union* 12 (1931): 189.

–. "Status, Scope, and Problems of the Section of Oceanography." *Transactions, American Geophysical Union* 4 (1923): 68-71.

Livingstone, D.N. *The Geographical Tradition: Episodes in the History of a Contested Enterprise.* Malden, MA: Blackwell, 1992.

–. *Putting Science in Its Place: Geographies of Scientific Knowledge.* Chicago: University of Chicago Press, 2003.

Lonergan, S., and D.B. Brooks. *Watershed: The Role of Fresh Water in the Israeli-Palestinian Conflict.* Ottawa: International Development Research Centre, 1994.

Lovejoy, A.O. *The Great Chain of Being: A Study of the History of an Idea.* 1936. First Harper Torchbook ed. New York: Harper and Row, 1960.

Lovins, A. *Soft Energy Paths: Toward a Durable Peace.* Cambridge, MA: Ballinger/Friends of the Earth, 1977.

L'vovitch, M.I. "World Water Balance (General Report)." In *World Water Balance: Proceedings of the Reading Symposium, July 1970,* 401-15. Gentbrugge, Belgium: IASH-UNESCO-WMO, 1972.

–. *World Water Resources and Their Future,* trans. R.L. Nace. Chelsea, MI: American Geophysical Union, 1979.

L'vovich, M., G.F. White, A.V. Belyaev, J. Kindler, N.I. Koronkevic, T.R. Lee, and G.V. Voropaev. "Use and Transformation of Terrestrial Water Systems." In *The Earth as Transformed by Human Action: Global and Regional Changes in the Biosphere over the Past 300 Years,* ed. B.L. Turner, W.C. Clark, R.W. Kates, J.F. Richards, J.T. Mathews, and W.B. Meyer, 235-52. Cambridge: Cambridge University Press with Clark University, 1990.

Lyon, A. "Yemen Sleepwalks into Water Nightmare." Reuters News Service, 2 March 2008. http://uk.reuters.com/article/environmentNews/idUKL2970383520080302.

Maass, A. "Public Investment Planning in the United States: Analysis and Critique." *Public Policy* 18, 2 (1970): 211-43.

Macagno, E.O. "Leonardo da Vinci: Engineer and Scientist." In *Hydraulics and Hydraulic Research: A Historical Review,* ed. G. Garbrecht, 33-53. Rotterdam: A.A. Balkema, 1987.

Maccagni, C. "Galileo, Castelli, Torricelli and Others: The Italian School of Hydraulics in the 16th and 17th Centuries." In *Hydraulics and Hydraulic Research: A Historical Review,* ed. G. Garbrecht, 55-79. Rotterdam: A.A. Balkema, 1987.

Mackenzie, F.D. "Contested Ground: Colonial Narratives and the Kenyan Environment, 1920-1945." *Journal of Southern African Studies* 26, 4 (2000): 697-718.

Maffioli, C.S. *Out of Galileo: The Science of Waters, 1628-1718.* Rotterdam: Erasmus Publishing, 1994.

Maidment, D.R., ed. *Handbook of Hydrology.* New York: McGraw-Hill, 1993.

Marsh, G.P. *Man and Nature: Or Physical Geography as Modified by Human Action.* New York: Charles Scribner, 1864.

Marvin, C.F. "The Status and Problems of Meteorology." *Transactions, American Geophysical Union* 1 (1920): 561-72.

–. "Status, Scope and Present-Day Problems of Meteorology." *Transactions, American Geophysical Union* 4 (1923): 54-60.

Marx, K. "The German Ideology." 1845-46. In *The Marx-Engels Reader,* ed. R.C. Tucker, 146-200. New York: W.W. Norton, 1978.

–. "The Grundrisse." 1857-58. In *The Marx-Engels Reader,* ed. R.C. Tucker, 221-93. New York: W.W. Norton, 1978.

Mather, J.D. "200 Years of British Hydrogeology: An Introduction and Overview." In *200 Years of British Hydrogeology,* ed. J.D. Mather, 1-14. London: Geological Society, 2004.

Matless, D. "A Modern Stream: Water, Landscape, Modernism, and Geography." *Environment and Planning D: Society and Space* 10 (1992): 569-88.

Mautner, T. *The Penguin Dictionary of Philosophy.* London: Penguin, 1996.

McCully, P. *Silenced Rivers: The Ecology and Politics of Large Dams.* London: Zed Books, 1996.

McCutcheon, S. *Electric Rivers: The Story of the James Bay Project.* Montreal: Black Rose Books, 1991.

McGee, W.J. "The Beginning of Agriculture." *The American Anthropologist* 8 (1895): 350-75.

–. "Our Great River: What It Is and May Be Made for Commerce, Agriculture, and Sanitation: The Largest Inland Water Project of Our Time." *The World's Work* 13 (1907): 8576-84.

–. "Principles of Water-Power Development." *Science* 34, 885 (1911): 813-25.

–. "Water as a Resource." *Annals of the American Academy of Political and Social Science* 33 (1909): 521-34.

McMahon, K., dir. *Waterlife.* National Film Board of Canada, 2009.

Mead, D.W. *Hydrology: The Fundamental Basis of Hydraulic Engineering.* New York: McGraw-Hill, 1919.

–. *Hydrology: The Fundamental Basis of Hydraulic Engineering.* 2nd ed. New York: McGraw-Hill, 1950.

Meadows, D., D.L. Meadows, J. Randers, and W.W. Berens. *The Limits to Growth.* New York: Signet Books, 1972.

Mehta, D., dir. *Water.* Deepa Mehta Films: 2005.

Meigs, P. "Water Problems in the United States." *Geographical Review* 42, 3 (1952): 346-66.

Meinzer, O.E. "Formation of the Section of Hydrology of the American Geophysical Union." *Transactions, American Geophysical Union* 12 (1931): 227-29.

–. "The History and Development of Ground-Water Hydrology." *Journal of the Washington Academy of Sciences* 24, 1 (1934): 6-32.

–, ed. *Hydrology.* New York: Dover, 1942.

–. "Introduction ." In *Hydrology,* ed. O.E. Meinzer, 1-31. New York: Dover, 1942.

Melville, H. *Moby-Dick.* London: Penguin, 1972.

Merchant, C. *The Death of Nature: Women, Ecology and the Scientific Revolution.* San Francisco: HarperCollins, 1980.

Meyer, A.F. *The Elements of Hydrology.* New York: John Wiley and Sons; London: Chapman and Hall, 1917.

–. *The Elements of Hydrology.* 2nd ed. New York: John Wiley and Sons; London: Chapman and Hall, 1928.

Miller, P. "John Wesley Powell: Vision for the West." *National Geographic* 85, 4 (1994): 89-115.

Millner, A. "Post-Colonialism." In *The New Fontana Dictionary of Modern Thought,* ed. A. Bullock and S. Trombley, 669-70. London: HarperCollins, 2000.

Mitchell, B. *Resource and Environmental Management.* Harlow: Longman, 2002.

Mitchell, B., and D. Shrubsole. *Canadian Water Management: Visions for Sustainability.* Cambridge, ON: Canadian Water Resources Association, 1994.

Mitchell, T. "Fixing the Economy." *Cultural Studies* 12, 1 (1998): 82-101.

Mohamed, Y.A. "Drought and the Need to Change: The Expansion of Water Harvesting in Central Darfur, Sudan." In *Sustaining the Soil: Indigenous Soil and Water Conservation in Africa,* ed. C. Reij, I. Scoones, and C. Toulmin, 35-43. London: Earthscan, 1996.

Mohun, S. "Labour Process." In *A Dictionary of Marxist Thought,* ed. Tom Bottomore, Laurence Harris, V.G. Kiernan, and Ralph Miliband, 297-301. Oxford: Blackwell, 1999.

Mol, A.P.J., and D.A. Sonnenfeld. "An Introduction." In *Ecological Modernisation around the World,* ed. A.P.J. Mol and D.A. Sonnenfeld, 3-16. London: Frank Cass, 2000.

Molle, F., P.P. Mollinga, and R. Meinzen-Dick. "Water, Politics and Development: Introducing Water Alternatives." *Water Alternatives* 1, 1 (2008): 1-16.

Morello, N. "La question des eaux douces au XVIIe siècle: Parcours des eaux, parcours des hypothèses." In *Colloque International OH₂, Dijon, 9-11 mai 2001.* Dijon: Université de Bourgogne, 2003.

Morse, B., and T. Berger. *Sardar Sarovar: Report of the Independent Review.* Ottawa: Resource Futures International, 1992.

Moser, P.H. "Hydrological Cycle." In *Encyclopedia of Climate and Weather,* ed. S.H. Schneider, 415-17. New York: Oxford University Press, 1996.

Moss, Senator F.E. *The Water Crisis.* New York: Praeger, 1967.

Mosse, D. "Local Institutions and Power: The History and Practice of Community Management of Tank Irrigation Systems in South India." In *Power and Participatory Development: Theory and Practice,* ed. N. Nelson and S. Wright, 144-56. London: Intermediate Technology Publications, 1995.

–. *The Rule of Water: Statecraft, Ecology and Collective Action in South India.* New Delhi: Oxford University Press, 2003.

Mumford, L. *Technics and Civilization.* New York: Harcourt, Brace and World, 1963.

Myler, J.L. "The Dirty Animal: Man." In *Eco-Crisis,* ed. C.E. Johnson, 116-48. New York: John Wiley and Sons, 1970.

Nace, R.L. "Development of Hydrology in North America." *Water International* 3, 3 (1978): 20-26.

–. "Editor's Preface." In *World Water Resources and Their Future,* by M.I. L'vovich, trans. R.L. Nace. Chelsea, MI: American Geophysical Union Translation Board, 1979.

–. "General Evolution of the Concept of the Hydrological Cycle." In *Three Centuries of Scientific Hydrology: Key Papers Submitted on the Occasion of the Celebration of the Tercentenary of Scientific Hydrology,* ed. UNESCO-WMO, 40-49. Paris: UNESCO-WMO, 1974.

–. "The Hydrological Cycle: Historical Evolution of the Concept." *Water International* 1, 1 (1975): 15-21.

–. "Water Resources: A Global Problem with Local Roots." *Environmental Science and Technology* 1, 7 (1967): 550-60.

–. "Water of the World: Distribution of Man's Liquid Assets is a Clue to Future Control." *Natural History* 73, 1 (1964): 10-19.

–. "World Hydrology: Status and Prospects." In *World Water Balance: Proceedings of the Reading Symposium, July 1970*, ed. IASH-UNESCO-WMO, 1-10. Gentbrugge, Belgium: IASH-UNESCO-WMO, 1972.

–. "World Water Inventory and Control." In *Water, Earth, and Man*, ed. R.J. Chorley, 31-42. London: Methuen, 1969.

Naranjo, M.F., and B. Bobee. "Le cycle hydrologique et les systèmes d'irrigation de la ville de Mexico à l'époque préhispanique." In *Colloque International OH₂, Dijon, 9-11 mai 2001*, Dijon: Université de Bourgogne, 2003.

Nash, J.E. "Hydrology and Hydrologists: Reflections." In *Advances in Theoretical Hydrology: A Tribute to James Dooge*, ed. J.P. O'Kane, 191-99. Amsterdam: Elsevier, 1992.

National Atlas of the United States. *Water of the United States.* http://nationalatlas.gov/water.html.

National Conservation Commission. *Report of the National Conservation Commission*, ed. under the direction of the Executive Committee by Henry Gannett. Washington DC: Government Printing Office, 1909.

National Research Council. Committee on Opportunities in the Hydrologic Sciences. *Opportunities in the Hydrologic Sciences.* Ed. Peter Eagleson. Washington, DC: National Academies Press, 1991.

National Resources Board. *A Report on National Planning and Public Works in Relation to Natural Resources Including Land Use and Water Resources with Findings and Recommendations.* Washington, DC: US Government Printing Office, 1934.

National Resources Committee. *Deficiencies in Basic Hydrologic Data: Report of the Special Advisory Committee on Standards and Specifications for Hydrologic Data of the Water Resources Committee.* Washington, DC: US Government Printing Office, 1937.

– (Water Resources Committee). *Drainage Basin Problems and Programs, 1937 Revision.* Washington, DC: US Government Printing Office, 1938.

Nelles, H.V. *The Politics of Development: Forests, Mines and Hydro-Electric Power in Ontario, 1849-1941.* Toronto: Macmillan, 1974.

Nemec, J. "International Aspects of Hydrology." In *Facets of Hydrology*, ed. John C. Rodda, 331-62. London: John Wiley and Sons, 1976.

Nevarez, L. "Just Wait Until There's a Drought: Mediating Environmental Crises for Urban Growth." *Antipode* 28, 3 (1996): 246-72.

Newell, F.H. *Water Resources: Present and Future Uses.* New Haven, CT: Yale University Press; London: Oxford University Press, 1920.

Nikolaieff, G.A. Editor's introduction to *The Water Crisis*, ed. G.A. Nikolaieff, 9. New York: H.W. Wilson, 1967.

–, ed. *The Water Crisis.* New York: H.W. Wilson, 1967.

Nuttle, W.K. "Is Ecohydrology One Idea or Many?" *Hydrological Sciences Journal* 45, 5 (2002): 805-8.

–. "Taking Stock of Water Resources." *EOS* 85, 513 (2002): 1-2.

Ohlsson, L., ed. *Hydropolitics: Conflicts over Water as a Development Constraint.* London: Zed Books, 1995.

Ohlsson, L., and A.R. Turton. *The Turning of a Screw: Social Resource Scarcity as a Bottle-Neck in Adaptation to Water Scarcity.* London: School of Oriental and Asian Studies, Water Study Group, University of London, 1999.

Ollman, B. *Dialectical Investigations.* New York: Routledge, 1993.

O'Neill, K.M. *Rivers by Design: State Power and the Origins of U.S. Flood Control.* Durham, NC: Duke University Press, 2006.

Padovani, F. "The Chinese Way of Harnessing Rivers: The Yangtze River." In *A History of Water,* vol. 1, *Water Control and River Biographies,* ed. T. Tvedt and E. Jakobsson, 120-43. London and New York: I.B. Tauris, 2006.

Palmer, B. *Descent into Discourse: The Reification of Language and the Writing of Social History.* Philadelphia: Temple University Press, 1990.

Parfit, M. "Sharing the Wealth of Water." In "Water," ed. M. Parfit, R. Conniff, J.G. Mitchell and W.S. Ellis. Special issue, *National Geographic* 184, 5A (1993): 20-37.

Parizek, R.R. "The Hydrologic Cycle Concept and Our Challenge in the 20th Century." *Mineral Industries* 32, 7 (1963): 1-13.

Patrick, R. "Is Water Our Next Crisis?" *Proceedings of the American Philosophical Society* 128, 3 (1994): 371-76.

Pawluk, R.R., J.A. Sandor, and J.A. Tabor. "The Role of Indigenous Soil Knowledge in Agricultural Development." *Journal of Soil and Water Conservation* 47 (July-August 1992): 298-302.

Pazwash, H. "Iran's Modes of Modernization: Greening the Desert, Deserting the Greenery?" *Civil Engineering* 53, 3 (March 1983): 48-51.

Pearce, F. *The Dammed: Rivers, Dams, and the Coming Water Crisis.* London: Bodley Head, 1992.

Pearse, P., F. Bertrand, and J.W. MacLaren. *Currents of Change: Final Report, Inquiry on Federal Water Policy.* Ottawa: Government of Canada, Inquiry on Federal Water Policy, 1985.

Peet, R. *Modern Geographical Thought.* Oxford: Blackwell, 1998.

Pentland, R., and A. Hurley. "Thirsty Neighbours: A Century of Canada-US Transboundary Water Governance." In *Eau Canada: The Future of Canada's Water,* ed. Karen Bakker, 163-82. Vancouver: UBC Press, 2007.

Pereira, H.C., "The Influence of Man on the Hydrological Cycle." In *World Water Balance: Proceedings of the Reading Symposium, July 1970,* ed. IASH-UNESCO-WMO, 41-55. Gentbrugge, Belgium: IASH-UNESCO-WMO, 1972.

Permutit Company, The. *Water Conditioning Handbook.* New York: The Permutit Company, 1943.

Perrault, P. *De l'origine des fontaines (Textes fondateurs de l'hydrologie N. 2).* 1674. Reprint, Asnières, France: Association Internationale des Sciences Hydrologiques, Comité National Français des Sciences Hydrologiques, 2001.

–. *On the Origin of Springs.* 1674. Reprint, New York: Hafner, 1967.

Petrella, R. "Blue Gold of the 21st Century." *Le Monde Diplomatique.* March 2000. Available as "The New Conquest of Water." http://www.goodplanet.info.

–. *Le Manifeste de l'eau: Pour un contrat mondial.* Brussels: Labor, 1998.

Pfannkuch, H.O. "Medieval Saint Barbara Worship and Professional Traditions in Early Mining and Applied Earth Sciences." *The History of Hydrology,* ed. E.R. Landa and S. Ince, 39-48. Washington, DC: American Geophysical Union, 1987.

Pietz, D.A. "Controlling the Waters in Twentieth-Century China: The Nationalist State and the Huai River." In *A History of Water.* Vol. 1, *Water Control and River Biographies,* ed. T. Tvedt and E. Jakobsson, 92-119. London and New York: I.B. Tauris, 2006.

Pinto, J.A. *The Trevi Fountain.* New Haven, CT: Yale University Press, 1986.

Pliny. *Natural History.* 10 vols. Cambridge, MA: Harvard University Press; London: William Heinemann, 1938-1963.

Polanyi, K. *The Great Transformation: The Political and Economic Origins of Our Time.* Boston: Beacon Press, 2001.

Postel, S. *Last Oasis: Facing Water Scarcity.* New York: W.W. Norton, 1992.

–. *Pillar of Sand: Can the Irrigation Miracle Last?* New York: W.W. Norton, 1999.

Postel, S.L., G.C. Daily, and P.R. Ehrlich. "Human Appropriation of Renewable Fresh Water." *Science* 271 (9 February 1996): 785-88.

Press, J. "Group Seeks Beach Cleanup: Richardson Site Was Once Popular Swimming Spot." *Kingston Whig Standard,* 18 July 2008.

Progress in Human Geography. "Classics in Geography Revisited: Gilbert White, Human Adjustment to Floods." *Progress in Human Geography* 21, 2 (1997): 243-50.

Rattue, J. *The Living Stream: Holy Wells in Historical Context.* Woodbridge, UK: Boydell Press, 1995.

Ray, J. 1692. *The Wisdom of God Manifested in the Works of the Creation.* 2nd ed. London: Samuel Smith.

Rees, J. *Natural Resources: Allocation, Economics and Policy.* London: Routledge, 1990.

Reij, C., I. Scoones, and C. Toulmin, eds. *Sustaining the Soil: Indigenous Soil and Water Conservation in Africa.* London: Earthscan, 1996.

Reisner, M. *Cadillac Desert: The American West and Its Disappearing Water.* New York: Penguin, 1986.

–. "Unleash the Rivers." *Time Magazine.* 26 April 2000. http://www.time.com.

Rescher, N. *Dialectics: A Controversy-Oriented Approach to the Theory of Knowledge.* Albany: State University of New York Press, 1977.

–. *Process Philosophy: A Survey of Basic Issues.* Pittsburgh: University of Pittsburgh Press, 2000.

–. "The Promise of Process Philosophy." In *Process and Analysis: Whitehead, Hartshorne, and the Analytic Tradition,* ed. G.W. Sheilds, 49-66. Albany: State University of New York Press, 2003.

Reuss, M. "The Development of American Water Resources: Planners, Politicians and Constitutional Interpretation." In *Managing Water Resources Past and Present,* ed. J. Trottier and P. Slack, 51-71. Oxford: Oxford University Press, 2004.

–. "Hydrology." In *The History of Science in the United States: An Encyclopedia,* ed. M. Rothenberg, 274-76. New York: Garland, 2001.

–. "Surface Water Hydrology and the 'Big Dam Era' in the United States." Paper presented at the Second International Water History Conference, University of Bergen, Norway, 10-12 August, 2001.

Rockstrom, J. "Managing Rain for the Future." In *Rethinking Water Management: Innovative Approaches to Contemporary Issues,* ed. C.M. Figuertes, C. Tortajada, and J. Rockstrom, 70-101. London: Earthscan, 2003.

Rodda, J.C. "Foreword." In *Global Hydrology: Processes, Resources and Environmental Management,* by J.A.A. Jones, xi. Essex: Longman, 1997.

–. "Guessing or Assessing the World's Water Resources?" *Journal of the Institute of Water and Environmental Management* 9, 4 (1995): 360-68.

–. "Whither World Water?" *Water Resources Bulletin* 31, 1 (1995): 1-7.

Rodda, J.C., and N.C. Matalas, eds. *Water for the Future: Hydrology in Perspective (Proceedings of the Rome Symposium, April 1987).* IAHS publication no. 164. Wallingford, UK: International Association of Hydrological Sciences, 1987.

Rodda, J.C., and L. Ubertini. *The Basis of Civilization: Water Science?* IAHS publication no. 286. Wallingford, UK: International Association of Hydrologic Sciences Press, 2004.

Rogers, P. *America's Water: Federal Roles and Responsibilities.* Cambridge, MA: MIT Press, 1996.

–. "Fresh Water." In *The Global Possible: Resources, Development and the New Century,* ed. R. Repetto, 255-98. New Haven, CT: Yale University Press, 1985.

–. "Hydrology and Water Quality." In *The Earth as Transformed by Human Action: Global and Regional Changes in the Biosphere over the Past 300 Years,* ed. B.L. Turner, W.C. Clark, R.W. Kates, J.F. Richards, J.T. Mathews, and W.B. Meyer, 231-57. Cambridge: Cambridge University Press with Clark University, 1990.

Ross, A. "The Lonely Hour of Scarcity." *Capitalism, Nature, Socialism* 7, 3 (1996): 3-26.

Rouse, H., and S. Ince. *History of Hydraulics.* Iowa City: Iowa Institute of Hydraulic Research, State University of Iowa, 1957.

Roy, A., and S. Lane. "Putting the Morphology Back into Fluvial Geomorphology: The Case of River Meanders and Tributary Junctions." In *Contemporary Meanings in Physical Geography: From What to Why?* ed. S. Trudgill and A. Roy, 103-25. London: Arnold, 2003.

Royte, E. *Bottlemania: How Water Went on Sale and Why We Bought It.* London: Bloomsbury, 2008.

Saberwal, V.K. "Science and the Desiccationist Discourse of the 20th Century." *Environment and History* 3 (1997): 309-43.

Sachs, W. *Planet Dialectics: Explorations in Environment and Development.* London: Fernwood, 1999.

Saeijs, H.L.F., and M.J. van Berkel. "Global Water Crisis: The Major Issue of the 21st Century, a Growing and Explosive Problem." *European Water Pollution Control* 5, 4 (1995): 26-40.

Said, E. "Yeats and Decolonization." In *The Edward Said Reader,* ed. M. Bayoumi and A. Rubin, 291-313. New York: Vintage, 2000.

Sambursky, S. *The Physical World of the Greeks.* New York: Collier, 1962.

Sammani, M.O.E., and S.M.A. Dabloub. "Making the Most of Local Knowledge: Water Harvesting in the Red Sea Hills of Northern Sudan." In *Sustaining the Soil: Indigenous Soil and Water Conservation in Africa,* eds. C. Reij, I. Scoones, and C. Toulmin, 28-43. London: Earthscan, 1996.

Sandbach, F. *Environment, Ideology and Policy.* Oxford: Basil Blackwell, 1980.

Sardar, Z. "Above, Beyond, and at the Center of the Science Wars: A Postcolonial Reading." In *After the Science Wars,* ed. K.M. Ashman and P. Baringer, 117-136. London: Routledge, 2001.

Sauer, C.O. "The Morphology of Landscape." 1925. In *Land and Life: A Selection from the Writings of Carl Ortwin Sauer,* ed. J. Leighly, 315-50. Berkeley: University of California Press, 1963.

Saunders, P. "Environmental Refugees: The Origins of a Construct." In *Political Ecology: Science, Myth and Power,* ed. P. Stott and S. Sullivan, 218-46. London: Arnold, 2000.

Saville, T. "Basic Principles of Water Behavior." In *Headwaters Control and Use: A Summary of Fundamental Principles and Their Application in the Conservation and Utilization of Waters and Soils throughout Headwater Areas.* Papers presented at the Upstream Engineering Conference held in Washington, D.C., 22-23 September 1936. Washington, DC: Soil Conservation Service and Forest Service of the United States Department of Agriculture, 1937.

Scarce, R. *Fishy Business: Salmon, Biology and the Social Construction of Nature*. Philadelphia, Temple University Press, 2000.

Schama, S. *Landscape and Memory*. Toronto: Random House, 1995.

Scheuerlein, H. "Fluvial Hydraulics throughout History." In *Hydraulics and Hydraulic Research: A Historical Review*, ed. G. Garbrecht, 185-90. Rotterdam: A.A. Balkema, 1987.

Schmit, B. "Water: H_2O = Life: Case Study of an Exhibition." ExhibitFiles, a community site for exhibit designers and developers. http://www.exhibitfiles.org/water_h2o__life.

Scientific American. "Managing Planet Earth." Special issue, *Scientific American*, September 1989.

Scoones, I., C. Reij, and C. Toulmin. "Sustaining the Soil: Indigenous Soil and Water Conservation in Africa." In *Sustaining the Soil: Indigenous Soil and Water Conservation in Africa*, ed. C. Reij, I. Scoones, and C. Toulmin, 1-27. London: Earthscan, 1996.

Scott, J.C. *Seeing Like a State: How Certain Schemes to Improve the Human Condition Have Failed*. New Haven, CT: Yale University Press, 1998.

Section Internationale d'Hydrologie Scientifique – Conseil International de Recherches – Union Géodésique et Géophysique International. " Bulletin N. 1, Première Réunion plénière de la Section. (Madrid, Octobre 1924)." Venezia: Premiate Officine Grafiche C. Ferrari, 1924. http://www.cig.ensmp.fr.

Seenivasan, R., ed. *Neerkattis: The Rural Water Managers*. Madurai, India: DHAN (Development of Humane Action) Foundation, 2003.

Seenivasan, R.P., and A. Kumar. *Vision for Village Tanks of Tamil Nadu*. Madurai, India: DHAN Foundation, 2004.

Seneca, L.A. *Naturales Quaestiones*. Cambridge, MA: Harvard University Press; London: William Heinemann, 1972.

Sennett, R. *Flesh and Stone: The Body and the City in Western Civilization*. New York: W.W. Norton, 1994.

Shapin, S. *The Scientific Revolution*. Chicago: University of Chicago Press, 1996.

Shapin, S., and S. Schaffer. *Leviathan and the Air-Pump: Hobbes, Boyle, and the Experimental Life*. Princeton, NJ: Princeton University Press, 1985.

Shiklomanov, I.A. "World Fresh Water Resources." In *Water in Crisis: A Guide to the World's Freshwater Resources*, ed. Peter H. Gleick, 13-24. New York: Oxford University Press, 1993.

Shiklomanov, I.A., and John C. Rodda, eds. *World Water Resources at the Beginning of the Twenty-First Century*. Cambridge: Cambridge University Press, 2003.

Shiva, V. *Staying Alive: Women, Ecology, and Development in India*. London: Zed Books, 1988.

–. *Water Wars: Privatization, Pollution, and Profit*. Toronto: Between the Lines, 2002.

Sims, P. "Previous Actors and Current Influences: Trends and Fashions in Physical Geography." In *Contemporary Meanings in Physical Geography: From What to Why?* ed. S. Trudgill and A. Roy, 3-23. London: Arnold, 2003.

Smith, N. "The Production of Nature." In *FutureNatural: Nature, Science, Culture*, ed. G. Robertson, M. Mash, L. Tickner, J. Bird, B. Curtis, and T. Putnam, 36-54. London: Routledge, 1996.

–. *Uneven Development: Nature, Capital and the Production of Space*. Oxford: Basil Blackwell, 1984.

Smith, N.A.F. *Man and Water: A History of Hydro-Technology*. New York: Scribner, 1975.

Soper, K. "Nature/'nature.'" In *FutureNatural: Nature, Science, Culture*, ed. G. Robertson, M. Mash, L. Tickner, J. Bird, B. Curtis, and T. Putnam, 22-34. London: Routledge, 1996.

–. *What Is Nature?* Oxford: Blackwell, 1995.

Soule, M.E., and G. Lease, eds. *Reinventing Nature? Responses to Postmodern Deconstruction.* Washington, DC: Island Press, 1995.

Soullard, É. "Les Académiciens des Sciences: Claude, Charles et Pierre Perrault, l'Abbé Picard, La Hire ... et le chantier des eaux de Versailles, sous Louis XIV." In *Colloque International OH₂, Dijon, 9-11 mai 2001.* Dijon: Université de Bourgogne, 2003.

Special Correspondent. Stockholm Water Prize for Sunita Narain. *The Hindu,* 27 August 2005. http://www.hindu.com.

Spidel, D.H., and A.F. Agnew. "The World Water Budget." In *Perspectives on Water: Uses and Abuses,* ed. D.H. Spidel, L.C. Ruedisili, and A.F. Agnew. New York: Oxford University Press, 1988.

Stehlik, D., Geoffrey Lawrence, and Ian Gray. "Gender and Drought: Experiences of Australian Women in the Drought of the 1990s." *Disasters* 24, 1 (2000): 38-53.

Stichweh, R. "The Sociology of Scientific Disciplines: On the Genesis and Stability of the Disciplinary Structure of Modern Science." *Science in Context* 5, 1 (1992): 3-15.

Stiglitz, J. *Globalization and Its Discontents.* New York: W.W. Norton, 2002.

Stott, P., and S. Sullivan, eds. *Political Ecology: Science, Myth and Power.* London: Arnold, 2000.

Strang, V. *The Meaning of Water.* Oxford: Berg, 2004.

Sutcliffe, R.C. "Introduction." In *World Water Balance: Proceedings of the Reading Symposium, July 1970; A Contribution to the International Hydrological Decade,* xii. Gentbrugge, Belgium: IASH-UNESCO-WMO, 1972.

Swain, H. "Report of the Expert Panel on Safe Drinking Water for First Nations." Ottawa: Minister of Public Works and Government Services Canada, 2006.

Swift, J. "Desertification: Narratives, Winners and Losers." In *The Lie of the Land: Challenging Received Wisdom on the African Environment,* ed. M. Leach and R. Mearns, 73-90. Oxford: International African Institute in association with James Currey and Heinemann, 1996.

Switzer, J. "Swimmers Head Back to the Beach: Young and Old Take the Plunge as Organizers Look to Revive Richardson." *Kingston Whig Standard,* 23 July 2008.

Swyngedouw, E. "The City as a Hybrid: On Nature, Society and Cyborg Urbanization." *Culture, Nature, Socialism* 7, 2 (1996): 65-80.

–. "Dispossessing H₂O: The Contested Terrain of Water Privatization." *Capitalism, Nature, Socialism* 16, 1 (2005): 81-98.

–. "The Marxian Alternative: Historical-Geographical Materialism and the Political Economy of Capitalism." In *A Companion to Economic Geography,* ed. E. Sheppard and T.J. Barnes, 41-59. Oxford: Blackwell, 2000.

–. "Metabolic Urbanization: The Making of Cyborg Cities." In *In the Nature of Cities: Urban Political Ecology and the Politics of Urban Metabolism,* ed. N. Heynen, M. Kaika, and E. Swyngedouw, 21-40. London: Routledge, 2006.

–. "Modernity and Hybridity: Nature, *Regeneracionismo,* and the Production of the Spanish Waterscape, 1890-1930." *Annals of the Association of American Geographers* 89, 3 (1999): 443-65.

–. "Neither Global nor Local: 'Glocalization' and the Politics of Scale." In *Spaces of Globalization: Reasserting the Power of the Local,* ed. K. Cox, 137-66. London: Guilford Press, 1997.

–. "Power, Nature and the City: The Conquest of Water and the Political Ecology of Urbanization in Guayaquil, Ecuador; 1880-1990." *Environment and Planning A* 29 (1997): 311-32.

–. *Social Power and the Urbanization of Water: Flows of Power.* Oxford: Oxford University Press, 2004.

Swyngedouw, E., and N.C. Heynen, eds. "Urban Political Ecology, Justice and the Politics of Scale." Special Issue, *Antipode* 35, 5 (2003).

Tanner, R.G. "Philosophical and Cultural Concepts Underlying Water Supply in Antiquity." In *Water for the Future: Hydrology in Perspective (Proceedings of the Rome Symposium, April 1987)*, ed. J.C. Rodda and N.C. Matalas, 27-36. IASH publication no. 164. Wallingford, UK: International Association of Hydrological Sciences, 1987.

Tarnas, R. *The Passion of the Western Mind.* New York: Ballantine Books, 1991.

Taylor, P.J., and F.H. Buttel. "How Do We Know We Have Global Environmental Problems? Science and the Globalization of Environmental Discourse." *Geoforum* 23, 3 (1992): 405-16.

Tempelhoff, J., H. Hoag, M. Ertsen, E. Arnold, M. Bender, K. Berry, C. Fort, D. Pietz, M. Musemwa, M. Nakawo, J. Ur, P. van Dam, M. Melosi, V. Winiwarter, and T. Wilkinson. "Where Has the Water Come From?" *Water History* 1, 1 (2009): 1-8.

Thomas, H.E. "Changes in Quantities and Qualities of Ground and Surface Water." In *Man's Role in Changing the Face of the Earth,* ed. W.L. Thomas, C.O. Sauer, M. Bates, and L. Mumford, 542-63. Chicago: University of Chicago Press, 1956.

Thomas, W.L., C.O. Sauer, M. Bates, and L. Mumford, eds. *Man's Role in Changing the Face of the Earth.* Chicago: University of Chicago Press, 1956.

Thornwaite, C.W. "The Hydrologic Cycle Re-Examined." *Soil Conservation* 3, 4 (1937-38): 85-91.

Tixeront, J. "L'Hydrologie en France au XVIIme siècle." In *Three Centuries of Scientific Hydrology: Key Papers Submitted on the Occasion of the Celebration of the Tercentenary of Scientific Hydrology,* ed. UNESCO-WMO, 24-36. Paris: UNESCO-WMO, 1974.

Todd, D.K. *The Water Encyclopedia: A Compendium of Useful Information on Water Resources.* Port Washington, NY: Water Information Center, 1970.

Tonini, D. "The Evolution of the Concept of the Hydrological Cycle in the Western World, with Special Regard to the Contributions of Italian Scholars." *Water International* 2, 4 (1977): 16-31.

Tozer, H.F. *A History of Ancient Geography.* 2nd ed. New York: Billo and Tannen, 1971.

Tuan, Yi-Fu. *The Hydrologic Cycle and the Wisdom of God: A Theme in Geoteleology.* Toronto: University of Toronto Press, 1968.

–. *Topophilia: A Study of Environmental Perception, Attitudes, and Values.* Englewood Cliffs, NJ: Prentice Hall, 1974.

Turner, B.L., W.C. Clark, R.W. Kates, J.F. Richards, J.T. Mathews, and W.B. Meyer, eds. *The Earth as Transformed by Human Action: Global and Regional Changes in the Biosphere over the Past 300 Years.* Cambridge: Cambridge University Press, 1990.

Tvedt, T., and E. Jakobsson, eds. *A History of Water.* Vol. 1, *Water Control and River Biographies.* London: I.B. Tauris, 2006.

–. "Introduction: Water History Is World History." In *A History of Water.* Vol. 1, *Water Control and River Biographies,* ed. T. Tvedt and E. Jakobsson, ix-xxiii. London: I.B. Tauris, 2006.

Tvedt, T., and T. Oestigaard, eds. *A History of Water.* Vol. 3, *The World of Water.* London: I.B. Tauris, 2006.

UNESCO (United Nations Educational, Scientific and Cultural Organization). *Scientific Framework of World Water Balance.* Paris: UNESCO, 1971.

–. *Textbooks on Hydrology.* Vol. 1, *Analyses and Synoptic Tables of Contents of Selected Textbooks.* Technical Papers in Hydrology 6. Paris: UNESCO, 1974.

–. *Textbooks on Hydrology.* Vol. 2, *Analyses of Selected Textbooks.* Technical Papers in Hydrology 6. Paris: UNESCO, 1974.

–. *Water for People, Water for Life: The United Nations World Water Development Report.* New York: UNESCO and Berghahn Books, 2003.

UNESCO/FAO Working Group on the International Hydrological Decade. *Man's Influence on the Hydrological Cycle: A Draft Report of the UNESCO/FAO Working Group on the International Hydrological Decade.* Rome: Water Resources and Development Service, Land and Water Development Division, Food and Agriculture Organization of the United Nations, 1973.

UNESCO-WMO-IAHS. "Foreword." In *Three Centuries of Scientific Hydrology: Key Papers Submitted on the Occasion of the Celebration of the Tercentenary of Scientific Hydrology,* ed. UNESCO-WMO, 11. Paris: UNESCO-WMO, 1974.

United Nations Commission on Sustainable Development. "Comprehensive Assessment of the Freshwater Resources of the World." United Nations document E/EN.17/1997/9, 1997.

United Nations Environment Programme. "Vital Water Graphics: An Overview of the State of the World's Fresh and Marine Waters." New York: UNEP/GRID Arendal, 2002.

United Nations Secretariat of UN-Water. *Water for Life Decade: 2005-2015.* New York: UN Department of Public Information, 2005.

United Nations World Commission on Environment and Development (Gro Harlem Brundtland, chair), *Our Common Future.* Oxford: Oxford University Press, 1987.

United States Federal Council for Science and Technology Ad Hoc Panel on Hydrology. *Scientific Hydrology.* Washington, DC: US Government Printing Office, 1962.

Urban, M., and B. Rhoads. "Conceptions of Nature: Implications for an Integrated Geography." In *Contemporary Meanings in Physical Geography: From What to Why?* ed. S. Trudgill and A. Roy, 211-31. London: Arnold, 2003.

van der Leeden, F. *Water Resources of the World: Selected Statistics.* Port Washington, WI: Water Information Center, 1975.

van der Leeden, F., F.L. Troise, and D.K. Todd. *The Water Encyclopedia.* 2nd ed. Chelsea, MI: Lewis Publishers, 1990.

Van Hylckama, T.E.A. *The Water Balance of the Earth.* Centerton, NJ: Drexel Institute of Technology, Laboratory of Climatology, 1956.

Varenius. *Cosmography and Geography.* London: Samuel Roycroft, for Richard Blome, 1693.

Veissman, W.J., and G.L. Lewis. *Introduction to Hydrology.* 4th ed. New York: HarperCollins College Publishers, 1996.

Visvanathan, Shiv. "Mrs Bruntland's Disenchanted Cosmos." *Alternatives* 16, 3 (1991): 377-384.

Vitruvius. *Ten Books on Architecture.* Cambridge: Cambridge University Press, 1999.

Volker, A. "International Cooperation in Hydrology and Water Resources Development." *Hydrological Sciences Journal* 28, 1 (1983): 49-56.

Vörösmarty, C., D. Lettenmaier, C. Leveque, M. Meybeck, C. Pahl-Wostl, J. Alcamo, W. Cosgrove, H. Grassi, H. Hoff, P. Kabat, F. Lansigan, R. Lawlord, and R. Naimann. "Humans Transforming the Global Water System." *EOS* 85, 48 (2004): 509-14.

Wainwright, J. "Politics of Nature: A Review of Three Recent Works by Bruno Latour." *Capitalism, Nature, Socialism* 16, 1 (2005): 115-22.

Walkem, A. "The Land Is Dry: Indigenous Peoples, Water, and Environmental Justice." In *Eau Canada: The Future of Canada's Water*, ed. Karen Bakker, 303-19. Vancouver: UBC Press, 2007.

Ward, C. *Reflected in Water: A Crisis of Social Responsibility*. London: Cassell, 1997.

Ward, R.C., and M. Robinson. *Principles of Hydrology*. 4th ed. London: McGraw-Hill, 2000.

Watts, M. "Political Ecology." In *A Companion to Economic Geography*, ed. E. Sheppard and T.J. Barnes, 257-74. Oxford: Blackwell, 2000.

Weiner, P.P., ed. *Dictionary of the History of Ideas: Studies of Selected Pivotal Ideas*. New York: Scribner, 1973-74.

Wescoat, J.L. Jr., and G.F. White. *Water for Life: Water Management and Environmental Policy*. Cambridge: Cambridge University Press, 2003.

Whatmore, S. "Hybrid Geographies: Rethinking the 'Human' in Human Geography." In *Human Geography Today*, ed. M. Doreen, John Allen, and Phillip Sarre, 22-39. Cambridge, UK: Polity Press, 1999.

–. "Introduction: More Than Human Geographies." In *Handbook of Cultural Geography*, ed. K. Anderson, M. Domosh, S. Pile, and N. Thrift, 165-67. London: Sage Publications, 2002.

Whitcombe, E. "The Environmental Costs of Irrigation in British India: Waterlogging, Salinity, Malaria." In *Nature, Culture and Imperialism: Essays on the Environmental History of South Asia*, ed. D. Arnold and R. Guha, 237-59. Delhi: Oxford University Press, 1995.

White, G.F. "Human Adjustment to Floods." *University of Chicago, Department of Geography Research Papers*. Research Paper 29. Chicago: University of Chicago, 1945.

–. *Strategies of American Water Management*. Ann Arbor: University of Michigan Press, 1969.

–. "Water Resource Adequacy: Illusion and Reality." In *The Resourceful Earth: A Response to Global 2000*, ed. J.L. Simon and H. Kahn, 250-66. Oxford: Blackwell, 1984.

White, R. *The Organic Machine: The Remaking of the Columbia River*. New York: Hill and Wang, 1995.

Whitehead, A.N. *Adventures of Ideas*. New York: Macmillan, 1933.

–. *Nature and Life*. Chicago: University of Chicago Press, 1934.

–. *A Philosopher Looks at Science*. New York: Philosophical Library, 1965.

–. *Process and Reality: An Essay in Cosmology*. New York: Macmillan, 1960.

–. *Science and the Modern World: Lowell Lectures, 1925*. New York: Macmillan, 1925.

Whitely, P., and V. Masayesva. "The Use and Abuse of Aquifers: Can the Hopi Indians Survive Multinational Mining?" In *Water, Culture, and Power: Local Struggles in a Global Context*, ed. J.M. Donahue and B.R. Johnston, 9-34. Washington, DC: Island Press, 1998.

Wikipedia. *Water*. (accessed 26 August 2007). http://en.wikipedia.org/wiki/Water#_note-0.

Williams G., and E. Mawdsley, "Postcolonial Environmental Justice: Government and Governance in India." *Geoforum* 37, 5 (2006): 660-70.

Wisler, C.O., and E.F. Brater. *Hydrology*. New York: John Wiley and Sons; London: Chapman and Hall, 1949.

Wittfogel, K.A. "The Hydraulic Civilizations." In *Man's Role in Changing the Face of the Earth*, ed. W.L. Thomas, C.O. Sauer, M. Bates, and L. Mumford, 152-64. Chicago: University of Chicago Press, 1956.

–. *Oriental Despotism: A Comparative Study of Total Power.* New Haven, CT: Yale University Press, 1957.

Wolf, A.T. "The Present and Future of Transboundary Water Management." In *Rethinking Water Management: Innovative Approaches to Contemporary Issues,* ed. C.M. Figuertes, C. Tortajada, and J. Rockstrom, 164-79. London: Earthscan, 2003.

Wolff, G., and P.H. Gleick. "The Soft Path for Water." In *The World's Water: 2002-2003; The Biennial Report on Freshwater Resources,* ed. P.H. Gleick, 1-32. Washington, DC: Island Press, 2002.

World Commission on Dams. *Dams and Development: A New Framework for Decision-Making; The Report of the World Commission on Dams.* London: Earthscan, 2000.

World Commission on Environment and Development. *Our Common Future.* Oxford: Oxford University Press, 1987.

World Resources Institute. *World Resources 1992-1993.* New York: Oxford University Press, 1992.

World Resources Institute, United Nations Development Programme, United Nations Environment Programme, and the World Bank. *World Resources 2000-2001: People and Ecosystems; The Fraying Web of Life.* Washington, DC: World Resources Institute, 2000.

Worster, D. *Nature's Economy: A History of Ecological Ideas.* 2nd ed. Cambridge: Cambridge University Press, 1994.

–. *Rivers of Empire: Water, Aridity, and the Growth of the American West.* New York: Pantheon, 1985.

–. "Water in the Age of Imperialism – and Beyond." In *A History of Water.* Vol. 3, *The World of Water,* ed. T. Tvedt and T. Oestigaard, 5-17. London: I.B. Tauris, 2006.

Wright, C.J. *The Coming Water Famine.* New York: Coward-McCann, 1966.

Wulff, H.E. "The Quanats of Iran." *Scientific American* 218, 4 (1968): 94-101.

Wynne, B. "Scientific Knowledge and the Global Environment." In *Social Theory and the Global Environment,* ed. M. Redclift and T. Benton, 169-89. New York: Routledge, 1994.

Xenos, N. *Scarcity and Modernity.* London: Routledge, 1989.

Yearley, S. *Sociology, Environmentalism, Globalization: Reinventing the Globe.* London: Sage Publications, 1996.

Young, G.J., J.C.I. Dooge, and J.C. Rodda. *Global Water Resource Issues.* Cambridge: Cambridge University Press, 1994.

Young, R.M. "Dialectics of Nature." In *A Dictionary of Marxist Thought,* ed. Tom Bottomore, Laurence Harris, V.G. Kiernan, and Ralph Miliband, 150. Oxford: Blackwell, 1999.

–. "Nature." In *A Dictionary of Marxist Thought,* ed. Tom Bottomore, Laurence Harris, V.G. Kiernan, and Ralph Miliband, 399. Oxford: Blackwell, 1999.

Zimmerman, M.E. *Heidegger's Confrontation with Modernity: Technology, Politics, Art.* Bloomington: Indiana State University Press, 1990.

Zimmermann, E.W. *World Resources and Industries: A Functional Appraisal of the Availability of Agricultural and Industrial Resources.* New York: Harper and Brothers, 1933.

–. *World Resources and Industries: A Functional Appraisal of the Availability of Agricultural and Industrial Materials.* Rev. ed. New York: Harper and Brothers, 1951.

Index

Note: "(f)"after a page number indicates a figure. "(t)"after a page number indicates a table.

abstraction: Cartesian, 18; as essence, 6-7, 8, 16(f); vs experience, 75; and hydrologic cycle, 149; metaphorical, 18, 249n41
actor-network theory (ANT), 176-77
Adams, Frank Dawson, 111-13, 112(f), 114, 115, 116, 144, 145, 146
agency: of humans, 187, 283n7; of ideas, 40; of non-human actants, 176-77
Agenda 21 (UN action plan), 217, 219
alien water, 225-28, 240, 293n7
American Association of Geographers, 232-33
American Geophysical Union (AGU), 128; Hydrology Section, 131, 134
American Society of Civil Engineers, 141, 142(f)
Anaxagoras, 112, 146
aqueducts, 81-83, 82(f), 83-84, 261n21, 264n101
aquifers, 241-46
argument from design. *See* intelligent design
aridity, 123-24, 270n65. *See also* drought
Aristotle, 76-77, 86, 113, 267n22

Bachelard, Gaston, 6
Baker, M.N., 144, 275n87
Bakker, Karen, 48-50, 67, 238
Bangladesh, 256n56
Berzelius, Jöns Jakob, 78-80
binaries, dialectic: history-consciousness, 31-32; humans-nature, 28, 64, 288n39; nature-society, 27-29, 35-36, 40, 67-68, 69
biosphere, 203, 288n39
Biswas, Asit K., 114, 145, 146, 217, 268n27
Blaikie, Piers, 70, 259n100
body public, 231, 232(f), 236
Braun, Bruce, 33, 246, 253n67
Brickner, E. Ya., 165, 272n29
Britain, 60-61, 91, 95-96, 97, 102, 117, 118, 123, 256n56
Brooks, David, 228, 239, 241, 245
Brundtland Commission, 203, 204, 288n39
Bureau of Reclamation, 19, 150, 152, 160, 161

calculation: dynamic vs. static, 164-66; of evaporation, 118; of global water, 164-67; of water balance, 115-19

Printed and bound in Canada

Set in Garamond by Artegraphica Design Co. Ltd.

Copy editor: Judy Phillips

Proofreader and indexer: Dianne Tiefensee